Algal Biotechnology and Environment

Algal Biotechnology and Environment

Editors

Dinabandhu Sahoo

Department of Botany
University of Delhi
Delhi-110007, India

B. D. Kaushik

Formerly Head
Division of Microbiology
Indian Agricultural Research Institute
New Delhi-110013, India

I.K. International Publishing House Pvt. Ltd.

NEW DELHI • BANGALORE

Published by
I.K. International Publishing House Pvt. Ltd.
S-25, Green Park Extension
Uphaar Cinema Market
New Delhi–110 016 (India)
E-mail: info@ikinternational.com
Website: www.ikbooks.com

ISBN 978-93-81141-71-7

10 9 8 7 6 5 4 3 2 1

Published by Krishan Makhijani for I.K. International Publishing House Pvt. Ltd., S-25, Green Park Extension, Uphaar Cinema Market, New Delhi–110 016. Printed by Rekha Printers Pvt. Ltd., Okhla Industrial Area, Phase II, New Delhi–110 020.

Preface

Algae are a large and diverse group of simple, autotrophic organisms, ranging from unicellular to multicellular forms, such as the giant kelps that grow upto 70 meters in length. The largest and most complex marine forms are called seaweeds. The prokaryotic cyanobacteria (commonly referred to as blue-green algae) were traditionally included as "algae" in older textbooks but they are now considered to be bacteria. However, still large numbers of plant biologists use both the terms frequently. Algae are found in the fossil records dating back to approximately 3 billion years in the Precambrian. Fossilized filamentous algae from the Vindhya basin have been dated back to 1.6 to 1.7 billion years ago. Algae occur widely in aquatic habitats (ocean, rivers, lakes and temporary water bodies) both in fresh and saline waters; and on land surface wherever moisture is available. They serve as primary food to a large number of small animals in ocean and other aquatic systems. They are photosynthetic hence enrich the environment with oxygen. Algae serve as a source of food, feed, fertilizer, pigments and their extracts are used in various types of industries such as ice cream, toothpaste, textile printing, pharmaceuticals, nutraceuticals, packaging, meat processing, chocolate making, tomato ketchup and several other industries. Some species of algae are extensively used for wastewater treatments. Recently algae have been found as an excellent source of biodiesel and bioethanol. Algae can be used for carbon dioxide capture from power plants and various other high emission industries. **Algal Biotechnology and Environment** includes some of the papers presented by leading phycologists of the world during the International Conference on Applied Phycology "Algae in Biotechnology and Environment" held at New Delhi, India. The major thrust is on community analysis, technological improvement in product harvesting, e.g., agar-agar and astaxanthin; phycoremediation, wastewater treatment, biofertilizer, algal diversity in diverse habitats and related topics. The publication of this book was delayed due to several factors which was beyond our control. This book provides a wealth of information for all those studying or working with algae. We are confident that this book would find widespread readership.

We thank Dr. Pooja Baweja, M. Arunjit Singh, E. Geetanjali, Savindra Kumar, Salam Sonia Devi, Gaurav Kumar, Nitin Pipralia, Arvind K. Bharadwaj, Priyanka Verma and Vivek Chopra—research students of Marine Biotechnology Laboratory, Department of Botany, University of Delhi—for their help during the compilation of this book.

Dinabandhu Sahoo

B.D. Kaushik

Contents

1

Effects of Temperature and Ammonium on Growth, Pigment Production and Nitrogen Uptake by Four Species of *Porphyra* (Bangiales, Rhodophyta) Native to the New England Coast

Jang K. Kim[1,4], George P. Kraemer[2], Christopher D. Neefus[3] Ik Kyo Chung[4,5] and Charles Yarish[4]

[1]Department of Ecology and Evolutionary Biology, University of Connecticut, Groton, CT 06340, USA.

[2]Department of Environmental Science, Purchase College, State University of New York Purchase, NY 10577 USA .

[3]Department of Plant Biology, University of New Hampshire, Durham, NH 03824, USA.

[4]Department of Ecology and Evolutionary Biology, University of Connecticut, CT 06901, USA.

[5]Department of Marine Science, Pusan National University, Busan 609-735, KOREA.

ABSTRACT

Porphyra is one of the world's most valued maricultured seaweeds. The effects of ammonium concentration and temperature on the growth, pigment content and tissue nitrogen content of *Porphyra* from coastal New England were studied in 50 liter tanks to produce critical information for the development of a land based aquaculture system. Four Northwest Atlantic *Porphyra* species: *P. leucosticta, P. amplissima, P. linearis* and *P. umbilicalis*, were cultivated for one and two weeks at saturating light intensities (100 – 150 $\mu Em^{-2}s^{-1}$) and six combinations of ammonium (25 and 250 µM) and temperature (10, 15 and 20°C). Specific growth rate (SGR) increased with decreasing temperature in *P. leucosticta, P. linearis* and *P. umbilicalis* and increased with increasing temperature in *P. amplissima*. The SGR of all species was greater at the higher ammonium concentration. *Porphyra linearis* had the highest SGR, increasing in biomass by about 16% d^{-1}. Phycoerythrin (PE) content was higher at 10°C and 250 µM in all species except *P. amplissima*. The PE content of *P. linearis* (29 mg g^{-1} fw^{-1}) and *P. umbilicalis* (26 mg g^{-1} fw^{-1}) was significantly higher than other two species. Tissue nitrogen content of all species measured in dry weight (DW) was on average 1.45% higher at 250 μM than at 25 μM ammonium concentration. This percentage represents the absolute difference between the two values. *Porphyra umbilicalis* had the highest tissue nitrogen contents (6.76%) at 10°C and 250 μM ammonium. Based on these results, *P. linearis* and *P. umbilicalis* should be considered the potential candidates for bioremediation with fin- and shellfish mariculture.

Keywords: *Porphyra*, nutrient uptake, temperature, ammonium, bioremediation, mariculture.

INTRODUCTION

An emerging problem of coastal fish mariculture is the significant loading of inorganic nutrients into local waters (Beveridge, 1987), which contributes to blooms of phytoplankton and weedy macroalgae (Cuomo et al., 1993). Reducing the net release of nutrients into the environment is, therefore, an important issue; bioremediation of fish mariculture effluent is an option that provides both ecological and economic incentives. Integrated aquaculture, in which seaweeds are grown downstream from the animals, is a process of bioremediation. It can reduce nutrient loading in coastal waters because the nitrogen (N) and phosphorus (P) in the animal effluent enable rapid growth of the seaweeds. Obviously, the best seaweed to integrate into an animal aquaculture operation is the one characterized by rapid growth, high accumulation of N and P in tissue (ecological value), and high commercial price (*economic value*; Troell et al., 2003; Neori et al., 2004). Therefore, knowledge of ecological and physiological aspects of seaweed growth is a very important part of the development of an integrated aquaculture system.

Nitrogen has traditionally been considered the limiting nutrient in temperate oceans both for phytoplankton (Ryther and Dunstan, 1971; Twomey and Thompson, 2001) and seaweed communities (Lobban and Harrison, 1994; Harrison and Hurd, 2001). Generally, nitrogen addition increases photosynthetic activity and growth of seaweeds (Chopin et al., 1990). Nitrogen is an indispensable element incorporated into many organic macromolecules (proteins, nucleic acids and pigments). Synthesis of chlorophyll *a* (Chl *a*) and phycoerythrin (PE) also requires N (Lobban and Harrison, 1994). On an average, proteins contain about 15% N, nucleic acids about 13% N and PE is comprised of up to 20% N (Lapointe and Duke, 1984; Harrison and Hurd, 2001). Depending upon the polysaccharide content, algae may contain between 3 – 10% N by dry weight (Ryther et al., 1981; Harrison and Hurd, 2001).

In general, algae can use a wide variety of nitrogenous compounds to fulfill their N requirements: ammonia, nitrate, urea, amino acids and nucleosides may be taken up from the growth medium (Rai and Gaur, 2001), with NO_3^- and NH_4^+ being the primary sources in most circumstances. Some seaweeds, such as *Gelidium nudifrons* N.L. Gardner (Bird 1976) and *Laminaria groenlandica* Rosenvinge (Harrison et al., 1986), take up NO_3^- and NH_4^+ simultaneously and at the same rate. Thus, these species have the potential to assimilate more N per unit time if provided both forms of N than if limited to only one N form for uptake. However, in many species, NH_4^+ inhibits the uptake of NO_3^- by up to 50% (Conway, 1977; DeBoer, 1981). For instance, NH_4^+ concentrations in the range 0.5 – 10.0 μM suppress NO_3^- uptake in phytoplankton cells by 50% (Conway, 1977). Thalli of *Gracilaria foliifera* (Forsskål) Børgesen and *Neoagardhiella baileyi* (Kützing) M.J. Wynne and W.R. Taylor grown with NH_4^+-N showed a greater biomass yield than did NO_3^--cultured thalli and appeared to be capable of storing more N (DeBoer et al., 1978, Bird et al., 1982). Exposure of *Gracilaria tikvahiae* to NH_4^+-enriched seawater also led to greater levels of phycoerythrin than did exposure to the same level of NO_3^- (Bird et al. 1982). *Porphyra yezoensis* also prefers NH_4^+ in preference over other nitrogen sources, such as NO_3^-, NO_2^-, amino acid and urea (Amano and Noda, 1987).

The genus *Porphyra* (Bangiales, Rhodophyta) is capable of rapid growth and is an efficient nutrient concentrator. Recently Kraemer et al. (2003) measured specific growth rates of *Porphyra* species exceeding 24% d^{-1}, and more recent studies suggest this genus is capable of 45% d^{-1} in the short-term (Kraemer, pers. comm.). Since it consists of one or two cell layers, *Porphyra* has an extremely high surface area to volume ratio. With all cells taking up nutrients, it is capable of rapid assimilation and growth (Neori et al., 2004). In addition, *Porphyra* is a valuable source of food and in Asia this seaweed is commercially maricultured to supply a billion dollar (U.S.) annual market (Yarish et al., 1999; FAO, 2003). *Porphyra* contains high levels of protein (25 – 50%), vitamins (higher vitamin C than in oranges), trace minerals and dietary fibers (Noda 1993). Eighteen types of free amino acids have been reported, including taurine, which has been found to reduce blood cholesterol levels (Tsujii et al., 1983; Noda, 1993). This alga is also a preferred source of the red pigment *r*-phycoerythrin, which is utilized as a fluorescent "tag" by the medical diagnostic industry (Mumford and Miura 1988).

Porphyra has been cultivated for the past several hundred years in Japan and has become one of the most successful aquaculture industries in Japan, Korea and China (Mumford and Miura, 1988; FAO, 2003). Several attempts at culturing commercial *Porphyra* have been made in North America. Commercial operations on the west coast of the U.S. and Canada were unsuccessful (Yarish et al., 1998) due to the inability of *Porphyra* farmers to obtain aquaculture lease permits (Yarish et al., 1998). In coastal New England, a commercially valuable Asian taxon, *P. yezoensis*, was selected for aquaculture trials in part because little was known of the biology of the native New England *Porphyra* species. *Porphyra yezoensis* was not grown successfully because the gametophyte was not well adapted for the temperature and nutrient regimes of northeastern Maine's coastal environment (Yarish et al., 1998). Therefore, selection of an appropriate local *Porphyra* cultivar is necessary for a successful integrated aquaculture system.

A wide variety of biological factors such as inter-individual variability, nutritional history, type of tissue, life history stage/age, surface area:volume ratio of the thallus, and blade morphology may influence the nutrient uptake and growth. Young tissues exhibit higher rates of uptake and growth than older tissues and therefore one must be careful when uptake rates are determined only on portions of the thallus (Thomas et al., 1985; Hurd and Dring, 1990; Kraemer and Yarish, 1999). For example, the kelp *Laminaria groenlandica* Rosenvinge is a perennial plant wherein the first year plants have higher uptake rates than the third year plants (Harrison et al. 1986). Younger blades and small size tissue of *Porphyra yezoensis* also grew at a significantly higher rate than mature blades and large size tissue (Hafting, 1999). To avoid these biological effects, materials size and age should be considered.

The primary objective of this study was to collect physiological measurements to guide the selection of an appropriate species of *Porphyra* (*P. leucosticta, P. linearis, P. umbilicalis* and *P. amplissima*) from coastal New England to be used in the development of an integrated land-based aquaculture system. This will enable us to compare the efficiency of different species of *Porphyra* as nutrient scrubbers to decide which candidate would be best suited for bioremediation.

We present here results that describe the influence of temperature and nutrient availability (ammonium concentration) on growth rate, phycoerythrin content, tissue N content and N removal from the media.

MATERIALS AND METHODS

Algal Material and Culture

Porphyra amplissima (Kjellman) Setchell et Hus (ME-32p) used in this study was in culture at the Marine Biotechnology Laboratory of the University of Connecticut at Stamford. The strain of *Porphyra amplissima* was originally collected from Gore Point, Cobscook Bay, ME, USA. *Porphyra leucosticta* Thuret in Le Jolis was collected in the mid intertidal zone at Groton and Waterford, Connecticut, USA in March and May 2002. *Porphyra umbilicalis* Kützing and *Porphyra linearis* Greville were collected in the upper intertidal zone at Rye, New Hampshire, USA in March 2003.

Experiments were carried out for one or two week(s) in 50L tanks at saturating light intensities (100 – 150 $\mu E\ m^{-2}s^{-1}$; Kraemer and Yarish 1999). Two 50L tanks were contained in each of 9 water baths. Temperature in each water bath was controlled independently via 1000 watt immersion heaters. Light was supplied by 400 watt Ceramalux lamps (Philips, Somerset, NJ) placed above each pair of tanks. Irradiance was measured by a light meter (LI-185A, Li-Cor Inc.) and adjusted with neutral density filters. Photoperiod was 12:12h L:D. The culture medium was filtered (0.45 μm) and UV-irradiated seawater with von Stosch's enrichment (Ott, 1965) without nitrogen and phosphorus. Nitrogen and phosphorus levels were regulated by addition of ammonium ($NaNO_3$) and phosphate (Na_2HPO_4) to cultures twice a week at a molar N:P ratio of 10:1. The initial stocking densities of tanks for *P. leucosicta, P. linearis P. umbilicalis* and *P. amplissima* were 0.08, 0.1, 0.14 and 0.08 g L^{-1} respectively.

Acclimation

Porphyra leucosticta, P. umbilicalis and *P. linearis,* were acclimated for 4 – 6 days in gently aerated, Avery Point (CT) filtered seawater at 10°C and 50 – 100 $\mu E\ m^{-2}s^{-1}$ under a 12:12 L:D photoperiod. Filtered seawater was renewed daily to maintain a stable nutrient status in the algal tissues during acclimation. After acclimation, discs (20 mm diameter) of *P. leucosticta* and *P. umbilicalis* were punched from different blades and cultivated for two days under the same conditions to allow recovery from wounding effects (Drew, 1983). Whole blades (5 × 1 cm) of *P. linearis* and *P. amplissima* were used. As the *Porphyra amplissima* had been grown in von Stosch enriched seawater, an acclimation period was considered unnecessary.

Experimental Design

The experiments were conducted using a split-plot, randomized complete block design, with temperature (3 levels) as main plots and ammonium (2 level) as subplots, resulting in a total of six treatment combinations, each with three replicates. The design accommodated practical limitations in temperature and ammonium control: three levels of temperature (10, 15 and 20°C) and two levels of ammonium concentration (25 and 250 μM). The ammonium and temperature levels reflect the range in marine aquaculture (Day, 2003).

Measurements

At one-week intervals, all of the biomass in each tank was weighed (fresh weight; FW) and samples were taken for dry weight and pigment analysis. FW were obtained after blotting the thalli dry with paper towels. Dry weight (DW) and moisture content were calculated by drying a sample of the biomass at 60°C to constant weight. Specific growth rate (SGR, expressed as % increase d^{-1}) was calculated as follows:

$$SGR = 100*(\ln S_2 - \ln S_1)/(T_2 - T_1)$$

where S_1 and S_2 are the fresh weights at days T_1 and T_2, respectively (D'Elia and DeBoer, 1978). PE was extracted using a modification of the method of Beer and Eshel (1985). Approximately 100 mg FW of tissue was ground in a mortar with pestle in 0.1M phosphate buffer (pH 6.5) and kept at 4°C before being centrifuged at 19,000g for 15 min. The supernatant was analyzed with a Spectronic Genesys 5 spectrophotometer. PE content was calculated according to the equations used in Beer and Eshel (1985). For the analysis of tissue total N and C content, samples were dried at 60°C before being ground. The powder was analyzed using a Perkin Elmer 2400 series II CHNS/O elemental analyzer. Specific growth rate, tissue PE, nitrogen and carbon content were determined weekly.

N removal was calculated using the equation,

N Removal = Growth rate × Tissue N content × DW/FW

The difference in nitrogen removal ability between species and between conditions could be compared with this method. For example, when comparing two species under the same or different conditions, similar or different growth rates but different tissue nitrogen content were obtained, therefore the nutrient uptake ability of these two species could be different.

Statistical Analysis

Two-way split-plot ANOVA ($\alpha = 0.05$) was used to analyze data. When ANOVA indicated treatment effect of temperature or an interaction between ammonium level and temperature, Tukey's HSD analysis ($\alpha = 0.05$) was used as a post hoc test to determine pairwise comparison probabilities between treatment level means. All statistic analyses were done using Minitab (release 13, Minitab Inc. State College, PA U.S.A.).

RESULTS

Growth

Temperature significantly influenced the growth rate of all species. The growth rate of *Porphyra leucosticta* was higher at 10 and 15°C than at 20°C (Fig. 1; $p = 0.01$). *Porphyra linearis* and *P. umbilicalis* grew fastest at 10°C (Fig. 1). However, *P. amplissima* showed the higher growth rate at 20°C than that at the lower temperatures (Fig. 1).

Nutrient availability also significantly influenced growth rate. The growth rates of *Porphyra linearis* and *P. umbilicalis* were significantly higher at 250 μM ammonium than that at 25 μM ammonium concentration (Fig. 1; $p \leq 0.022$). Of the four species studied, *Porphyra linearis* had

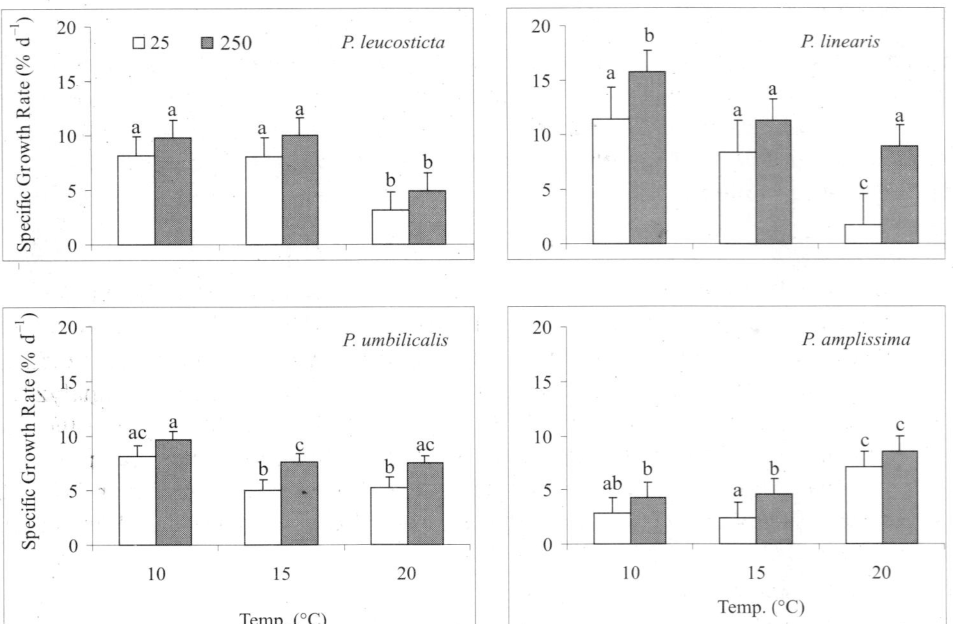

Fig. 1 Growth rate of *P. leucosticta**, *P. linearis***, *P. umbilicalis*** and *P. amplissima** grown at 10, 15 and 20°C and 25 and 250 μM ammonium. Error bars represent standard error. Bars with the same letter are not significantly different ($p > 0.05$; *first week results, **average results over first two weeks).

the highest growth rate, increasing in biomass by about 16 % d^{-1} over 14 days at 10°C, while other three species grew at about 10 % d^{-1} at optimum conditions.

Phycoerythrin Content

The phycoerythrin (PE) content was affected by temperature ($p \leq 0.027$) at high ammonium concentration in all species except *Porphyra amplissima* (Fig. 2). The PE contents of *P. leucosticta* were higher at 10°C than at other temperatures and *P. linearis* and *P. umbilicalis* showed significantly higher PE contents at 10 and 15°C. The PE content of *P. linearis* and *P. umbilicalis* was on average 90% higher at 250 μM than at 25 μM ammonium concentration at all temperature conditions (Fig. 2; $p \leq 0.012$). The PE content of *P. linearis* (29 mg g^{-1} FW^{-1}) and *P. umbilicalis* (26 mg g^{-1} FW^{-1}) was much higher than *P. leucosticta* and *P. amplissima*. For *P. linearis*, the effect of temperature on PE content was greater at high than at low ammonium levels; similar interactions were not found in the other species.

Nitrogen Contents and C:N Ratio in Tissue

Nitrogen values in tissue varied from 2.25% (*P. leucosticta* at 25 μM and 20°C) to 6.76% (*P. umbilicalis* at 250 μM and 10°C). Temperature alone had no effect on tissue nitrogen content of any *Porphyra* species ($p > 0.05$). Ammonium concentration significantly affected nitrogen content

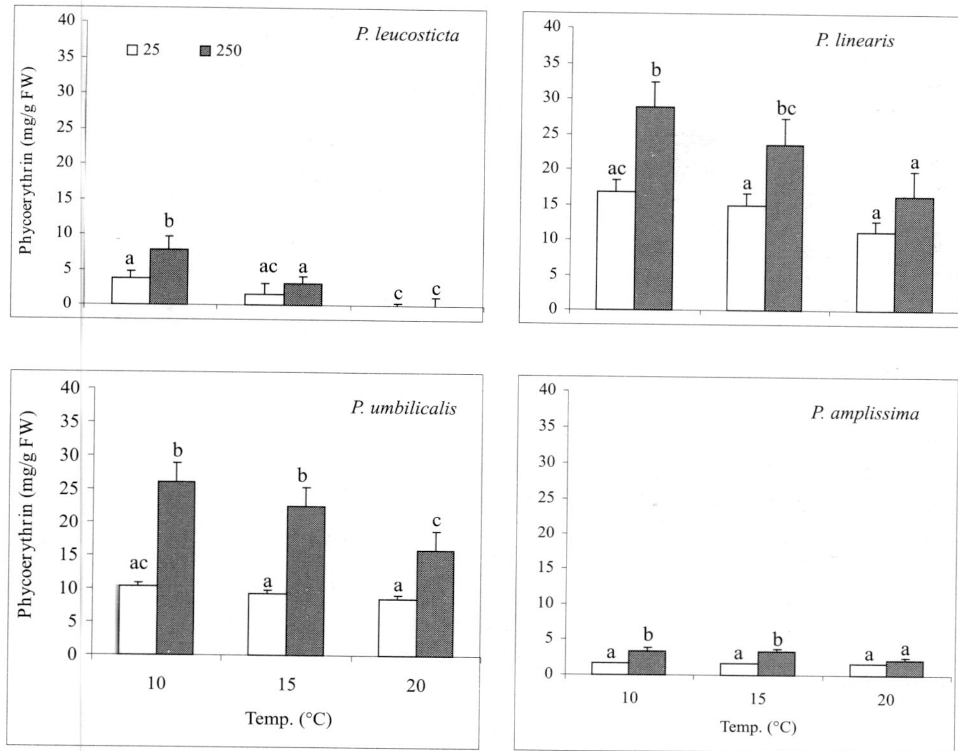

Fig. 2 Phycoerythrin contents of *P. leucosticta**, *P. linearis***, *P. umbilicalis*** and *P. amplissima** grown at 10, 15 and 20°C and 25 and 250 µM ammonium. Error bars represent standard error. Bars with the same letter are not significantly different ($p > 0.05$; *first week results, **average results over first two weeks).

(Fig 3; $p \leq 0.001$) in all species. The tissue nitrogen content of all species was on an average 1.45% higher in the DW at 250 µM than at 25 µM ammonium concentration. This percentage represents the absolute difference between two values. *Porphyra umbilicalis* had the highest tissue nitrogen contents (6.76%) at 10°C and at 250 µM ammonium (Fig. 3). The starting C:N ratio of the algae was between 7.2 and 9.5. Lower C:N ratios were observed in all *Porphyra* species grown at 250 µM than at 25 µM ammonium concentration over 14 days. *Porphyra leucosticta* showed the highest ratio (up to 16.98) under 25 µM ammonium.

Nitrogen Removal

The nitrogen removal ability of *Porphyra leucosticta, P. linearis* and *P. umbilicalis* was affected by temperature. In *P. leucosticta, P. linearis* and *P. umbilicalis*, N removal was higher at high ammonium concentration than at low ammonium concentration, and decreased with increasing temperature in all species except *P. amplissima* at 25 µM (Fig. 4). The highest N removal capability of *P. leucosticta* occurred at 10 and 15°C, for *P. linearis* and *P. umbilicalis,* it was at 10°C and for *P. amplissima,* it was at 20°C. The N removal of *P. umbilicalis* (0.14 mg d^{-1} g DW^{-1}) and *P.*

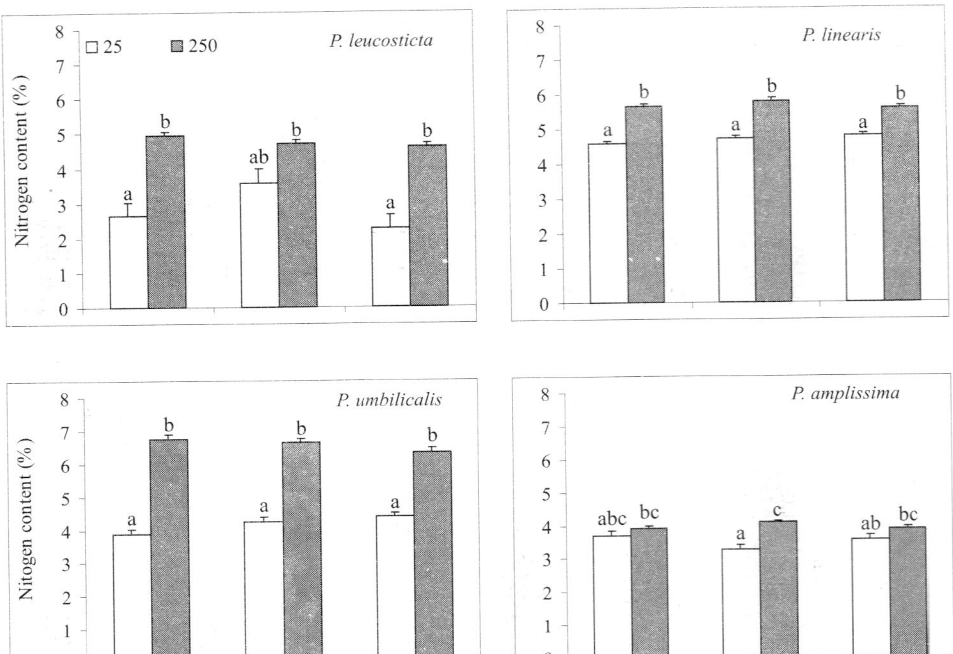

Fig. 3 Tissue N contents of *P. leucosticta**, *P. linearis***, *P. umbilicalis*** and *P. amplissima**. Error bars represent standard error. Bars with the same letter are not significantly different (p > 0.05; *first week results, **average results over first two weeks).

linearis (0.13 mg d^{-1} g DW^{-1}) was markedly higher than *P. leucosticta* at 10°C and *P. amplissima* at 20°C (0.05 and 0.02 mg d^{-1} g DW^{-1}, respectively) and 250 μM ammonium concentration.

DISCUSSION

Porphyra linearis grew at over 16% SGR, which was significantly higher than that of other seaweeds being used for integrated aquaculture, e.g., *Gracilaria parvispora* Abbott (10% d^{-1}; Nelson et al., 2001) and *Laminaria saccharina* (Linnaeus) J.V. Lamouroux (9% d^{-1}; Subandar et al., 1993). *Porphyra umbilicalis* has a high capacity for nitrogen accumulation (6.76% tissue nitrogen at optimal conditions). This value is also much higher than that of other maricultured seaweeds, including *Gracilaria tikvahiae* McLachlan (1.03 – 2.95%; Lapointe and Duke, 1984), *Gracilaria gracilis* (Stackhouse) Steentoft, L.M. Irvine and Farnham (3.68%; Hernandez et al. 2002), and *Gracilaria pacifica* I.A. Abbott (4.18 – 4.59%; Naldi and Wheeler, 1999), *Gracilaria parvispora* (< 4%; Nelson et al., 2001). It is even markedly higher than other efficient nutrient scrubbers such as *Ulva rotundata* Bliding, *Ulva*

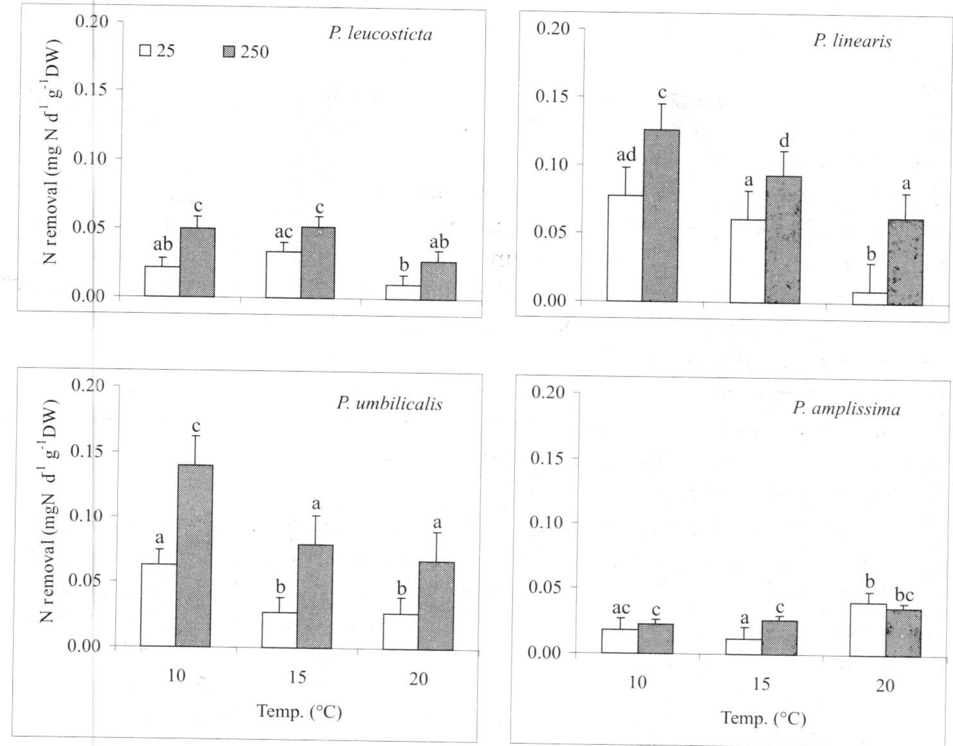

Fig. 4 Nitrogen removal by *P. leucosticta**, *P. linearis***, *P. umbilicalis*** and *P. amplissima** grown at 10, 15 and 20 °C and 25 and 250 μM ammonium. Error bars represent standard error. Bars with the same letter are not significantly different ($p > 0.05$; *first week results, **average results over first two weeks).

fenestrata Postels and Ruprecht and *Enteromorpha intestinalis* (Linnaeus) Link (3.06%, 4.71% and 3.35%, respectively; Hernández et al., 2002).

Some species grow well under conditions in laboratory that never occur in their native environment. For example, the distribution of the red alga, *Polyneura hilliae* (Greville) Kylin, indicates that it grows well between 11 and 15°C. However, Yarish et al. (1986) reported that the alga grew very well from 10 to 20°C under laboratory conditions. Below 5°C and above 25°C, temperatures never experienced in nature, *P. hilliae* died. Another red alga *Calliblepharis ciliata* (Hudson) Kützing showed similar tendencies (Yarish et al., 1986). The fundamental niche describes the environmental "space" within which a species can survive and reproduce in the absence of biotic interactions. Competition, predation, and parasitism restrict organisms to the realized niche, only a part of the fundamental niche (Hutchinson, 1958). Temperature is an important part of the ecological niche, a concept that is often used to describe the range of tolerance (Lampert and Sommer, 1997). Hutchinson (1958) emphasized that organisms have ranges of tolerance for many environmental factors, rather than only a single factor. Stress in one abiotic factor may reduce the tolerance range of another; by providing in aquaculture an optimal environment, tolerance ranges are likely to be as broad as possible. Therefore, when organisms are protected from biotic interactions and cultivated under an optimal condition, the physiological tolerance range can be wider. In the present study, *P. leucosticta, P. linearis* and *P. umbilicalis* grew well at low temperature similar to those of the natural habitat of its gametophytes. However, *P. amplissima* attained the highest growth rate at 20°C, a temperature above the highest reached in the field (Chopin et al., 1999).

Porphyra amplissima gametophyte is present from spring through autumn along the coast of Maine, where sea surface temperatures generally range between 0 – 15°C (Chopin et al., 1999; Yarish et al., 1999). *Porphyra leucosticta* is common in winter/spring in Connecticut, (USA) when water temperatures range between 1 – 16°C (Chopin et al., 1999). Therefore, *P. leucosticta* was expected to be at least as tolerant of higher temperatures as *P. amplissima* because *P. amplissima* is rarely found in south of New Hampshire (Chopin et al., 1999). However, *Porphyra amplissima* grew well at high temperature, which demonstrated higher temperature tolerance than the other three species. *P. amplissima* gametophytes came from conchospores grown in the lab without the effects of biotic interactions while other three species came from the field which was previously exposed to biotic interactions.

Recently, Day (2003) and Kraemer et al. (2003) reported that *P. amplissima* has very high growth capability; the specific growth rates of *P. amplissima* were over 32% d^{-1} at 12°C and 24% d^{-1} at 15°C respectively. However the growth rate of *P. amplissima* measured during the present study was much lower. The slower growth rates measured in this study were probably due the effect of tissue age. *Porphyra amplissima* used in this study was mature (over 10 cm), while Day (2003) and Kraemer et al. (2003) used younger blades. Young tissue has much higher growth and nutrient uptake rate than older tissue (Kraemer and Yarish, 1999; Harrison and Hurd, 2001).

The ability to store N has ramifications in management techniques in algal aquaculture. Previous studies have shown that *Gracilaria* is able to take up ambient N very rapidly and store it in organic form for later use during periods of N-limitation (Bird et al., 1982; Hwang et al., 1987). *Gracilaria* can obtain and store enough nitrogen for non-limited growth if given a single pulse of nitrogen every 2 weeks (Ryther et al., 1981). This storage is reflected in thallus N contents (3 – 5%), which can be substantially higher than those indicating N deficiency (1.5 – 2%; Ryther et al. 1981). Pigments are sensitive to the N status of the algae and probably decline due to growth and a lack of sufficient ambient N for continued synthesis of the new pigments. Increases in the chlorophyll *a* content with increases in cellular N are well known for algae (Antia et al., 1963; Fogg 1965). Studies of the red alga *Gracilaria tikvahiae* indicated that chlorophyll and carotenoid pigments did not contribute greatly to the overall N content (Bird et al., 1982).

Our results indicate a storage function for phycoerythrin (PE) pigments in *Porphyra* tissue (except for *P. leucosticta*; Fig. 5). At 250 µM ammonium level, pigment and tissue N contents were significantly greater than those at lower level of ammonium, and both contents did not show differences between temperatures. This supports other studies that arrived at the same conclusion. PE comprised 20% of total N in *G. tikvahiae*, but decreased markedly under N-limitation (Lapointe and Duke, 1984). PE as a ratio of total protein decreased when tissue N content fell below 1.8% (Bird et al., 1982). Perhaps at incipient N limitation these pigments are preferentially utilized for the N requirements of continued algal growth. Quantification in *Gracilaria foliifera* (Lapointe, 1981), *G. tikvahiae* (Bird et al., 1982), *G. gracilis* (Smit et al., 1997) and *G. pacifica* (Naldi and Wheeler,

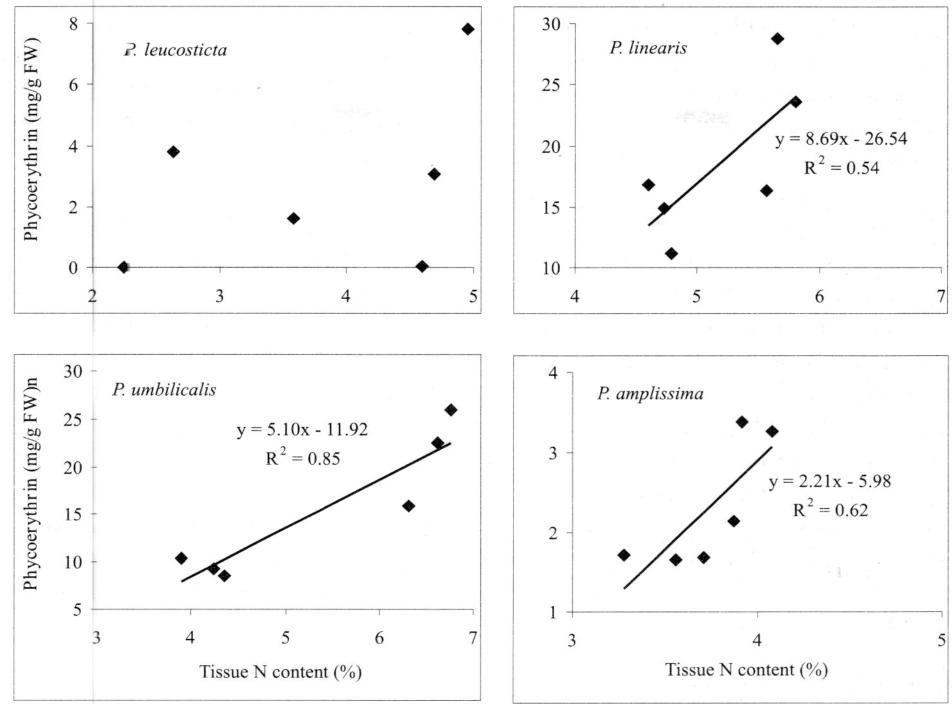

Fig. 5 Phycoerythrin vs. tissue N contents of *P. leucosticta* ($p > 0.05$), *P. linearis* ($p < 0.05$), *P. umbilicalis* ($p < 0.05$) and *P. amplissima* ($p < 0.05$).

1999) indicated PE levels of 7-11% of total nitrogen. In *Porphyra umbilicalis*, phycobiliproteins were found to represent 15-20% of soluble protein (Hernández et al., 1993), while in some red algae, like *Gracilaria pacifica*, PE pigment alone can comprise up to 60% of total soluble protein (Naldi and Wheeler, 1999). *Porphyra linearis* and *P. umbilicalis* exhibited much higher PE contents (276 and 205 mg g^{-1} DW^{-1} respectively; calculated from PE contents FW^{-1}) than that reported for *P. yezoensis* (40-60 mg g^{-1} DW^{-1}; Yan et al., 2000) and another maricultured seaweed, *G. tikvahiae* (15 mg g^{-1} DW^{-1}; Lapointe and Duke, 1984) grown under nutrient enriched conditions.

In many coastal areas, blooms of fast-growing opportunistic seaweeds have replaced the previously-dominant species (Fletcher, 1996; Raffaelli et al., 1998). Such blooms may be due to increased loading of inorganic nutrient into the seawater, decreased herbivore activity, global warming, or a combination of these effects (Smith et al., 1981; Lapointe et al., 1997). The former is the most likely, and potentially controllable. Fish farms are a major source of released nutrients (Ackefors and Enell, 1994). All species of *Porphyra* used in this study thrived in ammonium concentrations as high as those measured in fish farm effluent (Day, 2003), and had much higher growth rates (10-16 % d^{-1}) than under natural lower ammonium concentration.

The efficiency of the different species of *Porphyra* as nutrient scrubbers, as well as the economic potential, temperature range and growth ability may be examined to decide which candidate(s) would be best suited for bioremediation of aquaculture effluent. *Porphyra leucosticta* and *P. linearis* exhibit excellent gustatory properties (Chopin et al., 1999), while *P. umbilicalis* and *P. amplissima* may be marketable for a variety of industrial and biotechnological uses (taurine and r-phycoerythrin; Chopin et al., 1999). Our results indicate the possibility of inter-specific variation in growth, N accumulation and PE content as functions of temperature and ammonium availability. For year-round ammonium removal and production of *Porphyra* and phycobiliprotein, several species should be used in rotation according to their seasonality (Day, 2003). Day (2003) suggested that *P. linearis* and *P. umbilicalis* would be good sources of PE.

In our study, *P. linearis* and *P. umbilicalis* had high PE contents (Figs. 2 and 3), as well as high nitrogen uptake and fast growing capabilities. For these reasons, *Porphyra linearis* and *P. umbilicalis* should be considered potential candidates for bioremediation with fin- and shellfish mariculture. *Porphyra amplissima* grew well at the high temperature (20 °C) condition, but the apparent low nutrient uptake capability and low PE content seem to make this species less suitable for bioremediation. Early life history stage usually have higher nutrient uptake and growth rates than mature thalli of the same species (Thomas et al., 1985). Harrison et al. (1986) reported that *Laminaria groenlandica* was a perennial and first year plants have higher uptake rates than the third year plants. Therefore, further study of the effect of temperature on the bioremediatory performance of young *P. amplissima* tissue should be performed if this species is to be utilized at high temperature.

ACKNOWLEDGEMENTS

This study was supported by grants to C. Yarish from the State of Connecticut's Critical Technologies Program, Connecticut Sea Grant College Program and grants to C. Yarish, G.P. Kraemer and C.D. Neefus from NOAA's National Marine Aquaculture Initiative.

REFERENCES

Ackefors, H., and Enell M. 1994. The release of nutrients and organic matter from aquaculture systems in Nordic countries. *J. Appl. Ichthyol.* 10: 225-241.

Amano, H. and Noda H. 1987. Effect of nitrogenous fertilizers on the recovery of discolored fronds of *Porphyra yezoensis*. *Bot. Mar.* 30: 467-473.

Antia, N. J., McAllister, C. D., Parsons, T. R., Stephens K., and Strickland J. D. H. 1963. Further measurements of primary production using a large volume plastic sphere. *Limnol. Oceanogr.* 8: 166-183.

Beer, S. and Eshel, A. 1985. Determining phycoerythin and phycocyanin concentrations in aqueous crude extracts of red algae. *Aust. J. Mar. Freshwat. Res.* 36: 785-792.

Beveride, M. C. L. 1987. Cage Aquaculture. Fishing News Books Ltd., Farnham.

Bird, K. T. 1976. Simultaneous assimilation of ammonium and nitrate by Gelidium nudifrons (Gelidiales: Rhodophyta). *J. Phycol.* 12: 238-241.

Bird K. T., Habing C. and DeBusk, T. 1982. Nitrogen allocation and storage patterns in *Gracilaria tikvahiae* (Rhodophyta). *J. Phycol.* 18: 344 -348.

Chopin, T., Yarish, C., Wilkes, R., Belyea, E., Lu, S. and Mathieson A. 1999. Developing Porphyra/salmon integrated aquaculture for bioremediation and diversification of the aquaculture industry. *J. Appl. Phycol.* 11:3 463 - 472.

Chopin, T., Hanisak, M.D., Koehn, F.E., Million, J. and Moreau, S. 1990. Studies on carrageenans and effects of seawater phosphorus concentration on carrageenan content and growth of Agardhiella subulata (C. Agardh) Kraft and Wynne (Rhodophyceae, Solieriaceae). *J. Appl. Phycol.* 2: 3-16.

Conway, H.L. 1977. Interactions of inorganic initrogen in the uptake and assimilation by marine phytoplankton. *Mar. Biol.* 39: 221-232.

Cuomo, V., Merrill, J., Palomba, I. and Perretti, A. 1993. Systematic collection of *Ulva* and mariculture of *Porphyra*: Biotechnology against eutrophication in the Venice Laggon. *Int. J, Environ. Stud.* 43: 141-149.

D Elia , C.F. and DeBoer, J. A. 1978. Nutrition studies of two red algae. II. Kinetics of ammonium and nitrate uptake. *J. Phyol.* 14: 266-272.

Day. J. P. 2003. Effects of light and ammonium on growth, N uptake and pigmenatation of *P. umbilicalis* Kützing, *Porphyra linearis* Greville, *P. leucosticta* Thuret in Le Jolis, Porphyra amlissima Kjellman. MS thesis. Unisversity of New Hampshire.

DeBoer, J. A. 1981. Nutrients. In C.S. Lobban and M. J. Wynne (eds.), The Biolgy of Seaweeds, Blackwell Scientific Publications, Oxford. pp. 356-396.

DeBoer, J. A. Guigli, H. J., Israel, T. L. and D Elia, C. F. 1978. Nutritional studies of two red algae. I. Growth rate of a function of nitrogen source and connection. *J. Phycol.* 14: 261-266.

Drew, E. A. 1983. Physiology of *Laminaria* I. Use of excised lamina discs in short and long term experiments. P. S. Z. N. I. *Marine Ecology*. 4: 227-250.

FAO. 2003. Food and Agriculture Organization of the United Nations. http: // www. fao. org / unfao / bodies / conf / c 2003 - e. htm

Fletcher, R. L. 1996. The occurrence of "green tides": A review. In Schramn, W. and Nienhuis, P . H. (Eds.) *Marine Benthic Vegetation: Recent Changes and the Effects of Eutrophication*. Springer, Berlin, pp. 7-44.

Fogg, G. E. 1965. *Algal Cultures and Phytoplankton Ecology*. The University of Wisconsin Press. Madison.

Hafting, J. T. 1999. A noval technique for propagation of *Porphyra yezoensis* Ueda blades in suspension cultures via monospres

Harrison, P. J. and Hurd, C. L. 2001. Nutrient physiology of seaweeds: Application of concepts to aquaculture. *Ch. Biol. Mar.* 42: 71-82.

Harrison, P, J. Druejl, L. D., Lloyd K. E. and Thompsan, P. A. 1986. Nitrogen uptake kinetics in three year: Classes of *Laminaria groenladica* (Laminariales : Phaeophyta). Mar. Biol. 93: 29-35.

Hernandaz, I., Corzo, A., Gordillo, F. J. Robles, M. D. Saez, E., Fernandaz, J. A. and Niell, F. X. 1993. Seasonal cycle of the gametophytic from of *Porphyra umbilicalis*: Nitrogen and carbon. *Mar. Ecol. Progr. Ser.* 99: 301-311.

Hernández, I., Martínez-Aragón, J. F., Tovar, A., Pérez-Lloréns, J. L. and Vergara, J. J. 2002. Biofiltering efficiency in removal of dissolved nutrients by three species of estuarine macroalgae cultivated with sea bass (*Dicentrarchus labrax*) waste waters 2. Ammonium. *J. Appl. Phycol.* 14: 375-384.

Hurd, C. L. and Dring, M. J. 1990. Phosphate uptake by intertidal algae in relation to zonation and season *Mar. Biol.* 107:281-289.

Hutchinson, G. E. 1958. Homage to Santa Rsalia, or Why are there so many kinds of animals. *Amer. Nat.* 91: 145-159.

Hwang, S. P. L., Williams, S. L. and Brinkhuis, B. H. 1987. Changes in internal dissolved nitrogen pools as related to nitrate uptake and assimilation in *Gracilaria tikvahiae* McLachlan. *Bot. Mar.* 30: 11-19.

Kraemer, G. P. and Yarish, C. 1999. A preliminary comparison of the mariculture potential of *Porphyra purpurea* and *Porphyra umbilicalis*. *J. Appl. Phycol.* 11: 473-477.

Kraemer, G. P., Pereira, R., Snellgrove, D., Carmona, R., Neefus, C., Chopin, T. and Yarish, C. 2003. Algal bioremediation of eutrophic effluents in small scale integrated aquaculture system. *J. Phycol.* 39 (Suppl.): 30.

Lampert, W. and Sommer, U. 1997. Limnoecology : The Ecology of Lakes and Streams. Oxford University Press. pp. 48-49.

Lapointe, B. E. 1981. The effects of light and nitrogen on growth, pigment content, and biochemical composition of *Gracilaria foliifera* v. angustissima (Gigartinales, Rhodophyta). *J. Phycol.* 17: 90-95.

Lapointe, B. E. and Duke, C. S. 1984. Biochemical strategies for growth of *Gracilaria tikvahiae* (Rhodophyta) in relation to light intensity and nitrogen availability. *J. Phycol.* 20: 488-495.

Lapointe, B. E., Littler, M. M. and Littler, D. S. 1997. Macroalgal overgrowth of fringing coral reefs at Discovery Bay, Jamaica: Bottom-up versus top-down control. *Proc. 8th Coral Reef Symp.* 1: 927-932.

Lobban, C. S. and Harrison, P. J. 1994. Seaweed ecology and physiology. Cambridge University Press.

Mumford, T. F. and Miura, A. 1988. *Porphyra* as food: Cultivation and economics. In Lembi CA, Waaland JR (eds.), Algae and Human Affairs, Cambridge University Press, London, pp. 87-117.

Naldi, M. and Wheeler, P. A. 1999. Changes in nitrogen pools in *Ulva fenestrata* (Chlorophyta) and *Gracilaria pacifica* (Rhodophyta) under nitrate and ammonium enrichment. *J. Phycol.* 35: 70-77.

Nelson, S. G., Glen, E. P., Conn, J., Moore, D., Walsh, T. and Akutagawa, M., 2001. Cultivation of *Gracilaria parvispora* (Rhodophyta) shrimp-farm effluent ditches and floating cages in Hawaii: a two-phage polyculture system. *Aquaculture.* 193: 239-248.

Neori, A., Chopin, T., Troell, M., Buschmann, A. H., Kraemer, G. P., Halling, C., Shpigel, M. and Yarish, C. 2004. Integrated aquaculture: rationale, evolution and state of the art : Emphasizing seaweed biofiltration in modern mariculture. *Aquaculture.* In press.

Noda, H. 1993. Health benefits and nutritional properties of nori. *J. Appl. Phycol.* 5: 255-258.

Ott, F. D. 1965. Synthetic media and techniques for the xenic cultivation of marine algae and flagellate. *Virginia. J. Sci.* 16: 205-218.

Raffaelli, D. G., Raven, J. and Poole, L. 1998. Ecological impact of macroalgal blooms. Oceanogr. *Mar. Biol. Annu. Rev.* 36: 97-125

Rai, L. C. and Gaur, J. P. 2001. Algal adaptation to environmental stresses. Springer. New York.

Ryther, J. H., Corwin, N., DeBusk, T. A. and Williams, L. D. 1981. Nitrogen uptake and storage by the red alga *Gracilaria tikvahiae* (McLachlan, 1979). *Aquaculture* 26: 107-115.

Ryther, J. H. and Dunstan, W. M. 1971. Nitrogen, phosphorus and eutrophication in coastal marine environment. *Science*. 171 (3975): 1008-1013. Harrison P. J. and C. L. Hurd. 2001. Nutrient physiology of seaweed: Application of concept to aquaculture. *Cah. Biol. Mar.* 42: 71-82.

Smit, A. J., Robertson, B. L. and du Preez, D. R. 1997. Influence of ammonium-N pulse concentrations and frequency, tank condition and nitrogen starvation on growth rate and biochemical composition of *Gracilaria gracilis*. *J. Appl. Phycol.* 8: 473-481.

Smith, S. V., Kimmerer, W. J., Laws, E. A., Brock, R .E. and Walsh, T. W. 1981. Kaneohe Bay sewage diversion experiment: perspectives on ecosystem responses to nutritional perturbation. *Pac. Sci.* 35: 84-92.

Subandar, A., Petrell, R. J. and Harrison, P. J. 1993. *Laminaria* culture for reduction of dissolved inorganic nitrogen in salmon farm effluent. *J. Appl. Phycol.* 5: 455-463.

Thomas, T. E., Harrison, P. J. and Taylor, E. B. 1985. Nitrogen uptake and growth of the germlings and mature thallus of *Fucus distichus*. *Mar. Biol.* 84:267-274.

Troell, M., Halling, C., Neori, A., Chopin, T., Buschmann, A. H., Kautsky, N. and Yarish, C. 2003. Integrated mariculture: asking the right questions. *Aquaculture* 226: 69-90.

Tsujii, K., Ichikawa, T., Matusuura Y. and Kawamura, M. 1983. Hypercholesterolemic effect of taurocyamine or taurine on the cholesterol metabolism in white rats. *Sulfur Amino Acids* 6: 239-248.

Twomey, L. and Thompson, P. 2001. Nutrient limitation of phytoplankton in a seasonally open bar-built estuary: Wilson Inlet, Western Australia. *J. Phycol.* 37: 16-29.

Yan, X., Fujita, Y. and Aruga, Y. 2000. Induction and characterization of pigmentation mutants in *Porphyra yezoensis* (Bangiales, Rhodophyta). *J. Appl. Phycol.* 12: 69-81.

Yarish, C., Breeman, A. M. and van den Hoek, C. 1986. Survival strategies and temperature responses of seaweeds belonging to different biogeographic distribution groups. *Bot. Mar.* 29: 215-230.

Yarish, C., Wilkes, R., Chopin, T., Fei, X. G., Mathieson, A. C., Klein, A. S., Neefus, C. D., Mitman, G. G. and Levine, I. 1998. Domestication of indigenous *Porphyra* (nori) species for commercial cultivation in Noetheast America. *World Aquaculture* 29: 26-29, 55.

Yarish, C., Chopin, T., Wilkes, R., Mathieson, A. C., Fei, X. G. and Lu., S. 1999. Domestication of nori for Northeast America: The Asian experience. *Bull. Aquacul. Assoc. Can.* 1: 11-17.

2

Improvement of *Gracilaria* Agar Solubility by Acid Treatments

Hirotaka Kakita[1], Hiroshi Kamishima[1], Anong Chirapart[2] and Masao Ohno[3]
[1]Institute for Environmental Management Technology, AIST Shikoku, National Institute of Advanced Industrial Science and Technology, Hayashi, Takamatsu, Kagawa 761-0395, Japan.
[2]Department of Fishery Biology, Faculty of Fisheries, Kasetsart University, Chatuchak, Bangkok 10900, Thailand.
[3]Usa Marine Biological Institute, Kochi University, Tosa, Kochi 781-1164, Japan.

ABSTRACT

In order to increase the solubility of alkali-treated agar biopolymers extracted from *Gracilariopsis lemaneiformis*, sixteen hydrolysis methods with four kinds of acids (sulfuric acid, acetic acid, oxalic acid, and citric acid) were evaluated to determine a suitable procedure. The agar solubility against DMSO and water at 25°C and 95°C were tested. Among the procedures tested, partial hydrolysis with 0.001N acetic acid for 1 hour was more effective for increasing agar solubility while maintaining its gelling ability. The results indicated that partial hydrolysis of agar under effective conditions improves agar solubility and produces high grade agar from the Japanese agarophyte, especially *Gracilaria* (*Gracilariopsis*). The resulting high solubility agar may facilitate the spread of industrial applications of *Gracilaria agars*.

Keywords: Marine macroalgae, *Gracilaria*, *agar*, solubility; hydrolysis.

INTRODUCTION

Agar is the first phycocolloid to be used in the human food industry (Armisen and Galatas, 1987). Nowadays, the commercially valuable agarocolloids are principally extracted from red algae genera *Gelidium* and *Gracilaria*. To overcome the shortage of *Gelidium*, the utilization of *Gracilaria* as an agar resource began just before the beginning of this century (Armisen, 1995). There are basically three grades of agar; (1) food grade, (2) bacteriological (microbiological) grade, and (3) sugar-reactive grade (Armisen, 1995; Murano, 1995; Critchley, 1993). *Gracilaria* species represent the most important source of food and sugar-reactive agars. The bacteriological grade agar is used as the medium for growing bacteria and fungi in laboratories and is the most expensive and microbiologically pure. It is usually made from *Gelidium* and *Pterocladia* because *Gracilaria* agar has a gelling temperature above 40°C (Armisen, 1995). Thus, several species of *Gelidium* are still exploited because of the high quality of their agarocolloids.

Because the genus *Gracilaria* is more abundant in nature than *Gelidium* and several cultivations of *Gracilaria* have been successful (Critchley 1993; Martines et al. 1990; Santelices and Ugarte, 1987), the red algae *Gracilaria* seem to hold good potential as a source of agars. However, improved technology for processing *Gracilaria* agar is necessary to accelerate the use of *Gracilaria* as an agar source. Points for improvement to produce high grade agar are as follows: (1) improvement of gel formation (obtaining a high gelling ability), (2) obtaining a high gel strength, (3) obtaining high solubility, (4) reducing agar's viscosity. Numerous reports have been made on alkali treatment to develop suitable agars for industrial materials (Chirapart et al., 1995a; Hurtado-Ponce, 1992a, 1992b; Nishinari and Watase, 1983; Tagawa and Kojima, 1972; Watase & Nishinari, 1981 & 1986), but there have been very few reports of the effect of acid hydrolysis on the solubility of the *Gracilaria* agar.

In this study, we report on a procedure suitable for increasing the solubility of alkali-treated agar biopolymers extracted from *Gracilariopsis lemaneiformis* in Japan (Chirapart et al., 1995b).

MATERIALS AND METHODS

Samples of a red alga, *G. lemaneiformis*, were collected in May 1991 at Uranouchi Inlet in Tosa Bay (Kochi Prefecture) on Shikoku Island, southwestern Japan. The *Gracilaria* was washed in fresh water to remove epiphytes, sand and mud, then sun-dried, and kept in a freezer at –30°C until used.

The detailed preparation of dried samples and extraction of agar was performed as previously described (Chirapart et al., 1995a). Alkali-treated agar was obtained after treatment with 5% NaOH incubated at 80°C for 2 h.

The alkali-treated agar powders (1.5%w/v) were hydrolyzed in 0.1 N, 0.01 N, and 0.001 N solutions of sulfuric acid, acetic acid, oxalic acid, or citric acid, for 1, 2, and 3 h at 100°C (Chirapart et al., 1997). The hydrolysate was filtered at reduced pressure through a glass fiber filter (porosity 0.8 μm). Non-gelling hydrolysate was precipitated by two volumes of acetone. The insoluble precipitate was recovered by centrifugation, washed in acetone again for the complete removal of any remaining acid, then washed in diethyl ether and dried at reduced pressure in a desiccator. To evaluate the solubility of each dried acid-treated agar, 5 mg of agar powder was added to 10 ml of water and DMSO (dimethyl sulfoxide), and stirred at 25 or 95°C. For some soluble treated-agars, agar solution was injected into an HPLC column in order to estimate their molecular weight.

RESULTS

Partial Hydrolysis with Various Acids

The solubility of alkali-treated agar biopolymers with partial hydrolysis with various acids and at two different temperatures is summarized in Table 1. Partial acid hydrolysis of alkali-treated agar biopolymers had a drastic effect on solubility. Sulfuric acid hydrolysis was the strongest, increasing agar solubility rapidly (only 1 h), but concentrated acids (more than 0.01 N) caused treated agar to become insoluble. Among hydrolysis with a dilute acid solution (0.001 N), only sulfuric acid treatment was able to produce hot water-soluble agar biopolymers. The order of acid effects on agar solubility was as follows; sulfuric acid > oxalic acid > citric acid > acetic acid.

Table 1 Effects of partial hydrolysis on agar solubility

Acid	Conc. (N)	Time (hour)	Solubility			
			Water		DMSO	
			25°C	95°C	25°C	95°C
Alkali-treated agar	---	---	I	I	I	S
Sulfuric acid	0.001	1	I	S	I	S
		2	I	S	S	S
		3	I	S	S	S
	0.01	1	I	I	S	S
		2	I	I	I	I
		3	I	I	I	I
	0.1	1	I	I	I	I
		2	I	I	I	I
		3	I	I	I	I
Acetic acid	0.001	1	I	I	I	S
		2	I	I	S	S
		3	I	I	S	S
	0.01	1	I	S	I	S
		2	I	S	S	S
		3	I	S	S	S
	0.1	1	I	S	S	S
		2	I	S	S	S
		3	I	S	S	S
Oxalic acid	0.001	1	I	I	I	S
		2	I	I	S	S
		3	I	I	S	S
	0.01	1	I	I	I	I
		2	I	I	I	I
		3	I	I	I	I
	0.1	1	I	I	I	I
		2	I	I	I	I
		3	I	I	I	I

Contd.

Citric acid	0.001	1	I	I	I	S
		2	I	I	S	S
		3	I	I	S	S
	0.01	1	I	S	S	S
		2	I	S	S	S
		3	I	S	S	S
	0.1	1	S	S	S	S
		2	S	S	S	S
		3	S	S	S	S

I: insoluble or partially soluble. S: soluble.

Molecular Weight

Molecular weights of alkali-treated agar biopolymer without acid treatment and after acid treatment are listed in Table 2. Sulfuric acid hydrolysis produced strong effects on the molecular weight of treated-agars. The order of acid effects on the degradation of agar polymers was as follows; sulfuric acid > oxalic acid > citric acid > acetic acid. This order was similar to the order of effect on solubility.

Table 2 Effects of partial hydrolysis on agar molecular weight

Treatment	Molecular weight	Sulphate content (%)	Agar property
Control (alkali-treated agar)	433,000	1.2	gelling polymer
0.001N sulfuric acid (100°C, 3hours)	3,590	0.4	non-gelling
0.001N oxalic acid (100°C, 3hours)	7,620	0.7	gelling polymer
0.001N citric acid (100°C, 3hours)	28,000	0.8	non-gelling
0.001N acetic acid (100°C, 3hours)	33,700	0.8	gelling polymer
0.01N acetic acid (100 °C 1hour)	7,790	0.7	gelling polymer
0.01N acetic acid (100°C, 2hours)	5,230	0.7	gelling polymer
0.01N acetic acid (100°C, 3hours)	3,080	0.6	non-gelling

From these results, we inferred that boiling in 0.001N acetic acid for 1 h prevented excess hydrolysis resulting in unnecessarily degraded agar. However the gelling ability and molecular weight were decreased with increase of the hydrolysis period and acid concentration (Table 2). At the same concentration of 0.001 N oxalic acid and citric acid hydrolysis had a greater effect on the decrease of agar molecular weight than did acetic acid hydrolysis.

DISCUSSION

Different kinds of acid used for hydrolysis gave different solubility, molecular weight, and gelling agar properties. From these results, we inferred that hydrolyzing by boiling in 0.001N acetic acid for 1 h produced an agar with gelling ability and high solubility. High grade, high solubility agar may be achieved by the simple partial acid hydrolysis method. While the detailed mechanism of acid hydrolysis of agar polymers has not yet been confirmed, these results give sufficient information on agar properties to justify further investigation on new applications of the agar of this Japanese agarophyte.

ACKNOWLEDGEMENTS

We are grateful to Dr Hirotoshi Yamamoto of Hokkaido University for identifying *G. lemaneformis*.

REFERENCES

Armisen, R and Galatas, F. 1987. Production, properties and uses of agar, in: *Production and utilization of Products from commercial seaweeds*, edited by D.J. McHugh, (FAO Fish. Tech. Pap. (288) pp. 189.

Armisen, R. 1995. World-wide use and importance of *Gracilaria*, *J. Appl. Phycol.* 7: 231-243.

Chirapart, A., Ohno, M., Sawamura, M. and Kusunose, H. 1996. Phenology and Morphology on a new member of Japanese *Gracilaria*, in Tosa Bay, Southern Japan, *Fish. Sci.* 61: 450-454.

Chirapart A., Ohno, M., Uketa, H., Sawamura, M. and Kusunose, H. 1997. Effects of partial acid hydrolysis on physical and chemical properties of agar from a newly reported Japanese agarophyte (*Gracilariopsis lemaneiformis*), *J. Appl. Phycol.* 9: 73-76.

Chirapart, A., Katou, Y., Ukeda, H., Sawamura, M. and Kusunose, H. 1995. Physical and chemical properties of agar from a new member of *Gracilaria*, *G. lemaneiformis* (Gracilariales, Rhodophyta) in Japan. *Fish. Sci.* 61: 450-454.

Critchley, A. T. 1993. *Gracilaria* (Gracilariales Rhodophyta): An economically important agarophyte, in: *Seaweed Cultivation and Marine Ranching*, edited by M. Ohno, and A.T. Critchley, (Japan International Cooperation Agency, Yokosuka, Japan) pp. 89-112.

Hurtado-Ponce, A.Q. 1992. Rheological properties of agar from *Gracilariopsis heteroclada* (Zhang et Xia) Zhang et Xia (Gracilariales, Rhodophyta) treated with powdered commercial lime and aqueuos alkaline solution, *Bot. Mar.* 35: 365-369.

Hurtado-Ponce, A.Q. 1992a. Influence of extraction time on the rheological properties of agar from some *Gracilaria* species from the Philippines. *Bot. Mar.* 35: 441-445.

Martines, C., Barrales, H. and Molina, M. 1990. Economic assessment of a successful method for mass field cultivation of *Gracilaria* in Chile, *Aquaculture*. 84: 101-116.

Murano, E. 1995. Chemical structure and quality of agars from *Gracilaria*, *J. Appl. Phycol.* 7: 245-254.

Nishinari, K., and Watase, M. 1983. Effect of alkali treatment on the rheological properties of concentrated agar-agar gels, *Carbohyd. Polymers*. 3: 39-52.

Santelices, B. and Ugarte, R. 1987. Production of Chilean *Gracilaria:* problems and perspectives, in: *Proc. Int. Seaweed Symp.* pp. 295-299.

Tagawa, S. and Y. Kojima, 1972. The alkali-treatment of the mucilage of *Gracilaria verrucosa*. in: *Proc.* 7th *Int. Seaweed Symp.* pp. 447-450.

Watase, M., and Nishinari, K. 1981. Effect of sodium hydroxide pretreatment on the relaxation spectrum of concentrated agar-agar gels, *Rheol. Acta*. 20: 155-162

Watase, M., and Nishinari, K. 1986. The effect of alkali pretreatment on the thermal and rheological properties of polysaccharide gels prepared from red seaweeds, in: *Gums and Stabilisers for the Food Industry* (vol. 3), edited by G. O. Phillips, D. J. Wedlock, and P. A. Williams, (Elsevier Applied Science Publishers, London) pp. 535-544.

3

Seaweed Community Analysis of a Rocky Shore for the Sustainable Seaweed Integrated Aquaculture System (SSIAS) in Korea

Ik Kyo Chung*[1], Yun Hee Kang[1], Si Jung Ryu[1], Myung Sook Kim[2], Charles Yarish[3] and Jin Ae Lee[4]

[1]Department of Marine Science Pusan National University, Busan 609-735, KOREA
[2]Research Institute for Basic Sciences, Pusan National University, Busan 609-735, KOREA
[3]Department of Ecology and Evolutionary Biology, University of Connecticut, Stamford, CT 06901, USA
[4]School of Environmental Science and Engineering, Inje University, Gimhae 621-749, KOREA

ABSTRACT

The ecological traits of the intertidal rocky shore seaweed communities were investigated for the site-specific development of the Sustainable Seaweed Integrated Aquaculture System (SSIAS) in Korea. Monthly phenological surveys of the intertidal communities were conducted in 2002 to obtain information on the wild seaweed species for their application to the SSIAS. The changes of the dominant species composition and their fresh weight and coverage were prominent among the seasons and sampling sites. Generally, brown algae, *Laminaria japonica* and *Sargassum* sp., were dominant in the winter; red algae were abundant in the spring and the autumn. Green alga, *Ulva pertusa*, showed the broadest coverage year around. The C/N ratios of seaweeds and the ambient nitrogen concentrations were positively correlated.

Keywords: C/N ratio, seaweed, sustainable seaweed integrated aquaculture system.

INTRODUCTION

Fed mariculture including intensive finfish aquaculture has caused many environmental problems. Modern intensive monoculture requires high levels of inputs such as seawater, feeds, fertilizers, chemicals, and inevitably produces considerable wastes. Therefore, many aquaculture operations put enormous pressure on coastal habitats. To reduce the nutrient burden of the fish farm effluents, the integration of seaweed cultivation with fish aquaculture has been proposed (Neori et al., 2004). The integrated aquaculture system is not a new concept; however, there have been significantly fewer examples of aquatic animals being polycultured with seaweeds than of aquatic animals polycultured with other aquatic animals (Stickney, 2000).

Studies on the methods of treating effluents from enclosed mariculture systems with macroalgae date back to the work of Ryther and colleagues in the mid-1970s, and this approach has gained new interest in the 1990s (Troell et al., 1999). Various strategies for integrating seaweed cultivation with fish culture have been successful (Brzeski and Newkirk, 1997).

Korea has some experience with polyculture. The National Fisheries Research and Development Institute (NFRDI) of Korea has already established guidelines for polyculture (Anonymous, 1994). Examples are *Porphyra tenera* and short necked clams or surf clams; *Hizikia fusiformis* (now considered a species of *Sargassum*) or *Laminaria japonica* and sea squirt; *Undaria pinnatifida* or *L. japonica* and abalone. The NFRDI's previous project did not provide the ecological basis for the present polyculture concept. They were simple combinations of several species with similar culture techniques and facilities or sharing the same space. In any case, based on these results, the legal basis of polyculture licensing was established. Soon after the finfish and shrimp aquacultures increased unexpectedly in the 1990s, a changed concept and strategy of an integrated culture to manage the effluent from these organisms was required. The study of environmentally sound integrated systems has been in progress under the Sustainable Seaweed Integrated Aquaculture System (SSIAS) project to reduce the environmental impact of marine farmed finfish effluent in the coastal zone (Chung et al., 2002). Under the auspices of the SSIAS, we investigated the rocky shore seaweed community to evaluate wild seaweed species as a nutrient sink and, accordingly, as candidates for the SSIAS.

MATERIALS AND METHODS

An ecological survey on the rocky shore seaweed community was conducted in the Tongyong area, known as 'the Home of Korean Aquaculture', located in the South Sea of Korea (Chung et al., 2002). Three sites were chosen based on the different environmental characteristics. Station (St.) 1 was close to the dock area near the waste treatment plant of Tongyong City, St. 2 was near the public beach area in between St. 1 and St. 3, and St. 3 was facing to the open channel of the South Sea (Fig. 1).

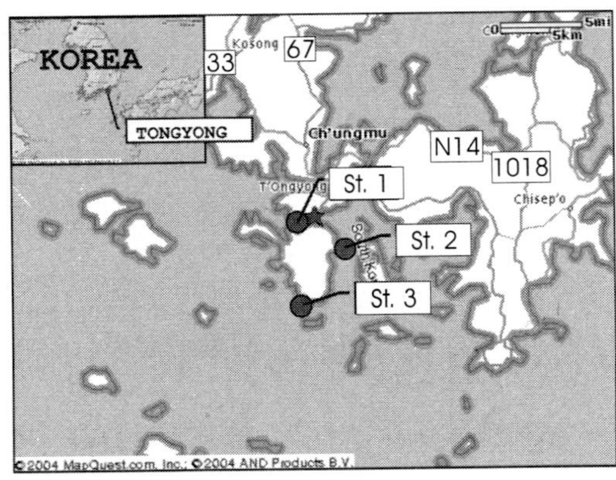

Fig. 1 Map of the study area (Tongyong area) showing the three study sites.

The basic ecological aspects of seaweed community structure were surveyed on monthly basis using the quadrat method (50 cm × 50 cm). The seawater nutrient contents were analyzed by standard methods (Parsons et al., 1984). The tissue elemental contents were analyzed by a CHNS/O elemental analyzer (Perkin Elmer 2400 Ser. II). Seawater and air temperature were referenced by the NFRDI's statistics for the fixed site observation data (Anonymous, 2002).

RESULTS AND DISCUSSION

As the aim of this study was to establish the background of the application of wild seaweed species for the SSIAS, for one year, in 2002, the seaweed community was studied using a different methodology from that of the conventional rocky shore ecological studies. Only several factors were considered.

The changes of seawater and air temperatures are shown in Fig. 2. Unusually low temperatures were recorded in May, which may be speculated to be the result of unusual atmospheric cooling. This cooling did not significantly affect the nutrient concentrations. The concentrations of dissolved inorganic nitrogen (NO_3^- and NH_4^+) and phosphate (PO_4^{3-}) varied among the sampling sites and seasons and did not show any significant seasonal trends. However, spatial variation was prominent; the concentrations of inorganic nutrients were highest at St. 1, second highest at St. 2, and relatively low at St. 3 (Fig. 3). The changes in the fresh weight of the three seaweed groups and dominant intertidal seaweed species at the three rocky shore sites during the study period are shown in Figs. 4 to 9. Generally, the total biomass was highest in late winter and early spring, when brown algae, such as *Undaria pinnatifida* and *Sargassum* sp., were dominant at all three sites. Generally, there were decreasing trends of biomass during late spring and summer.

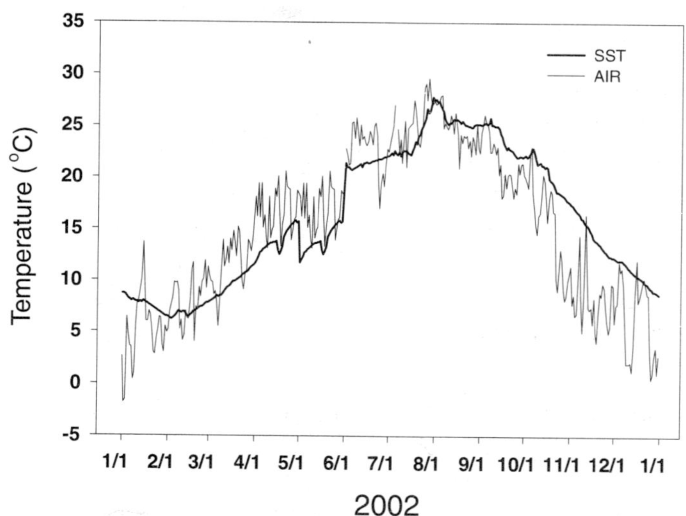

Fig. 2 Variation of seawater and air temperature at the fixed observation site at Tongyong (Choongmoo) Station in 2002.

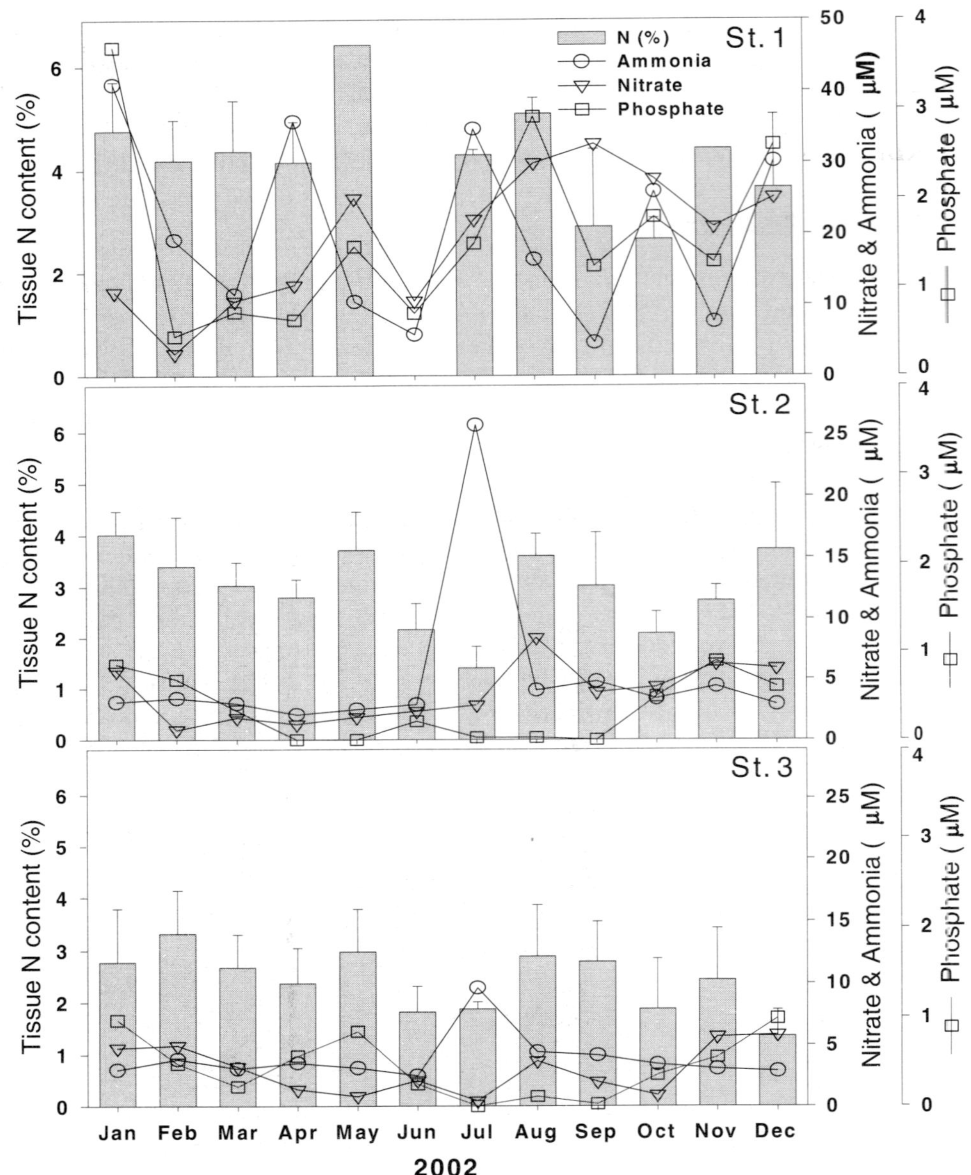

Fig. 3 The changes of concentrations of the dissolved inorganic nutrients and tissue nitrogen contents (%) in Tongyong area.

The patterns of dominant species composition, in coverage, are drawn in Figs. 10-12. The green alga *Ulva pertusa* showed the highest coverage throughout the whole year at all sites. Four dominant species were recorded at St. 1 and St. 2, but ten species appeared with high coverages at St. 3. At St. 1, other than *U. pertusa*, *Pachymeniopsis*, *Lomentaria catenata* and *Undaria pinnatifida* appeared in the winter and spring seasons only (Fig. 10). At St. 2, P. *elliptica*,

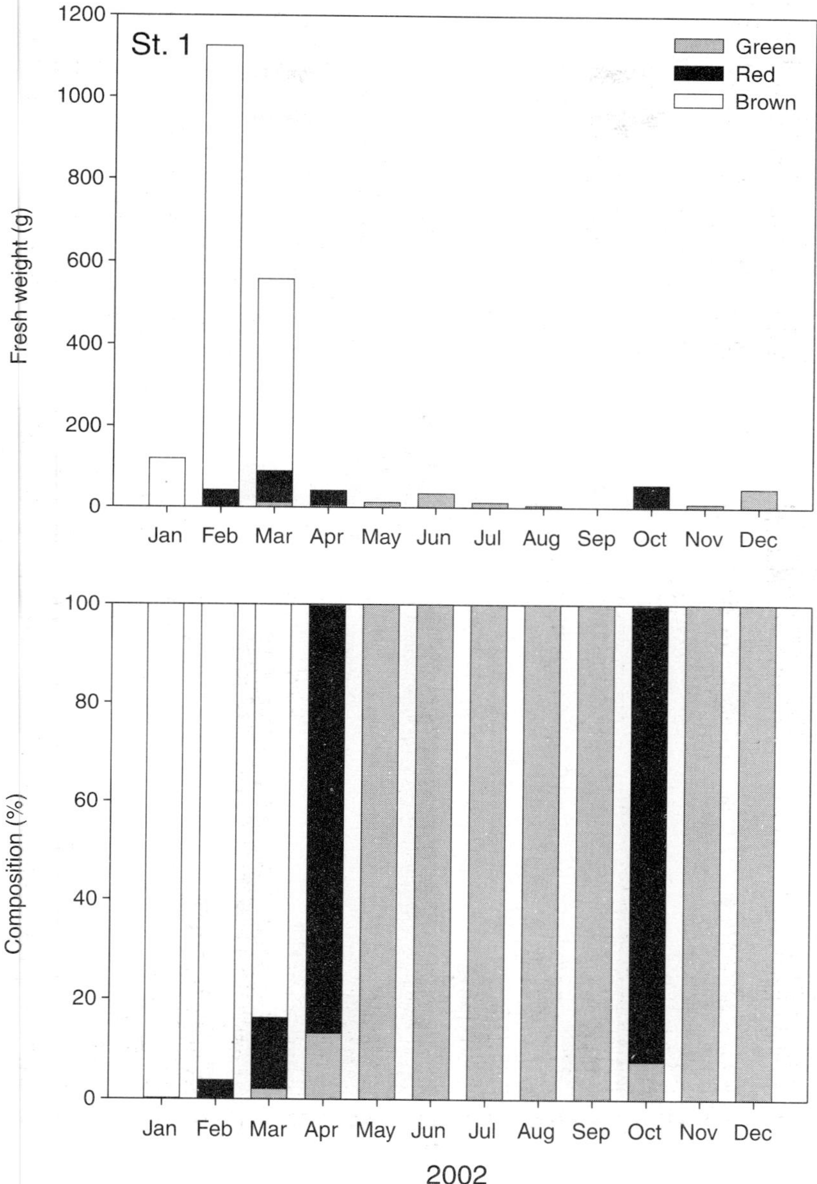

Fig. 4 The change of fresh weight and algal composition at St. 1 in Tongyong area.

Sargassum sp. and *U. pinnatifida* appeared, except in the summer (Fig. 11). There were more diverse species present at St. 3, and they showed species specific coverage patterns throughout the year, which were different from those at other sites (Fig. 12). An unusually high value was recorded at St. 2 in June due to the contribution of *Ulva* sp. and *Codium* sp.

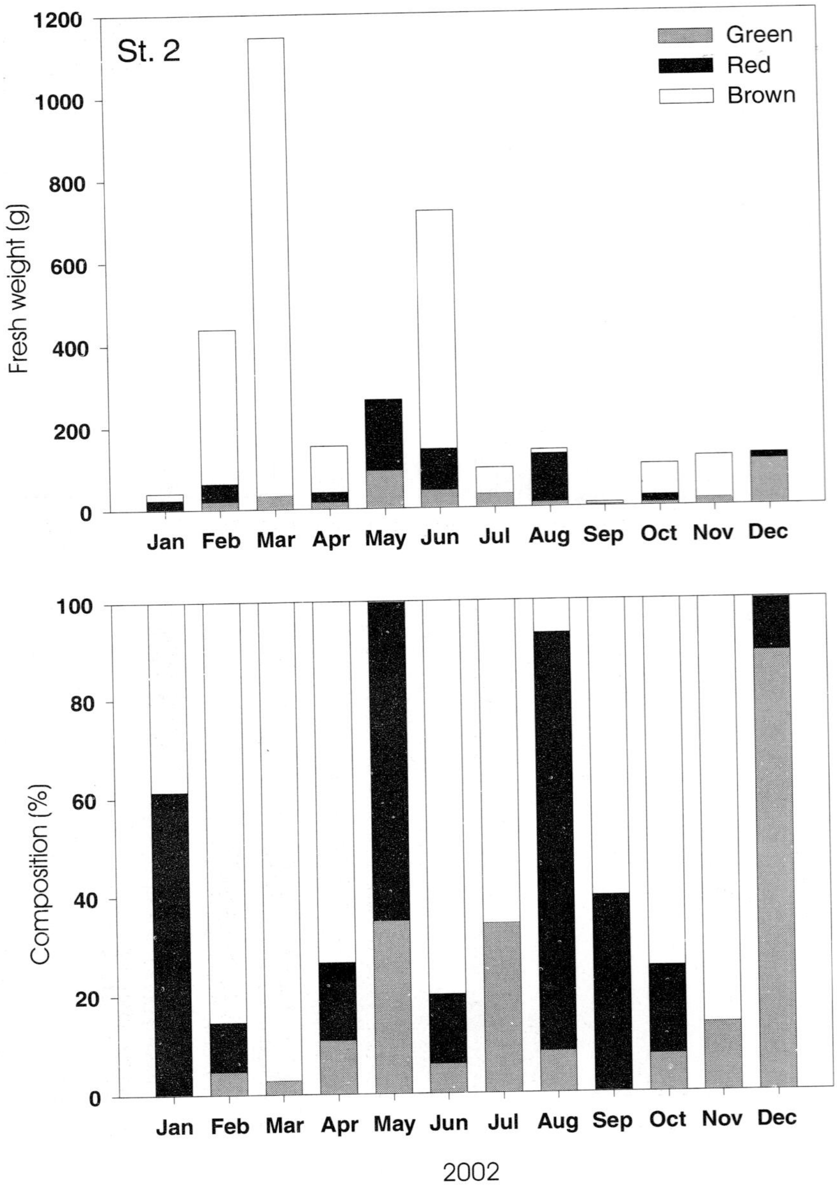

Fig. 5 The change of fresh weight and algal composition at St. 2 in Tongyong area.

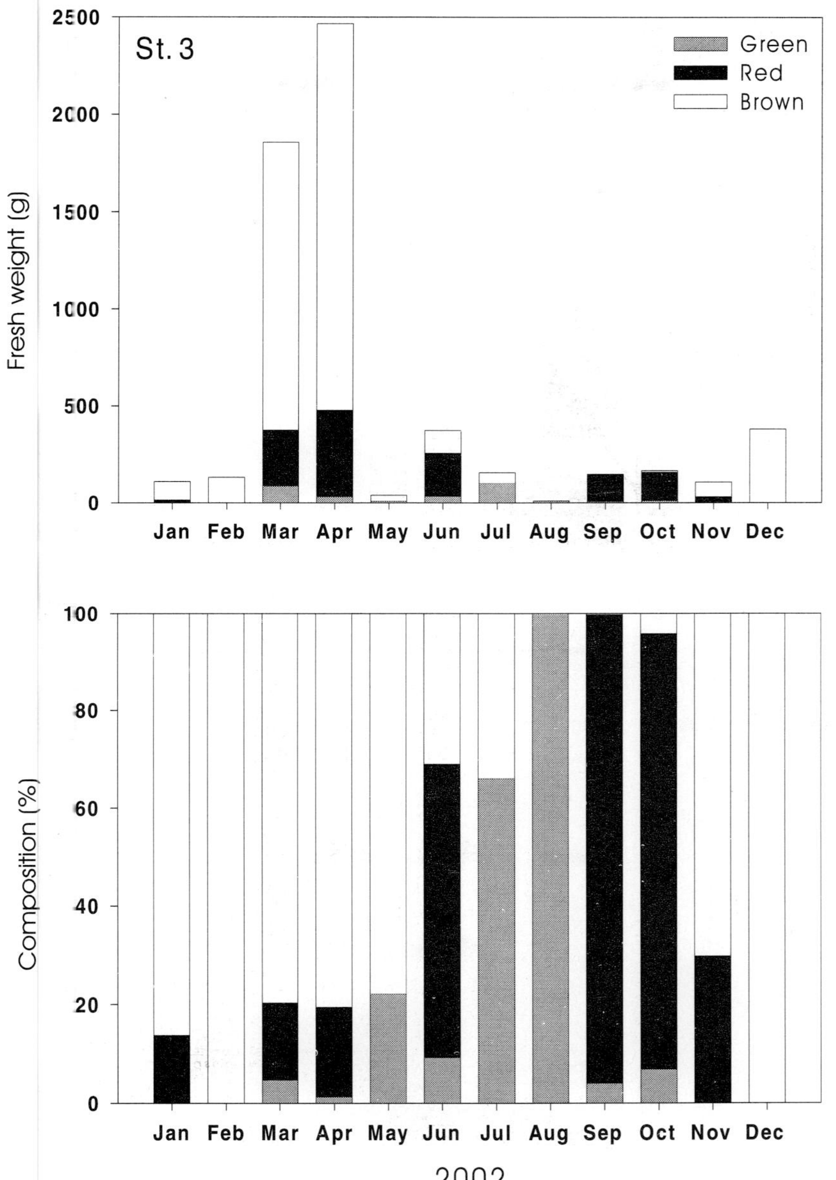

Fig. 6 The change of fresh weight and algal composition at St. 3 in Tongyong area.

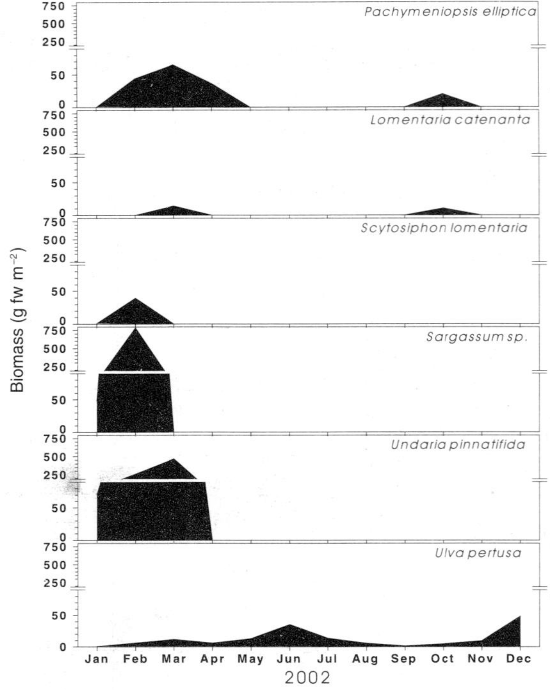

Fig. 7 The change of fresh weight of dominant intertidal seaweed species at St. 1 in Tongyong area.

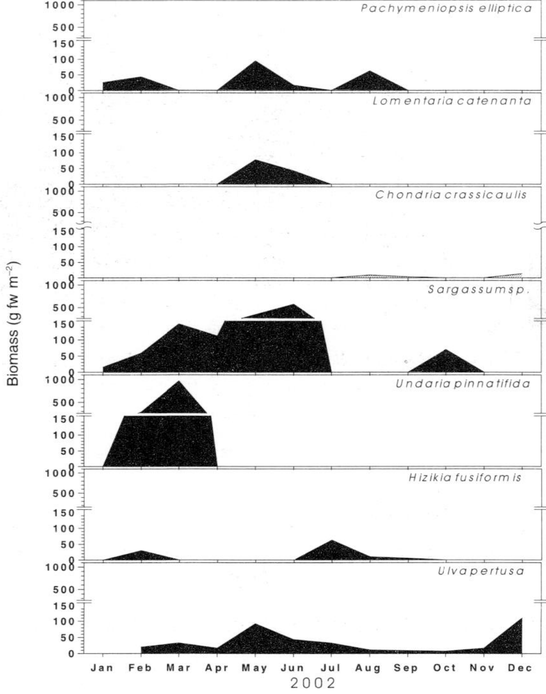

Fig. 8 The change of fresh weight of dominant intertidal seaweed species at. St. 2 in Tongyong area.

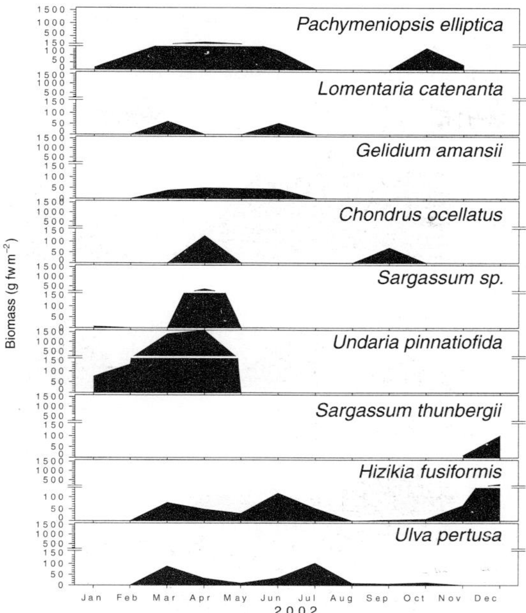

Fig. 9 The change of fresh weight of dominant intertidal seaweed species at. St. 3 in Tongyong area.

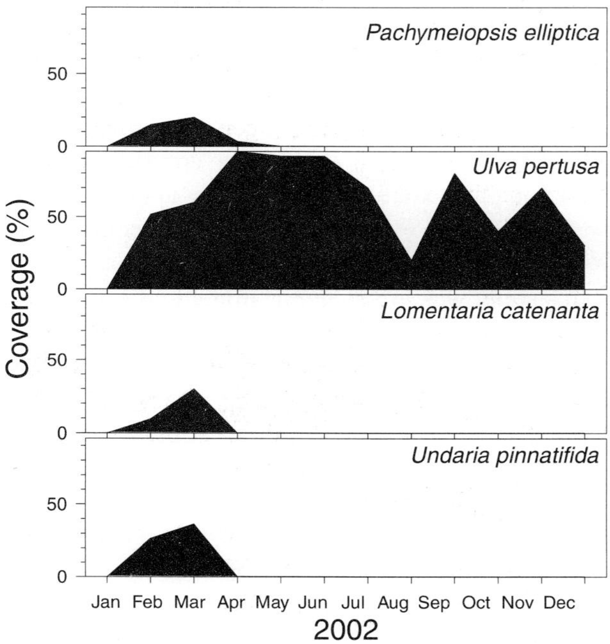

Fig. 10 The change of coverage of intertidal seaweed species at St. 1 in Tongyong area.

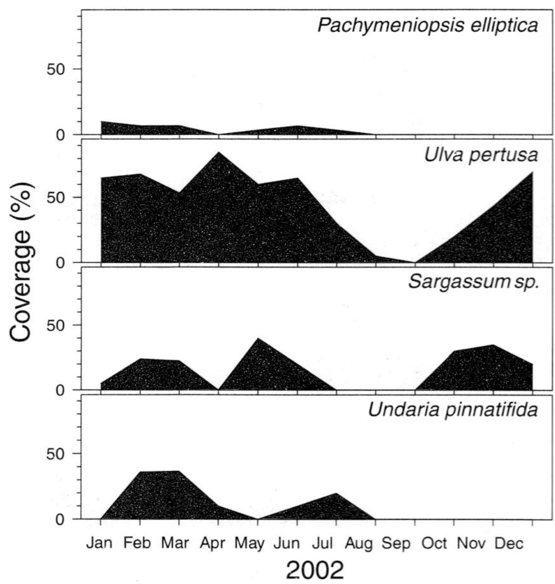

Fig. 11 The change of coverage of intertidal seaweed species at St. 2 in Tongyong area.

The tissue nitrogen contents of all seaweed species specimens of St. 1, where the amount of ambient inorganic nutrients was highest, were higher, with an average of 4.26% than those at St. 2 and St. 3, which averaged 2.97% and 2.43%, respectively (Fig. 3). The nitrogen contents of green, brown and red algae averaged 3.39%, 2.73% and 3.16%, respectively. The averaged values of tissue nitrogen contents from several species, which showed a coverage greater than 10%, revealed that red algae had higher average tissue nitrogen than brown algae (though the difference was not significant). Although the internal tissue nitrogen contents are thought to be the complex results of the interactions among the external environments, uptake kinetics, assimilation, growth and reproductive processes (Hanisak, 1983), the tissue nitrogen contents increased with increased concentration of dissolved nitrogen in seawater (Fig. 13).

The ultimate goal of this study was to develop a strategy to achieve the sustainability of the coastal ecosystem based on sound ecological and economic principles (Newkirk, 1996). There are many conflicts among the users of coastal waters, and aquaculture should be conducted under an integrated coastal zone management which maintains a balance between 'extractive' and 'fed' aquaculture (McVey et al., 2002; Rawson et al. 2002). Although the present local cultivation seaweed species are very promising (Noori et al., 2004; Chung et al., 2002), the results of this field study could be used as a basis for the management of selected seaweed species, because field phenology and nitrogen contents information on the seaweeds is required to confirm and preserve these seaweeds as probable local candidates for other systems in other locations.

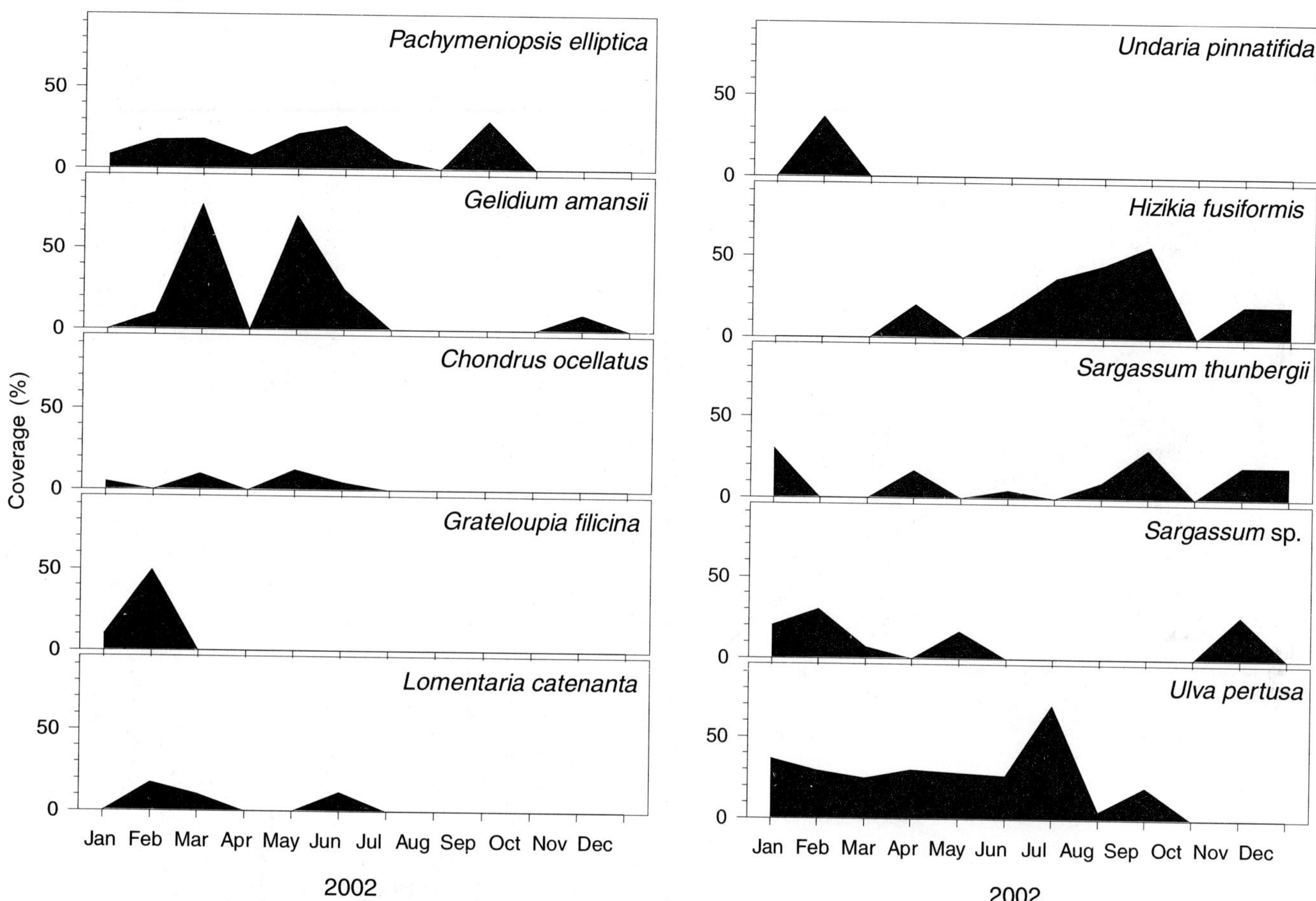

Fig. 12 The change of coverage of intertidal seaweed species at St. 3 in Tongyong area.

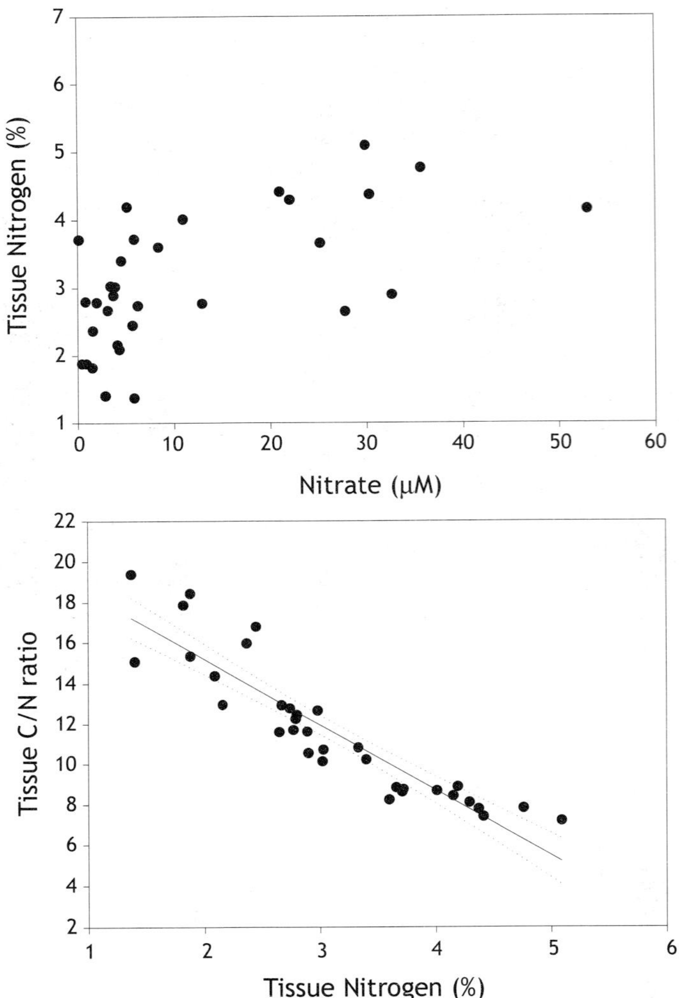

Fig. 13 A plot of total dissolved inorganic nitrogen vs. tissue nitrogen contents of seaweeds in Tongyong area.

REFERENCES

Anonymous, 1994. Studies on the development of polyculture system in thee coastal waters of Korea. Final report (National Fisheries Research and Development Institute) (in Korean with English summary).

Anonymous, 2002. National Fisheries Research and Development Institute. (http://www.nfrda.re.kr)

Brzeski, V. and Newkirk, G. 1997. Integrated coastal food production systems: A review of current literature. *Ocean Coastal Management.* 34: 55-71.

Chung, I. K., Kang, Y. H., Yarish, C., Kraemer, G. P. and Lee, J. A. 2002. Application of seaweed cultivation to the bioremediation of nutrient-rich effluent. *Algae.* 17: 187-194.

Hanisak, M. D. 1983. The nitrogen relationships of marine macroalgae, in: *Nitrogen in the marine environment*, edited by E. J. Carpenter and D. G. Capone (Academic Press, London), pp. 699-730.

McVey, J. P., Stickney, R. R., Yarish, C. and Chopin, T. 2002. Aquatic polyculture and balanced ecosystem management: New paradigms for seafood production, in: *Responsible marine aquaculture*, edited by R. R. Stickney and J. P. McVey, (CABI Publishing, New York), pp. 91-104.

Neori, A., Chopin, T., Troell, M., Buschmann, A. H., Kraemer, G. P., Halling, C. M., Shpigel, M. and Yarish, C. 2004. Integrated aquaculture: Rationale, evolution and state-of-the-art emphasizing seaweed biofiltration in modern mariculture. *Aquaculture* 231: 361-391.

Newkirk, G. 1996. Sustainable coastal production systems: A model for integrating aquaculture and fisheries under community management. *Ocean Coastal Management* 32: 69-83.

Parsons, T. R., Maita, Y. and Lalli, C. M. 1984. A manual of chemical and biological methods for seawater analysis, (Pergamon, Oxford).

Rawson, Jr. M. V., Chen, C., Ji, R., Zhu, M., Wang, D., Wang, L., Yarish, C., Sullivan, J. B., Chopin, T. and Carmona, R. 2002. Understanding the interaction of extractive and fed aquaculture using ecosystem modeling, in: *Responsible marine aquaculture*, edited by R. R. Stickney and J. P. McVey, (CABI Publishing, New York), pp. 263-296.

Stickney, R.R. 2000. Polyculture in: *Encyclopedia of aquaculture*. edited by R. R. Stickney (John Wiley & Sons, Inc., New York), pp. 658-660.

Troell, M., Rönnbäck, P., Halling, P., Kautshy, C. and Buschmann, A.H. 1999. Ecological engineering in aquaculture: Use of seaweeds for removing nutrients from intense mariculture. *J. Appl. Phycol.* 11: 89-97.

4

Succession of Seaweed Communities on Artificial Reefs and Marine Ranching in Japan

Masao Ohno[1], Chang Geun Choi[2] and Dinabandhu Sahoo[3]

[1]Usa Marine Biological Institute of Kochi University, Usa-cho, Tosa, Kochi, 781-1164 Japan

[2]Department of Aquaculture, Pukyong National University, Pusan, 608-737, Korea

[3]Marine Biotechnology Laboratory, Department of Botany, University of Delhi, Delhi-110007, India

ABSTRACT

Artificial reefs for the development of valuable fishery resources were placed on the sandy bottom at 8, 10 and 13 m depths on the Ikata of the southern Japan. They were observed monthly or bimonthly from February 1999 to June 2003. *Enteromorpha intestinalis* and *Colpomenia sinuosa* dominated the reef in spring, and several species of Melobesioideae dominated during summer and autumn. *E. intestinalis* and *C. sinuosa* were found to be typical primary colonists. The number of seaweed species at all the sites gradually increased in winter. In total, 38 seaweed species were found from all the reefs, over the study period. Each reef was dominated by *Sargassum* sp., *Ecklonia kurome* and *Padina arborescens* within 18 months of the placement. The seasonal changes of appearances of top shells have shown on the concrete reefs. Maximum 10-15 individuals of top shell have been observed on one artificial reef during the periods, when much seaweed fronds have grown from May to June. About 20 species of fishes appeared around and inside these artificial reefs through the survey period. Young and smaller fish, *Apogon natatus* (spotted cardinal fish) and *Charomisfumea* always appeared as dominate species around and inside of artificial reefs through the year.

Keywords: Artificial reef, marine ranching, gathering of animals, seaweed community, succession.

INTRODUCTION

Seaweed acts as a habitat, shelter, nursery and spawning bed for some fishes, abalone and spiny lobster, which are valuable fishery resources. The construction of artificial reefs has become an important and popular technique in the coastal areas for enhancement of fishery resources. The experimental construction of artificial seaweed beds has also been tried on many coasts in Japan (Ohno et al., 1990; Ohno 1993; Serisawa and Ohno, 1995 a, b, Sahoo and Ohno, 2000; Choi, et al., 2002) The artificial reef for nursery, are set on the coast of shallow waters. It is more effective to construct seaweeds bed on the artificial reefs for gathering of shell fish and fishes, because there are plenty of small animals which are available as feed for fishes in the seaweeds bed. We introduce seaweed bed and marine ranching with the steel and concrete artificial reefs by different substrata on the Ikata of southern Japan.

MATERIAL AND METHODS

The study was conducted at depths of 8, 10 and 13 m on the coast of Ikata, Shikoku Island, Japan. A large number of perennial seaweeds such as *Sargassum horneri*, *S. macrocarpum* and *Ecklonia kurome* grow on natural substrata around the study site. Artificial reefs were placed on the sandy bottom on February 16, 1999.

Two artificial concrete reefs of different sizes were placed on the rocky and sandy bottom of depths of 8 m (station A), 10 m (station B) and 13 m (station C) at the study site (Fig. 1, a, b. c). The "M type" artificial reef for gathering shellfishes were placed at stations A and B. This artificial reef was of the volume of 3.47 m^3 (2.5 × 1.5 × 1.25 m, Fig. 1d). The "RF type" reefs for gathering of lobster were placed at stations A and C. The artificial reefs for lobster and shell are created with many spaces on the side and inside by concrete. This artificial reef was of the volume of 4.8 m^3 (2 × 2 × 1.2 m, Fig. 1.e). The artificial reef with steel is set with three kinds of structures. The experimental steel structures were set with different blocks made of concrete, steel, wood on the roof of the structure. Another steel structures were the cages with natural stones on the roof of the structure. One of steel structures was the roof of steel plates.

The succession and growth of various marine algae on the reefs were observed by scuba diving monthly or bimonthly from March 1999 to June 2004. Photographs were taken by digital video camera, video camera and 35 mm camera. Photographs of seaweeds growing on the natural substrata around the artificial reefs were also taken.

RESULTS

Seaweed Succession

A total of 38 species of seaweed (6 Chlorophyta, 18 Phaeophyta and 14 Rhodophyta) were identified during the study period. Figure 2(a-d) shows seaweed beds and gathering fishes on the artificial reefs. Figure 3 shows changes in the seaweed succession of the dominant algal species on all of the reefs at study site during the study period. Diatom colonization on all reefs achieved 100% coverage within one month of placement. Within three months after placement, *Enteromorpha intestinalis* and *Colpomenia sinuosa* appeared luxuriously on the reefs, although they did not grow on the surrounding rocks in the study area. Thalli of *E. intestinalis* and *C. sinuosa* decayed over the summer season, at which time coralline red algal species started to grow on the reefs. From August to October 1999, coralline algae covered approximately 100% of the reefs at the study site. Diatoms and coralline algae had greatly increased their colonization of the reefs from October to December 1999.

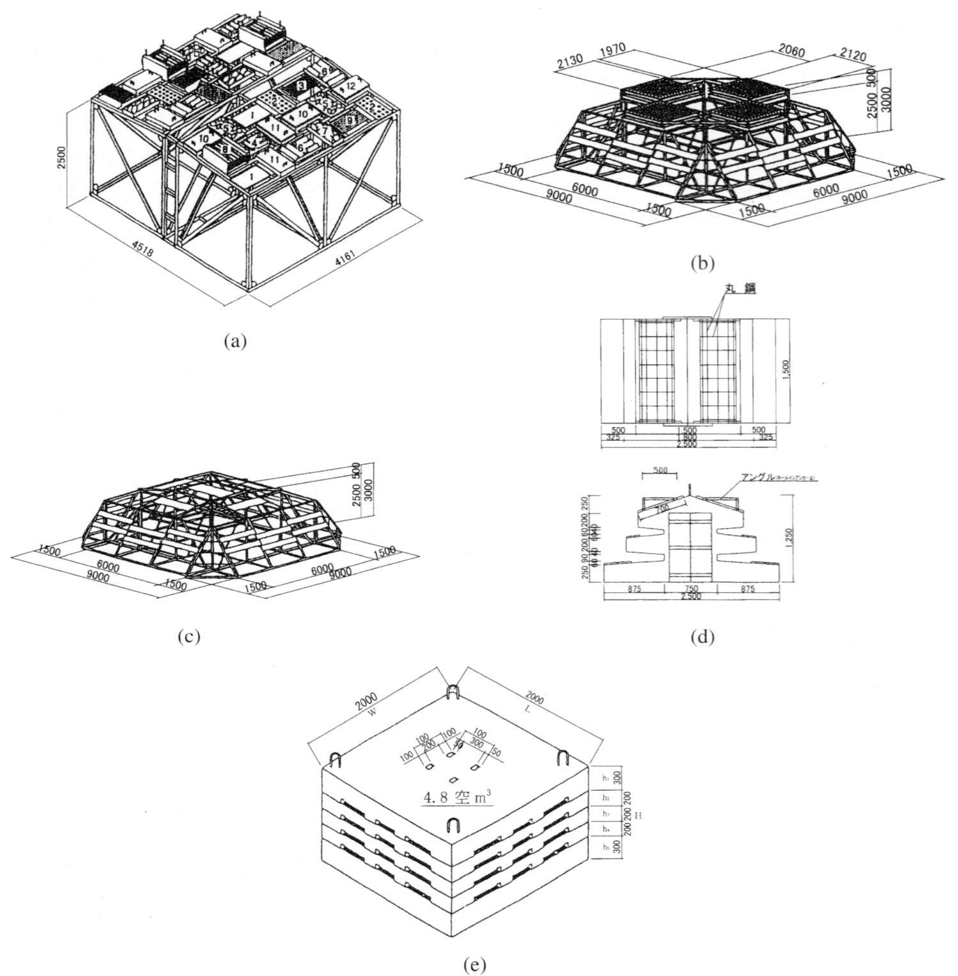

Fig. 1 Artificial reefs set on the Ikata coast. a, b, c: Steel Artificial reefs, d: M type artificial reefs for aba one, e: RF type artificial reef for lobster.

Brown algal coverages (such as *Sargassum* sp. and *Ecklonia kurome*) were recorded as over 20% of the reef during the winter season. The number of species tended to be comparatively greatest on the reefs during the winter and spring season (February to June 2000). In February 2000, *C. sinuosa* reappeared approximately covered 20-60% of the artificial reefs until June 2000. *Sargassum* sp. and *Jania adhaerens* from June 2000 to June 2002. *E. kurome* and *P. arborescens* dominated in late succession (climax) stages within 18 months of placement.

In the early stages of colonization, i.e. within 5 months of the placement of the reefs, green algae attached at station A tended to be comparatively more abundant than at stations B and C (both on the M type and RF type reefs). Furthermore, brown algae on station A appeared faster, and became more abundant than the reefs placed at other sites. However, at station A, the coverage ratio of red algae did not increase as fast and to such an extent as at stations B and C.

Fig. 2 a: three types of steel artificial reeds, b: seaweed bed on the steel artificial reef. c: seaweed bed on the concrete reef, d: gathering fishes around artificial reef.

Seaweed communities of all the reefs attained the climax condition after three years and became a similar with comparing of natural seaweed beds border on the site the artificial reef.

Figure 4 shows the number of species of seaweeds attached to the artificial reefs and natural seaweed bed border on the site of artificial reefs in 2002. The number of species in the artificial reefs were 7-10 species, and became almost the same number of natural seaweed beds.

The biomass of seaweed beds on the artificial reefs is shown in Fig. 5. The seaweed biomass on the reefs had a biomass of 1,800 g wet wt/m^2 at station A in June 2000. Stations B and C, had a biomass of 630g wet wt/m^2 and 650 g wet wt/m^2 in June, 2002, respectively. The biomass of seaweed beds on the reefs was found to decay and decrease significantly at all study areas, during the summer season (July to August 2001). The biomass of seaweeds on all artificial reefs at Station A was higher than that of the reefs at Stations B and C.

Shellfish

There are abalone and top shell on the coast in this study site. Many top shells were found on the rocks of study site, but a few abalones have also been found on the rocks there. The seasonal changes of appearances of top shells on two kinds is concrete reefs are showed in Table 1. The top shells found the abundant appearance of about 10-15 individuals per one artificial reef in the spring season, while top shell has been observed during the periods of much *Enteromorpha* fronds growing from May to June. In this season, there is not much *Enteromorpha* fronds on the natural rocks around artificial reefs. Top shell has shown to gather at the narrow space or at the

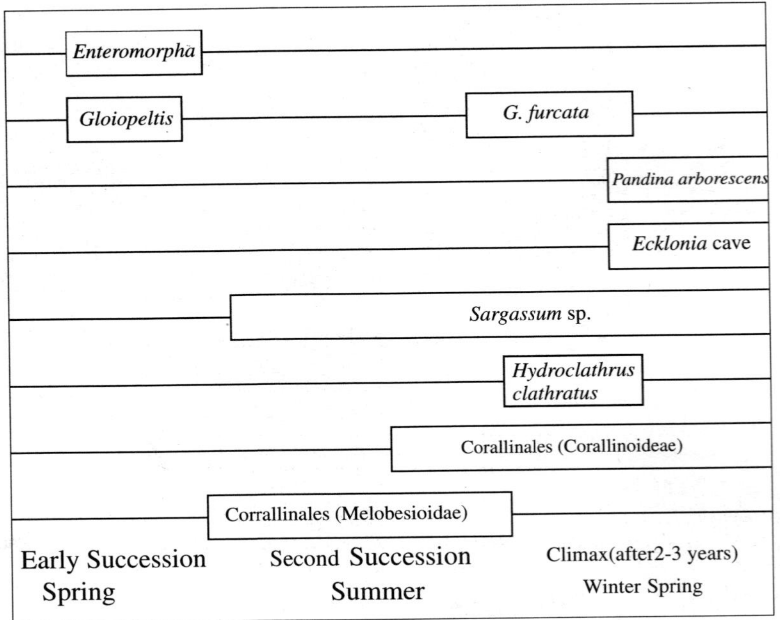

Fig. 3 Seaweed succession on the artificial reefs set on the coast in Ikata.

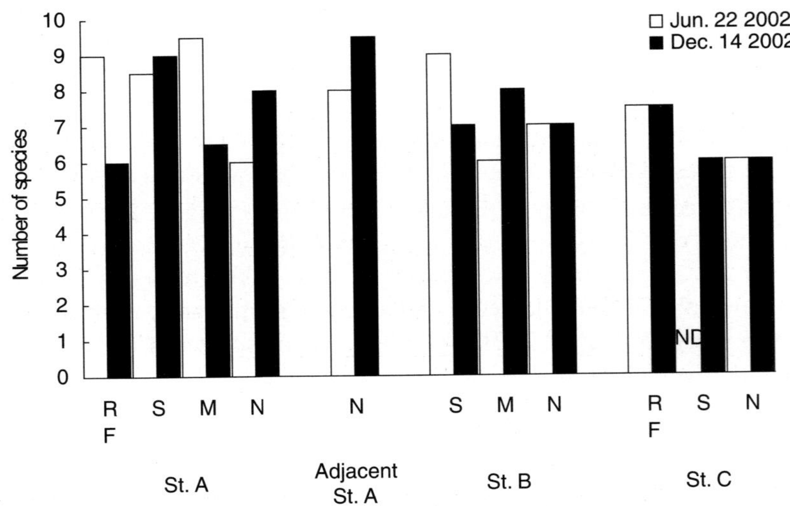

Fig. 4 Comparative appearance of species among the artificial reefs and natural coast

corner of 3 concrete blocks. Only two individuals of abalone were found at the narrow spaces during the survey period. They might attach at such a space. It is an important factor to have such optimum spaces of the artificial reef for top shells and abalone.

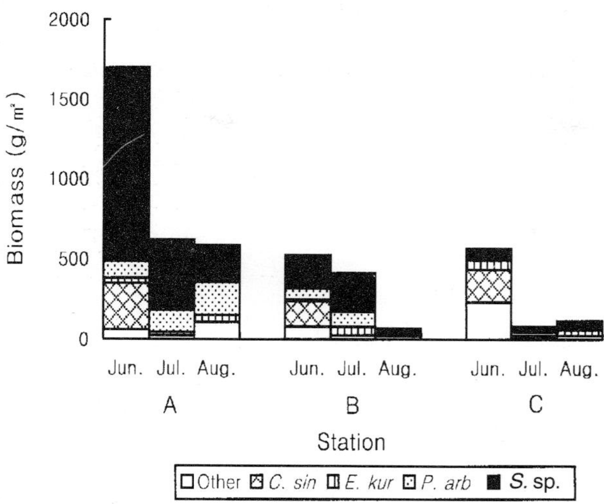

Fig. 5 Mean biomass of seaweed attached to the different types of artificial reefs at the study site during June to August 2000. Other: Other algae, *C. sin*: *Colpomenia sinuosa*, *E. kur*: *Ecklonia kurome*, *P. arb*: *Padina arborescens*, *S.* sp.: *Sargassum* sp.

Table 1 Seasonal changes of appearance number of top shell on the experimental artificial reef sites

		1999								2000					
St.	Type	M	M	J	J	A	O	N	D	F	M	A	J	J	A
St. A	M	1 – 3	5 – 15	3 – 10	0	0	0	0	0	1 – 2	3 – 5	1 – 3	1 – 2	1 – 2	3 – 4
	RF	1 – 3	8 – 20	5 – 15	0	0	0	0	3 – 5	1 – 2	3 – 5	1 – 3	1 – 2	1 – 2	3 – 4
St. B	M	1 – 2	3 – 10	3 – 5	0	0	0	0	0	1 – 2	3 – 5	1 – 3	1 – 2	1 – 2	3 – 4
St.C	RF	1 – 2	3 – 8	3 – 5	0	0	0	0	2 – 3	1 – 2	3 – 5	1 – 3	1 – 2	1 – 2	3 – 4

M:M type artificial reef RF:RF type artificial reef

Fishes

The seasonal changing of gathering fishes on and around artificial reefs is shown in Fig. 6. About 20 species of fishes appeared around and inside these artificial reefs through the survey period. Young and smaller fish, *Apogon natatus* (spotted cardinal fish) and *Charomisfumea* have always appeared as dominate species around and inside of artificial reefs through the year, but these species are not useful fishes. *Sebastes inermis* (black rock fish) cached, has been always several individuals inside artificial reefs, especially gathering inside the lobster reef. *Trachurus japonicus*

(Japanese horse mackel) cached by fishermen gathered in August and disappeared in September. There were several rockfishes in and around of steel artificial reef. They have gathered under the roof or shade spaces of artificial reef. The artificial reef with seaweed beds has gathered many kinds fishes though the year.

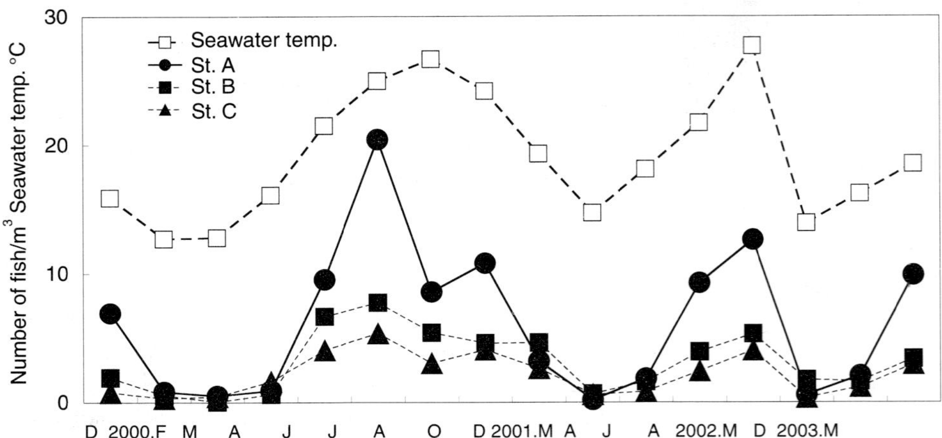

Fig. 6 Seasonal changes of appearance number of fishes in the experimental artificial reef sites

There are many kinds of juvenile fish associated with seaweed beds, which leave for the security of the underwater forest to deeper water or sandy coastal areas in adulthood.

DISCUSSION

The seaweed succession on the artificial reefs described here shows a similar pattern to that of other artificial reefs (Ohno et al., 1990; Ohno 1993; Serisawa and Ohno, 1995a,b; Choi, 2001). In this study, *E. intestinalis* and *C. sinuosa* were found to be typical primary colonizers. Early colonization by these species might be due to the annual release cycle of their reproductive cells. (Choi et al., 2000). In the initial stage, diatoms were colonizing all the reefs at approximately 100% coverage. Commercially valuable shellfish such as turban shell and abalone appeared on the artificial reefs after one month of establishment, because the shellfish grazed on the diatom and seaweed communities on the reefs.

Several species of the Melobesioideae dominated in summer and autumn when water temperatures were highest. During autumn and winter, when water temperatures were lower, the coverage of brown algae was comparatively higher. The course of succession was clearly affected by season and temperature. Seaweed occurring on the artificial reefs in winter, matured in May or June (Choi, 2001).

In our studies, the alga *Sargassum* sp., *E. kurome* and *P. arborescens* dominated in late succession (climax) stages within 2 years of placement. The reports of seaweed succession show to attain succession climax on the Japanese coast within 3 years (Serisawa et al. 1998). From this report. The seaweed succession might attain within 2-3 years by the periods the setting season.

In general, stability of the substratum is very important for maintaining seaweed beds (Watanuki and Yamamoto, 1990). During the investigation, all reefs were stable against waves and currents. Grazers, such as sea urchins, did not colonize the study site during the study period.

Foster and Schiel (1985) reported that *Macrocystis pyrifera* colonization was greatest near the upper horizontal edges of blocks and concrete prisms. He mentioned that this 'edge effect' may result from a combination of increased spore and larval settlement as well as from enhanced growth of plants associated with turbulent eddies formed around such obstructions. For similar reasons, *E. kurome* colonization might also be facilitated. We observed little sand cover on the edges and side the reefs as a result of turbulence. The above results suggest that the settlement of kelp species, such as *E. kurome*, on artificial reefs is the desired outcome, large-scale roughness and/or kelp knob could be important in maintaining the communities after initial colonization.

It is common opinion among fishermen that abalone eats as feed *Ecklonia* fronds by choice and the top shell eats any seaweed in the study site. But it supposes that their top shells had gathered to artificial reefs for feeding *Enteromorpha* fronds. Top shells have disappeared from the blocks, after biomass of seaweeds has decreased on the artificial reefs made by concrete. Top shell has showed to gather at the narrow space or at corner of concrete blocks. Abalone has found only two individuals at the narrow spaces during the survey period. They might attach at such a space. It is an important factor to have such optimum spaces of the artificial reef for top shells and abalone.

Artificial reefs gather many kind fishes, specially, juvenile fishes, as there is much feed for them in the artificial reef with seaweeds.

It is very difficult to calculate the productivity and value of this "ecosystem service" of providing a habitat to juvenile fish. From our project, the value of fish collected, food seaweeds and abalones from the seaweed beds has been estimated at approximately 2,000/m^2/year. We suggest that the value of the direct economic benefits provided by almost of seaweed beds in Japan is in the range of US$15-20/m^2/year.

REFERENCES

Choi, C. G., Serisawa, Y., Ohno M. and Sohn, C., H. 2000a. Construction of artificial seaweed beds: Using the spore bag method. *Algae.* 15(3) : 179-182.

Choi, C. G., Takayama, H, Segawa, S, Ohno, M and Sohn C.H. 2000b. Early stage of algal succession on artificial reefs at Muronohana, Ikata, Japan. 2000a; *J. Fish. Sci. Tech.* 3: 1-7.

Choi, C. G. 2001. Marine communities around the experimental artificial reefs. Ph.D. Thesis, Pukyong National University, Pusan, Korea.

Choi, C. G., Takeuchi, Y., Ohno, M. and Sohn C. H. 2002. Succession of seaweed communities on artificial reefs at Muronohana, Ikata, Japan, *Fisheries Science.* 68: 1755-1758.

Foster, M. S., and Schiel, D. S. 1985. The ecology of giant kelp forest in California: A community profile. U. S. Fish Wildlife Serv. *Boil.* 85. 1-52.

Ohno, M., Arai, S., Watanabe, M. 1990. Seaweed succession of artificial reefs on different bottom substrata. *J. Appl. Phycol.* 2: 327-332.

Ohno, M. 1993. Succession of seaweed communities on artificial reefs in Ashizuri, Tosa Bay, Japan. *Kor. J. Phycol.* 8: 191-198.

Sahoo, D., Ohno. M., 2000. Rebuilding the Ocean floor: Construction of artificial reefs around the Japanese coast. *Curr. Sci.* 78(3) 228-230.

Serisawa, Y., Ohno, M. 1995a. Succession of seaweed communities on artificial reefs in the inlet of Tosa Bay, Japan. Suisanzoshoku. 43: 437-443.

Serisawa, Y., Ohno, M. 1995b. Succession of seaweed communities on artificial reefs in Tei, Tosa Bay, Japan. Nippon Suisan Gakkaishi. 61: 854-859.

Serisawa, Y., Taino, S., Ohno M., and Aruga, Y. 1998. Succession of seaweeds on experimental plates immersed during different seasons in Tosa Bay, Japan. *Bot. Mar.* 41: 321-328.

Watanuki, A., Yamamoto, H. 1990. Settlement of seaweed on coastal structures. *Hydrobiologia* 204/205: 275-280.

5

Seaweeds: A Potential Source of Biofertilizer

Gaurav Kumar, Pooja Baweja and Dinabandhu Sahoo
Marine Biotechnology Laboratory, Department of Botany, University of Delhi, Delhi-110007, India.

ABSTRACT

The world is on the verge of global food crisis as fast growing population is creating high pressure on food production. To meet this increasing food demand, farmers are using chemical fertilizers to enhance the yield. The chemical fertilizers along with pesticides are accumulated in plants and soil which cause health problems in humans due to biomagnification. These adverse effects on soil and environment have pulled the attention to look for an alternative biofertilizer. Seaweeds are one of the important marine living renewable and economically valuable resources. In recent years, seaweeds are being used as biofertilizer, which can substitute the chemical fertilizer. Applications of seaweeds in agriculture as soil conditioners, fertilizers and green manure, are due to the presence of high amount of micronutrients, macronutrients, vitamins, amino acids and growth regulators like auxins, gibberellins, cytokinins and betaines. Application to the soil and foliar spray to the plants of different concentrations of aqueous seaweed extract enhanced percentage of seed germination, growth, and yield. Thus, seaweeds could serve as an alternative biofertilizer.

Keywords: Seaweed, biofertilizer, growth regulators, yield.

INTRODUCTION

The booming population of the world is placing a high pressure on food production. To sort out this increasing food demand, farmers are using chemical fertilizers along with pesticides. The adverse effects of pesticides on human health have been reported (Hansra, 1993). So there is a great demand to look for an alternative biofertilizer. Seaweeds which were traditionally used as food, feed, fodder, agar, alginate and carrageenan (Sahoo, 2000) are now being looked upon as potential source of biofertilizer (Hong et al., 2007; Kumar, 2008). There are about 9,000 seaweed species worldwide, mainly divided into three main groups on the basis of their pigment such as Chlorophyta (green), Phaeophyta (brown) and Rhodophyta (red) (Khan *et al.,* 2009). Brown seaweeds are the most commonly used biofertilizers due to their chelating property, *Ascophyllum nodosum* is the most commonly used one (Ugarte *et al.*, 2006). Besides *A. nodosum,* there are so many other brown seaweeds such as *Dictyopteris australis, Durvillaea potatorum, Ecklonia maxima, Fucus, Laminaria, Macrocystis pyrifera, Padina tetrastromatica, Sargassum* and *Turbinaria* are used as biofertilizers. In recent years, seaweeds are being used as fertilizer, which can substitute to the synthetic fertilizers (Crouch and van Staden, 1993), due to the presence of high amount of nutrients and growth regulators, chiefly cytokinins. Seaweeds extracts are marketed as liquid fertilizers (Table 1) since they contain many growth regulators, vitamins, amino acids, micronutrients and macronutrients (Tay et al., 1985; Khan et al., 2006). Several studies showed a wide range of beneficial effects of seaweed extract on plants, such as early seed germination, better growth and development, resistance to abiotic and biotic stress, and improve crop yield (Beckett and van Staden, 1989; Masny et al., 2004; Norrie and Keathley, 2006; Khan et al., 2009).

Table 1 List of selected commercial seaweed extract products available in market

Seaweed Used	Product Name	Company Name	Country Name
Ascophyllum nodosum	Acadian®	Acadian Agritech	Canada
Ascophyllum nodosum	Kelp Meal	Acadian Seaplants Ltd.	Canada
Ascophyllum nodosum	Biovita	PI Industries Ltd.	India
Ascophyllum nodosum	Maxicrop	Maxicrop (UK) Ltd.	UK
Ascophyllum nodosum	Alg-A-Mic	Bio Bizz Worldwide N.V.	The Netherlands
Ascophyllum nodosum	Bio-Genesis High Tide™	Green Air Products, Inc.	USA
Ascophyllum nodosum	Agri-Gro Ultra	Agri Gro Marketing Inc.	USA
Ascophyllum nodosum	Espoma	The Espoma Company	USA
Ascophyllum nodosum	Ocean Potion	Goji Berries. us	USA
Ascophyllum nodosum	Goëmar GA 14	Goemar Le Laboratoire de la Mer Asco products	France
Durvillaea potatorum	Natrakelp	Natrakelp Pty. Ltd.	Australia
Durvillaea potatorum	Seasol®	Seasol International Pty Ltd.	Australia
Ecklonia maxima	Kelpak®	Caltec Agri Marketing Services	South Africa
Macrocystis pyrifera	AgroKelp	Algas y Bioderivados Marinos, S.A. de C.V.	Mexico

Seaweed as Biofertilizer

The use of seaweeds as manure in the coastal areas dates back centuries. Nowadays seaweeds are used as biofertilizer or manure in several coastal countries such as China, France, Ireland, Japan, Korea, Norway, the Philippines, Scotland and Vietnam. Applications of seaweeds in agriculture as soil conditioners, fertilizers and green manure are due to the presence of high amount of micronutrients (Fe, Cu, Zn, B, Mn, Co and Mo), macronutrients (Ca, K and P), vitamins, amino acids and growth regulators such as auxins (Crouch et al., 1992; Stirk et al., 2004), betaines (Blunden et al., 1991; Wu et al., 1997), cytokinins (Durand et al., 2003; Stirk et al., 2003) and gibberellins (Taylor and Wilkinson, 1977) for the plant growth and development (Fletcher et al., 1982; Khan et al., 2006; Kumar and Sahoo, 2011). Generally, seaweed biofertilizers are found to be superior to farmyard manure and chemical fertilizers because of the presence of high amount of organic matter, which helps in maintaining the moisture retaining capacity of soil and enhance the availability of minerals at the soil surface.

Seaweed biofertilizers showed various advantages such as:

- Easy to handle and convenient for use
- Relatively cheaper than chemical fertilizer
- Relatively longer shelf life
- Increased aeration and soil moisture retaining capacity
- Increased nutrient uptake
- Increased soil fertility by humus formation
- Improved seed germination
- Enhanced vegetative and reproductive growth of treated plants
- Increased chlorophyll and carotinoid content of treated plants
- Improved nutritional quality and yield
- Improved plant resistance against diseases, insects, nematodes, pests and various stresses like frost, drought and salinity etc.

CHEMICAL CONTENTS OF SEAWEEDS

Micro- and Macronutrients of Seaweeds

There are several studies, which showed that seaweeds are the source of high amount of micronutrients (Fe, Cu, Zn, B, Mn, Co and Mo) and macronutrients (Ca, Mg, K and P) (Sivasankari et al., 2006; Rioux et al., 2007; Kumar, 2008).

PLANT GROWTH REGULATORS

Several plant growth bioassays clearly indicate that seaweeds and their extracts have plant growth regulators. Furthermore, seaweed extracts showed a wide range of growth responses that indicates the presence of more than one type of plant growth regulators (Tay *et al.,* 1985; Crouch and van Staden, 1993).

Auxin

Seaweeds contain auxins and auxin-like compounds (Crouch et al., 1992; Stirk et al., 2004). Extract prepared from *Ascophyllum nodosum* had as high as 50 mg Indole Acetic Acid (IAA) per gram of dry extract (Kingman and Moore, 1982; Khan et al., 2009). Crouch and van Staden (1991), observed the root promoting activity of the seaweed extract of *Ecklonia maxima* on mung bean, may be due to auxins. Stirk et al. (2004) also reported conjugates of IAA in seaweed extracts. Besides IAA, auxins-like compounds were also detected in alkaline hydrolyzates of *A. nodosum* and other marine algae (Buggeln and Craigie, 1971).

Betaines

Betaines and betaine-like substances have also been reported in seaweeds and their extracts (Blunden et al., 1991; Wu et al., 1997; Li-Jun et al., 2002). Extract prepared from *Ascophyllum nodosum* contains betaines and betaine-like compounds (Blunden et al., 1986). Betaines also enhance leaf chlorophyll content of plants (Blunden et al., 1997) and the increase in chlorophyll content may be due to a decrease in chlorophyll degradation (Whapham et al., 1993).

Cytokinins

Cytokinins have been reported in fresh seaweeds and seaweed extracts (Hussein and Boney, 1969; Ördög et al., 2004). The cytokinins present in seaweed extracts include trans-zeatin (Stirk and van Staden, 1997). Seaweed extracts also contain aromatic cytokinins like benzyl amino purine (BAP) and topolin (Stirk et al., 2004).

Gibberellins

Seaweeds also contain gibberellins and gibberellin-like compounds. Freshly prepared commercial extracts of *Ascophyllum* showed gibberellin-like bioactivity in a lettuce hypocotyl elongation assay (Williams et al., 1981). Paper chromatography and a dwarf pea bioassay indicate the presence of gibberellins and gibberellin-like substances in *Fucus vesiculosus* (Craigie, 2010).

Types of Seaweed Biofertilizer

Seaweeds are used as biofertilizers in different forms. Some important types are listed below:

- Seaweed manure
- Seaweed liquid extract or seaweed liquid fertilizer or seaweed extract
- Seaweed powder
- Seaweed granules

Growth Regulators: Mode of Action

Seaweed and their extracts showed growth promoting activities, which improve crop production. Seaweed contents such as amino acids, micro- and macronutrients, vitamins, growth regulators like auxins, betaines, cytokinins and gibberellins affect cellular metabolism to enhanced plant growth, development and yield (Crouch et al., 1992; Stirk et al., 2003; Ördög et al., 2004). Auxins and auxin-like substances stimulates cell elongation by stimulating wall loosening factors, such as elastins, to loosen cell walls. This effect became more efficient in the presence of gibberellins. Auxin along with cytokinins also stimulates cell division. On the cellular level, auxin is essential for cell growth and promotes axial elongation (as in shoots), lateral expansion (as in root swelling), or isodiametric expansion (as in fruit growth) (Campbell et al., 2008). Cytokinins have also been known to slow the aging of plant organs. This process occurs by preventing protein breakdown, activating protein synthesis, and assembling nutrients from nearby tissues (Campbell et al., 2008). Gibberellins stimulate cell elongation, breaking and budding, seedless fruits and seed germination. Seed germination done by breaking the seed's dormancy and acting as a chemical messenger. Betaines increase leaf chlorophyll content, may be due to decrease in chlorophyll degradation (Whapham et al., 1993 Blunden et al., 1997).

BENEFITS OF SEAWEED BIOFERTILIZER

Effect on Soil

Seaweeds and their extracts affect the physical, chemical, and biological properties of soil, which improves plant growth and development (Temple and Bomke 1988 1990). Application of seaweeds and their extracts improve soil structure by increasing moisture retention capacity and by promoting the growth of useful soil microbes. Brown seaweeds are a rich source of alginates and fucoidans thus, the gelling and chelating properties make these compounds important in making soil health (Cardozo et al., 2007). Alginates are also used as pesticides (Khan et al., 2009; Kumbar et al., 2003) due to its biodegradable and non-toxic nature. Alginates combine with the metallic ions of the soil to form chelates that absorb moisture, swell and improve soil structure that cause better soil aeration, which promotes the soil microbial activity for better plant growth (Moore, 2004 Khan et al., 2009).

Effect on Plant Growth

Application of seaweed liquid extracts or their commercial products promote plant growth and development (Jeannin et al., 1991; Crouch and van Staden, 1994; Kumar, 2008). This effect was more efficient when extracts were applied at an early growth stage to the roots or as a foliar spray to the plants (Jeannin et al., 1991). Seaweed extracts affect root development by promoting lateral root formation (Atzmon and van Staden, 1994) and increasing total volume of the root system (Slàvik, 2005). Application of seaweed extract enhanced root:shoot ratios and biomass in tomato seedlings (Crouch and van Staden, 1992). Root system may be improved due to endogenous auxins and other compounds in the seaweed extracts (Crouch et al., 1992). Seaweed extracts also influence the rate of nutrient uptake (Crouch et al., 1990), which promotes plant growth. Plants treated with *A. nodosum* extracts showed enhancement in growth of plants (Khan et al., 2009). Seaweeds and seaweed liquid extracts enhance plant chlorophyll content (Blunden et al., 1997; Sivasankari et al., 2006). Application of lower concentrations of *Ascophyllum nodosum* to soil or foliar spray to the leaves of tomato plants produced more green leaves than control. This increase in chlorophyll content may be due to the reduction in chlorophyll degradation, which might be due to betaines in the seaweed extract (Whapham et al., 1993). Genard et al., (1991) observed that glycine betaines delay the loss of photosynthetic activity.

Effect on Crop Yield

In crop plants yield enhancement effects may be due to improved chlorophyll content in leaves, or due to the betaines present in the seaweeds (Genard et al., 1991; Whapham et al., 1993; Blunden et al., 1997). Seaweed extracts accelerates early flowering and fruit set in crop plants, which results in enhancement in yield (Arthur et al., 2003). According to Crouch and van Staden (1992), seaweed extracts treated tomato seedlings set more flowers than control. Enhancement in yield is associated with the number of flowers at maturity. Yield increases in treated plants may be due to the presence of cytokinins in seaweed extracts (Blunden and Wildgoose, 1976; Featonby-Smith and van Staden, 1983). In vegetative plant organs, cytokinins are associated with nutrient partitioning, whereas high levels of cytokinins may be linked with nutrient mobilization in reproductive plant organs (Davey and van Staden, 1978; Gersani and Kende, 1982 Khan et al., 2009). Seaweed extract treated fruits reported higher concentrations of cytokinins as compared to control (Featonby-Smith and van Staden, 1984). Seaweed extract sprayed at vegetative stage on tomato plants, produced large sized fruits with superior quality (Crouch and van Staden, 1992). Application of seaweed extract (Kelpak) showed increased yield in barley (Featonby-Smith and van Staden, 1987) and peppers (Arthur et al., 2003). Simillarly, the number of flowers and seeds in marigold increased upto 50% over control plants (van Staden et al., 1994). Enhancement in beans yield was also observed by 24% after application of Kelpak 66 (Nelson and van Staden, 1984). Application of *A. nodosum* also increased the yield of 'Thompson seedless' grape (*Vitis vinifera* L.) Norrie and Keathley (2006).

IMPROVE RESISTANCE AGAINST STRESSES

Seaweeds and their extracts could be valuable to alleviate environmental (biotic and abiotic) stresses.

To Alleviate Biotic Stress

Seaweeds and their extracts have been reported to protect plants against various types of pest and diseases (Allen et al., 2001). Tomato plants treated with seaweed extracts showed a reduction in nematode infestation (Wu et al., 1997). These results imply that seaweed extracts impart nematode resistance. Plants improve pathogen resistance by elicitors (Patier et al., 1993; Boller, 1995; Khan et al., 2009). Laminaran introduce multiple defense responses in alfalfa and tobacco (Kobayashi et al., 1993; Klarzynski et al., 2003), while carrageenans are effective elicitors in tobacco (Mercier et al., 2001). The application of *Ascophyllum nodosum* extract with humic acid to bentgrass increased resistance to dollar spot disease (Zhang et al., 2003). Liquid seaweed extracts sprayed on plants reduced red spider mite populations (Terriere and Rajadhyaksha, 1964). Aphids also avoid plants after treatment with seaweed extracts. Similar findings were also observed in *Fragaria* when, treated with Maxicrop (Hankins and Hockey, 1990).

To Alleviate Abiotic Stress

The polyanionic properties of seaweeds have proved valuable in remediation of soils, especially those contaminated with heavy metals. Abiotic stresses such as drought, salinity and high temperature reduces the yield (Wang et al., 2003). Plants treated with seaweed extracts showed improved salt and freezing tolerance (Mancuso et al., 2006). Seaweed extracts of *Ascophyllum nodosum* increase drought tolerance in creeping bent grass (Zhang and Ervin, 2004). Similar findings were also observed by Zhang and Schmidt, (2000). Seaweed extracts induced drought and heat tolerance in creeping bent grass might be due to the cytokinin in seaweed or seaweed extracts (Ervin et al., 2004; Zhang and Ervin, 2008). Finally, these results suggest that seaweed products elicit abiotic stress tolerance in plants due to the presence of bioactive substances, which impart stress tolerance and enhance plant performance.

CONCLUSION

Seaweeds and their extracts are using to increasing crop production as they are efficient against various types of abiotic and biotic stresses. The recent challenges to food production due to the increasing environmental stresses is likely due to climate change hence, reduce crop yield. Not only this but also booming population is the key factor for the food crisis. Finally, the application of lower concentrations of seaweed extracts may increase plant growth and yield. Thus, the present review showed that seaweed and their extracts could serve as an alternative biofertilizer as they are eco-friendly, cheaper, more convenient to use and deliver substantial economic and environmental benefits.

REFERENCES

Allen, V. G., Pond, K. R., Saker, K. E., Fontenot, J. P., Bagley, C. P., Ivy, R. L., Evans, R. R., Schmidt, R. E., Fike, J. H., Zhang, X., Ayad, J. Y., Brown, C. P, Miller, M. F., Montgomery, J. L., Mahan, J., Wester, D. B. and Melton, C. 2001. Tasco: Influence of a brown seaweed on antioxidants in forages and livestock—A review. *J. Anim. Sci.* 79 (E Suppl): E21-E31.

Arthur, G. D., Stirk, W. A. and van Staden, J. 2003. Effect of a seaweed concentrate on the growth and yield of three varieties of *Capsicum annuum*. *S. Afr. J. Bot.* 69: 207-211.

Atzmon, N. and van Staden, J. 1994. The effect of seaweed concentrate on the growth of *Pinus pinea* seedlings. *New Forests*. 8: 279-288.

Beckett, R. P. and van Staden, J. 1989. The effect of seaweed concentrate on the growth and yield of potassium stressed wheat. *Plant Soil* 116: 29-36.

Blunden, G. and Wildgoose, P. B. 1976. The effects of aqueous seaweed extract and kinetin on potato. *J. Sci. Food Agric.* 28: 121-125.

Blunden, G. Jenkins, T. and Liu, Y. W. 1997. Enhanced leaf chlorophyll levels in plants treated with seaweed extract. *J. Appl. Phycol.* 8: 535-543.

Blunden, G., Cripps, A. L., Gordon, S. M., Mason, T. G. and Turner, C. H. 1986. The characterization and quantitative estimation of betaines in commercial seaweed extracts. *Bot. Mar.* 29:155-160.

Blunden, G., Smith, B. E., Mason, T. G., Ying, Xu., Yvin, J. C. and Chabot, R. 1991. Betaines in seaweed extracts. *J. Eur. Cosmet.* 13: 113-118.

Boller, T. 1995. Chemoperception of microbial signals in plant cells. *Annu. Rev. Plant. Physiol. Plant Mol. Biol.* 46:189-214.

Buggeln, R. G. and Craigie, J. S. 1971. Evaluation of evidence for the presence of indole-3-acetic acid in marine algae. *Planta* 97:173-178.

Campbell, N. A , Jane, B. R., Lisa, A. U., Michael, L. C., Steven, A. W., Peter, V. M. and Robert, B. J. 2008. Biology. 8th ed. San Francisco: Pearson, Benjamin Cummings, pp. 827-830.

Cardozo, K. H. M., Guaratini, T., Barros, M. P., Falcão, V. R., Tonon, A. P., Lopes, N. P., Campos, S., Torres, M. A., Souza, A. O., Colepicolo, P. and Pinto, E. 2007. Metabolites from algae with economical impact. *Comp. Biochem. Physiol. C Toxicol. Pharmacol.* 146:60-78.

Craigie, J.S. 2010. Seaweed extract stimuli in plant science and agriculture. *J. Appl. Phycol.* DOI 10.1007/s10811-010-9560-4.

Crouch, I. J. and van Staden, J. 1991. Evidence for rooting factors in a seaweed concentrate prepared from *Ecklonia maxima*. *J. Plant Physiol.* 137: 319-322.

Crouch, I. J. and van Staden, J. 1992. Effect of seaweed concentrate on the establishment and yield of greenhouse tomato plants. *J. App. Phycol.* 4(4): 291-296.

Crouch, I. J. and van Staden, J. 1993. Evidence for the presence of plant growth regulators in commercial seaweed products. *Plant Growth Regul.* 13: 21-29.

Crouch, I. J. and van Staden, J. 1994. Commercial seaweed products as biostimulants in horticulture. *J. Home. Consum. Hort.* 1: 19-76.

Crouch, I. J., Beckett, R. P. and van Staden, J. 1990. Effect of seaweeds concentrate on the growth and mineral nutrition of nutrient-stressed lettuce. *J. Appl. Biol.* 2: 269-272.

Crouch, I. J., Smith, M. T., van Staden, J., Lewis, M. J. and Hoad, G. V. 1992. Identification of auxins in a commercial seaweed concentrate. *J. Plant Physiol.* 139: 590-594.

Davey, J. E. and van Staden, J. 1978. Cytokinin activity in *Lupinus albus*. III. Distribution in fruits. *Physiol. Plant* 43: 87-93.

Durand, N., Briand, X. and Meyer, C. 2003. The effect of marine bioactive substances (NPRO) and exogenous cytokinins on nitrate reductase activity in *Arabidopsis thaliana. Physiol. Plant* 119: 489-493.

Ervin, E. H., Zhang, X. and Fike, J. 2004. Alleviating ultraviolet radiation damage on *Poa pratensis*: II. Hormone and hormone-containing substance treatments. *Hort. Sci.* 39: 1471-1474.

Featonby-Smith, B. C. and van Staden, J. 1983. The effect of seaweed concentrate and fertilizer on the growth of *Beta vulgaris. Z. Pflanzenphysiol.* 112:155-162.

Featonby-Smith, B. C. and van Staden, J. 1984. The effect of seaweed concentrate and fertilizer on growth and the endogenous cytokinin content of *Phaseolus vulgaris. S. Afr. J. Bot.* 3: 375-379.

Featonby-Smith, B. C. and van Staden, J. 1987. Effects of seaweed concentrate on grain yield in barley. *S. Afr. J. Bot.* 53: 125-128.

Fletcher, R. A., Kallidumbil, V. and Steele, P. 1982. An improved bioassay for cytokinins using cucumber cotyledons. *Plant Physiol.* 69: 675-677.

Genard, H., Le Saos, J., Billard, J. P., Tremolieres, A. and Boucaud, J. 1991. Effect of salinity on lipid composition, glycine betaine content and photosynthetic activity in chloroplasts of *Suaeda maritima. Plant Physiol. Biochem.* 29: 421-427.

Gersani, M. and Kende, H. 1982. Studies on cytokinin-stimulated translocation in isolated bean leaves. *J. Plant Growth Regul.* 1: 161-171.

Hankins, S. D. and Hockey, H. P. 1990. The effect of a liquid seaweed extract from *Ascophyllum nodosum* (Fucales, Phaeophyta) on the two spotted red spider mite *Tetranychus urticae*. *Hydrobiologia* 204(205): 555-559.

Hansra, B. S. 1993. Transfer of agricultural technology on irrigated agriculture. *Fer. News.* 38(4): 31-33.

Hong, D. D., Hien, H. M. and Son, P.N. 2007. Seaweeds from Vietnam used for functional food, medicine and fertilizer. *J. Appl. Phycol.* 19: 817-826.

Hussain, A. and Boney, A.D. 1969. Isolation of kinin-like substances from *Laminaria digitata. Nature.* 223: 504-505.

Jeannin, I., Lescure, J. C. and Morot-Gaudry, J. F. 1991. The effects of aqueous seaweed sprays on the growth of maize. *Bot. Mar.* 34:469-473.

Khan, M. M. A., Gautam, C., Mohammad, F., Siddiqui, M. H., Naeem, M. and Khan, M. N. 2006 Effect of gibberellic acid spray on performance of tomato. *Turk. J. Biol.* 30: 11-16.

Khan, W., Rayirath, U. P., Subramanian, S., Jithesh, M. N., Rayorath, P., Hodges, D. M., Critchley, A. T., Craigie, J. S., Norrie, J. and Prithiviraj, B. 2009. Seaweed extracts as biostimulants of plant growth and development. *Plant Growth Regul.* 28: 386-399.

Kingman, A. R. and Moore, J. 1982. Isolation, purification and quantification of several growth regulating substances in *Ascophyllum nodosum* (Phaeophyta). *Bot. Mar.* 25: 149-153.

Klarzynski, O., Descamps, V., Plesse, B., Yvin, J. C., Kloareg, B. and Fritig, B. 2003. Sulfated fucan oligosaccharides elicit defense responses in tobacco and local and systemic resistance against tobacco mosaic virus. *Mol. Plant Microbe Interact.* 16: 115-122.

Kobayashi, A , Tai, A., Kanzaki, H. and Kawazu, K. 1993. Elicitor-active oligosaccharides from algal laminaran stimulate the production of antifungal compounds in alfalfa. *Z. Naturforsch.* 48c: 575-579.

Kumar, G. 2008. Effect of seaweed extract on growth and development of wheat. M.Phil. Thesis. University of Delhi, New Delhi.

Kumar, G. and Sahoo, D. 2011. Effect of seaweed liquid extract on growth and yield of *Triticum aestivum* var. Pusa Gold. *J. Appl. Phycol.*

Kumbar, S. G., Dave, A. M. and Aminabhavi, T. M. 2003. Release kinetics and diffusion coefficients of solid and liquid pesticides through interpenetrating polymer network beads of polyacrylamide-g-guar gum with sodium alginate. *J. Appl. Polym. Sci.* 90: 451-457.

Li-Jun, H., Xiao, F. and Xiao-Jun, Y. 2002. The betaines from Chinese seaweeds. *Chinese J. Oceanol. Limnol.* 20(1): 97-100.

Mancuso, S., Azzarello, E., Mugnai, S. and Briand, X. 2006. Marine bioactive substances (IPA extract) improve ion fluxes and water stress tolerance in potted *Vitis vinifera* plants. *Adv. Hortic. Sci.* 20: 156-161.

Masny, A., Basak, A. and Żurawicz, E. 2004. Effects of foliar applications of KELPAK SL and GOËMAR BM 86® preparation on yield and fruit quality in two strawberry. *J. Fruit Ornamen. Plant Res.* 12: 23-27.

Mercier, L., Lafitte, C., Borderies, G., Briand, X., Esquerre-Tugaye, M. T. and Fournier, J. 2001. The algal polysaccharide carrageenans can act as an elicitor of plant defence. *New Phytol.* 149: 43-51.

Moore, K. K. 2004. Using seaweed compost to grow bedding plants. *BioCycle* 45: 43-44.

Nelson, W. R. and van Staden, J. 1984. The effect of seaweed concentrate on wheat culms. *J. Plant Physiol.* 115: 433-437.

Norrie, J. and Keathley, J. P. 2006. Benefits of *Ascophyllum nodosum* marine-plant extract applications to 'Thompson seedless' grape production. (Proceedings of the Xth International Symposium on Plant Bioregulators in Fruit Production, 2005). *Acta. Hortic.* 727:243-247.

Ördög, V., Stirk, W. A., van Staden, J., Novak, O. and Strnad, M. 2004. Endogenous cytokinins in the three genera of microalgae from the Chlorophyta. *J. Phycol.* 40:88-95.

Patier, P., Yvin, J. C., Kloareg, B., Lienart, Y. and Rochas, C. 1993. Seaweed liquid fertilizer from *Ascophyllum nodosum* contains elicitors of plant D-glycanases. *J. App. Phycol.* 5: 343-349.

Rioux, L. E., Turgeon, S. L. and Beaulieu, M. 2007. Characterization of polysaccharides extracted from brown seaweeds. *Carbohydrate Polym.* 69: 530-537.

Sahoo, D. 2000. *Farming the Ocean: Seaweeds Cultivation and Utilization.* Aravali Publication Corporation, New Delhi.

Sivasankari, S., Venkatesalu, V., Anatharaj, M. and Chandrasekaran M. 2006. Effect of seaweed extract on the growth and biochemical constitutes of *Vigna sinensis. Biores. Techn.* 97: 1745-1751.

Slàvik, M. 2005. Production of Norway spruce (*Picea abies*) seedlings on substrate mixes using growth stimulants. *J. For. Sci.* 51: 15-23.

Stirk, W. A. and van Staden, J. 1997. Comparision of cytokinin and auxin-like activity in some commercially used seaweed extracts. *J. Appl. Phycol.* 8: 503-508.

Stirk, W. A., Arthur, G. D., Lourens, A. F., Novak, O., Strnad, M. and van Staden, J. 2004. Changes in cytokinin and auxin concentrations in seaweed concentrates when stored at an elevated temperature. *J. Appl. Phycol.* 16: 31-39.

Stirk, W. A., Novak, M. S. and van Staden, J. 2003. Cytokinins in macroalgae. *Plant Growth Regul.* 41: 13-24.

Tay, S. A. B., MacLead, J. K., Palni, L. M. S. and Letham, D. S. 1985. Detection of cytokinins in a seaweeds extract. *Phytochem.* 24(11): 2611-2614.

Taylor, I. E. P. and Wilkinson, A. J. 1977. The occurrence of gibberellins and gibberellin-like substances in algae. *Phycologia* 16(1): 37-42.

Temple, W. D. and Bomke, A. A. 1988. Effects of kelp (*Macrocystis integrifolia*) on soil chemical properties and crop response. *Plant Soil.* 105: 213-222.

Temple, W. D. and Bomke, A. A. 1990. The short-term effects of fresh kelp (*Macrocystis integrifolia*) on physical properties of a fine-textured soil. *Plant Soil.* 125: 293-295.

Terriere, L. C. and Rajadhyaksha, N. 1964. Reduced fecundity of the two spotted spider mite on metal-chelate treated leaves. *J. Econ. Entomol.* 57: 95-99.

Ugarte, R. A., Sharp, G. and Moore, B. 2006. Changes in the brown seaweed *Ascophyllum nodosum* (L.) Le Jol. Plant morphology and biomass produced by cutter rake harvests in southern New Brunswick, Canada. *J. Appl. Phycol.* 18: 351-359.

van Staden, J., Upfold, J. and Dewes, F. E. 1994. Effect of seaweed concentrate on growth and development of the marigold *Tagetes patula. J. Appl. Phycol.* 6: 427-428.

Wang, Z., Pote, J. and Huang, B. 2003. Responses of cytokinins, antioxidant enzymes, and lipid peroxidation in shoots of creeping bentgrass to high root-zone temperatures. *J. Am. Soc. Hortic. Sci.* 128: 648-655.

Whapham, C. A., Blunden, G., Jenkins, T. and Hankins, S. D. 1993. Significance of betaines in the increased chlorophyll content of plants treated with seaweed extracts. *J. Appl. Phycol.* 5: 231-234.

Williams, D. C., Brain, K. R., Blunden, G., Wildgoose, P. B. and Jewers, K. 1981. Plant growth regulatory substances in commercial seaweed extracts. *Proc Int Seaweed Symp* 8: 760-763.

Wu, Y., Jenkins, T., Blunden, G. Whaphan, C. and Hankin, S. D. 1997. The role of betaines in alkaline extracts of *Ascophyllum nodosum* in the reduction of *Meloidogyne javanica* and *M. incognita* infestations of tomato plants. *Fundam. Appl. Nematol.* 20: 99-102.

Zhang, X. and Ervin, E. H. 2004. Cytokinin containing seaweed and humic acid extracts associated with creeping bentgrass leaf cytokinins and drought resistance. *Crop Sci.* 44: 1737-1745.

Zhang, X. and Ervin, E. H. 2008. Impact of seaweed extract-based cytokinins and zeatin riboside on creeping bentgrass heat tolerance. *Crop Sci.* 48: 364-370.

Zhang, X., Ervin, E. H. and Schmidt R.E. 2003. Physiological effects of liquid applications of a seaweed extract and a humic acid on creeping bentgrass. *J. Amer. Hortic. Sci.* 128(4): 492-496.

Zhang, X. and Schmidt, R. E. 2000. Hormone containing products impact on antioxidant status of tall fescue and creeping bentgrass subjected to drought. *Crop Sci.* 40: 1344-1349.

6

On the Occurrence of *Dichotomosiphon tuberosus* (A. Br.) Ernst from Ahmednagar, Maharashtra, India

A.K. Mohite, R.R. Sanap* and S.D. Pingle*
Dept. of Botany, R.B.N.B. College, Shrirampur, Dist. Ahmednagar.
*P.G. Dept. of Botany, P.V.P. College Pravaranagar, Loni, Tal. Rahata, Dist. Ahmednagar, M.S., India.

ABSTRACT

The present investigation deals with occurrence of rare green alga *Dichotomosiphon tuberosus* (A. Br.) Ernst collected from Ozar Dam, Ahmednagar district, Maharashtra. The alga is reported for the first time from Maharashtra. The benthic alga was found to grow attached to substratum in shallow, slow flowing percolated water stream of Ozar Dam. Morphological studies revealed that the species shows almost all identical characters as described by Venkataraman (1961) except slight larger size of vegetative as well as reproductive structures.

Keywords: *Dichotomosiphon tuberosus* (A. Br.) Ernst, Ozar dam, Ahmednagar, Maharashtra.

INTRODUCTION

Ernst (1902) first described the genus *Dichotomosiphon*. The only freshwater species *Dichotomosiphon tuberosus* (A. Br.) Ernst is being described for the first time from Ahmednagar district, Maharashtra. *Dichotomosiphon* (family Dichotomosiphonaceae) is considered to be one of the rare freshwater green algae. Biswas (1949) mentioned that the genus has not been reported from India. However there are earlier reports of its occurrence at a few places as Rajpur, Allahabad and Pinahat, district Agra by Randhawa (1942). Alga was later on also observed growing abundantly in the stream of Peeplyupala lake near Indore city by Sharma and Moghe (1957). Rao and Yalvigi (1973) reported growth of alga at Amruthur, Mysore state, South India. While Untawale et al. (1980) described a new species of alga as *Dichotomosiphon salina* sp. nov. from Goa estuary (Nom. Invalid.). Mukhopadhyay and Pal (2004) reported a new species *D.*

erecta sp. nov. from brackish water region of West Bengal. Henson (1984) recorded rare growth of alga in Michigan and Huron great lakes. The work on the life cycle of this alga is in progress. However, the present investigation deals only the with report of its occurrence and important morphological features of both vegetative and reproductive structures. The present study provides morphological variations in these structures.

MATERIAL AND METHODS

Alga was collected from Ozar Dam on 28th April 2005. Algal material was hand picked and brought to laboratory in polythene bag. Material was stored in 4% formaldehyde for morphological studies and deposited in Dept. of Botany, P. V. P. College Pravaranagar Dist. Ahmednagar.

For morphological observations, the material was mounted in glycerine and observed under research microscope (Focus- Biosupreme, BPM-3). Microphotographs were taken by using JSP Digital Color Camera (DF 2018) and Aver Media EZ Capture software.

RESULTS AND DISCUSSION

Morphological Observations (See Plate 1)

Alga was forming cushion-like mat with horizontal, downwardly projecting and erect branches. Thallus siphonaceous i.e. tubular without cross walls, characteristically dichotomously branched with constrictions at and between the dichotomies, cross walls formed only to cut off reproductive structures. Filaments measure up to 45 mm in length and 141 µm in diameter.

Akinetes were not observed. Fertile thalli with erect branches bearing both sex organs were observed i.e. monoecious. Sex organs are born on di, tri, or tetra-chotomously branched, strongly recurved branch lets, which may form corymb like clusters at the ends of vegetative filaments. Usually one spherical oogonium and several cylindrical antheridia were present in corymb like

Table 1 Comparative account of size of siphon and reproductive structures

Character	D. erecta	D. tuberosus		
	Mukhopadhyay and Pal, 2004	Sharma and Moghe, 1957	Venkataraman, 1961	Present study
Vegetative Siphon Diameter	42.7 - 45.0 µm	42 - 105 µm	50 - 100 µm	85 - 141 µm
Length	—	5 - 35 mm	—	45 mm
Oogonium (Diameter)	75.0 - 137.5 µm & 100.0 - 187.5 µm long	276 - 428 µm	250 - 300 µm	260 - 330 µm
Antheridium Diameter	42.7 - 45 µm	47 -85 µm	35 - 50 µm	40 - 71 µm
Length	80.0 µm - 1 mm	142 -190 µm	2 - 4 times than diameter	2 - 4 times than diameter

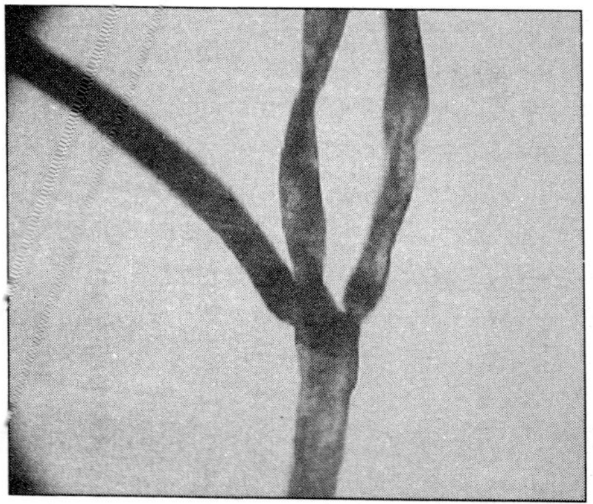

Fig. 1: Trichotomous Branching × 100

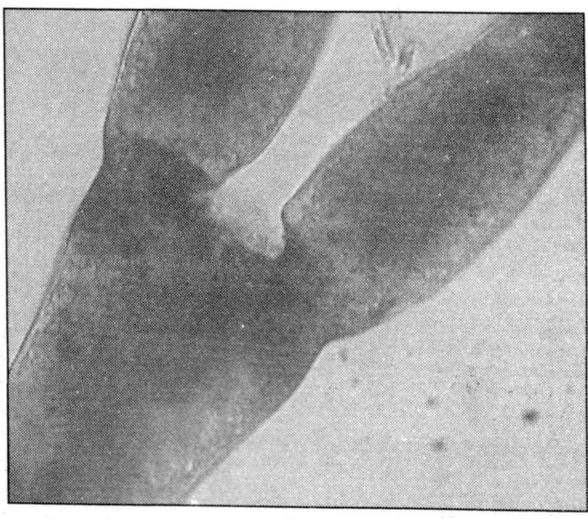

Fig. 2: Dichotomy with constriction × 400

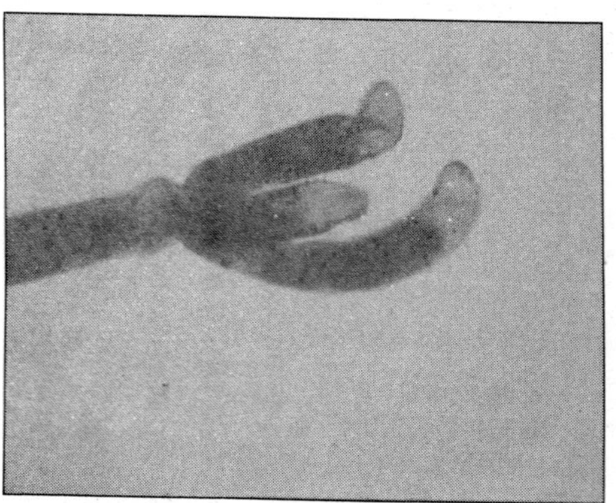

Fig. 3: Apical Tip × 100

Fig. 4: Apical Tip × 100

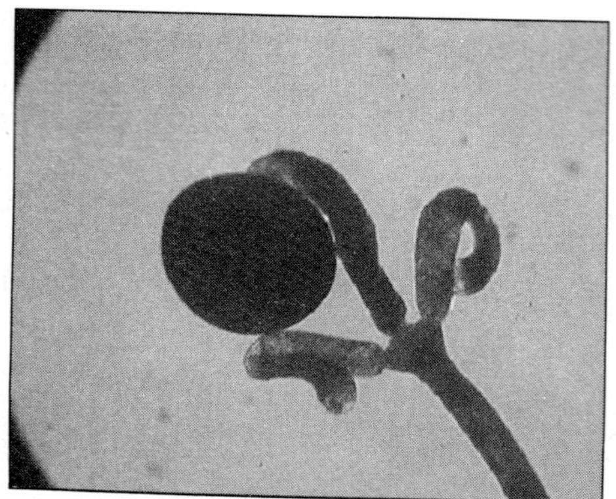

Fig. 5: Oogonium and Antheridia × 100

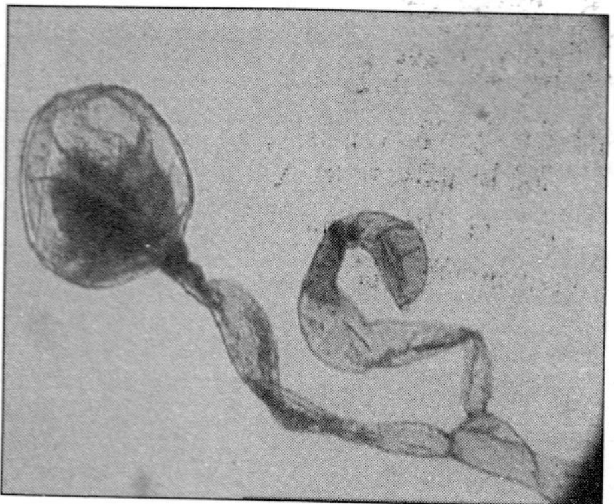

Fig. 6: Oogonium and Antheridia × 100

Plate 1-*Dichotomosiphon tuberosus* (A. Br.) Ernst.

cluster. Oogonia spherical 260 μm to 330 μm in diameter becomes beaked towards maturity with single globular or spherical oospore of nearly same size. Oospore becomes yellow at the maturity, oospore wall three layered thick. Antheridia 40 μm to 71 μm in diameter and 2 – 4 times the diameter in length. Table No. 1 gives compiled account of variation in the size of vegetative and reproductive structures of alga studied with forms described from other localities. Thus the present alga showed large size of vegetative and reproductive structures than that of described by Venkataraman (1961). However it exhibited large size of vegetative structures and small size of reproductive structure than the alga growing in Peeplyupala lake, Indore as described by Sharma and Moghe (1957).

ACKNOWLEDGEMENTS

A.K. Mohite expresses sincere thanks to Principal R.B.N.B. College, Shrirampur, Dist. Ahmednagar for his constant encouragement and Principal P.V.P. College Pravaranagar Tal. Rahata Dist. Ahmednagar for providing necessary facilities. Thanks are also due to UGC, Western Regional Office, Pune for granting fellowship.

REFERENCES

Biswas, K. 1949. Records of the Bot. Survey of India 15: Pt. I

Ernst, A. *1902. *Dichotomosiphon tuberosus* (A. Br.) Ernst Einneeue Oogame Susswasser-Siphonee, Beih. Bot. Central bl. 13: 115.

Henson, E. 1984. Notes on the Benthic Alga *Dichotomosiphon* from the Straits of Mackinac Area of Lakes Michigan and Huron. *J. Great Lakes Res.* 10(1): 85-89.

Mukhopadhyay, A. and Pal, R. 2004. *Dichotomosiphon erecta* sp. nova. A new species from brackish water region of West Bengal. Indian *Hydrobiology*. 7(1 and 2): 7-11.

ndhawa, M. 1942. Further Observations on Vaucheriaceae from Northern India *J. Ind. Bot. Soc.* 21:263.

, A. and Yalvigi, V. 1973. Occurrence of *Dichotomosiphon* (A. Br.) Ernst at Amruthur (Mysore state) South India. *Curr. Sci.* 42: 140-141.

a, R. and Moghe, S. 1957. On the Occurrence of *Dichotomosiphon* (A. Br.) Ernst fom Indore (M. P.). *rr. Sci.* 26: 254-256.

, A., Jagtap, T. and Dhargalkar, V. 1980. *Dichotomosiphon salina* sp. nov. A New Marine Alga Goa Estuary, India. Mahasagar 13(1): 73-76.

an, G. 1961. Vaucheriaceae. Published by Indian Council of Agricultural Research, New Delhi.

rint not seen.

7

Biochemical Composition of Some Freshwater Green Algae from Pravara Basin Region, Ahmednagar, Maharashtra, India

A. K. Mohite and S. D. Pingle

Department of Botany, P.V.P. College Pravaranagar, Loni Kd. Tal. Rahata Dist. Ahmednagar. (Maharashtra) 413 713.

ABSTRACT

In the present investigation some commonly occurring freshwater green algae have been collected from Pravara river basin, Ahmednagar district (Maharashtra) for estimation of their biochemical constituents such as proteins, total carbohydrates, lipids and phenols. The algae used for biochemical studies were *Hydrodictyon reticulatum* (L.) Lagerheim, *Enteromorpha intermedia* Bliding, *Spirogyra crassa* Kuetzing, *Spirogyra rhizoids* Randhawa and *Dichotomosiphon tuberosus* (A. Br.) Ernst. The order of the concentration in all algae under investigation was carbohydrates > lipids > proteins > phenols. Protein content ranged from 2.14 % (*Hydrodictyon reticulatum*) to 4.69% (*Dichotomosiphon tuberosus*), total carbohydrate content ranged from 39.5% (*Hydrodictyon reticulatum*) to 77.0% (*Spirogyra rhizoides*). While lipid concentration varied from minimum of 9.8% (*Spirogyra rhizoides*) to a maximum 23.2% (*Enteromorpha intermedia*). Total phenol composition observed was 0.32% (*Dichotomosiphon tuberosus*) to 1.77% (*Spirogyra rhizoides*) on dry weight basis. The biochemical composition of these forms is also compared with that of other green algal members, compiled from published data.

Keywords: Biochemical composition, freshwater green algae, Pravara, Ahmednagar district.

INTRODUCTION

Chemical composition plays a key role in the metabolism of algae. The exploitation of marine algae and freshwater algae for nutritional and industrial purpose is primarily based on the biochemical constituents of these forms. Studies on the biochemical constituents of several marine algae have been carried out from Indian coast as well as abroad by many workers. Majority of freshwater algae are morphologically and physiologically distinct from both marine algae and land plants. Although about 10% of green algae are marine in habitat, they were analyzed for their biochemical composition throughout India by Rao and Tipnis, 1963; Dhargalkar et al., 1980; Parekh et al. 1985; Jayshankar et al., 1990; Venkataraman Kumar, 1993; Devi et al., 1996; Sarojini and Subbarangaih, 1999. However very few freshwater algal forms like *Chlorella vulgaris* (in culture) was studied by Rathore and Kumar (1993), Chara corollina Willd by Jacob and Venkataraman (1962). No information is available on the biochemical composition of freshwater algae. Hence the present investigation was undertaken to study biochemical composition of freshwater algae from Pravara basin region of Ahmednagar district.

MATERIALS AND METHODS

In all, five freshwater green algal forms were collected from various sites of Pravara river basin of Ahmednagar district. The collected mass of material was brought to laboratory in separate polythene bags. Some portion of each alga was stored in 4% formaldehyde for morphological observations and species identification. Remaining bulk of each alga was repeatedly washed with tap water to remove other debris and carefully screened under binocular (Olympus SZ-30) to remove filaments of disastrous algae. Water drained material was then shade dried in the laboratory. Completely dried material was crushed in mortar and pestle to powder form. The fine powder of each alga was used for biochemical analysis.

Total protein content was determined with folin- ciocalteu reagent, with bovine albumen serum V as the standard by Lowry's method (Lowry et al., 1951). The total carbohydrate content was estimated by using anthrone reagent as per the method described in Sadashivam and Manickam (1996) using D-glucose as standard. Lipids were extracted and estimated by the procedure of Folch et al. (1957). While total phenols were determined by the method of Farkas and Kirlay (1962). Species identification was done with the help of monographs and standard literature (Joshi and Krishnamurthy, 1972; Randhawa, 1959; Prescott, 1951).

RESULTS AND DISCUSSION

The order of concentration in all algae under investigation was carbohydrates, lipids, proteins and phenols. The estimated concentrations of biochemical constituents were given in Table 1.

All five algal members under investigation are rich in total carbohydrate content. The range of total carbohydrate obtained was 39.5% to 77.0%. The minimum concentration of total carbohydrate, 39.5 % was recorded in *Hydrodictyon reticulatum* (L.) Lagerheim while maximum 77.0 % was observed in *Spirogyra rhizoides*. This obtained range appeared too much higher than the previously reported range of carbohydrate content i.e. 0.91% in *Cladophora fascicularis* (Devi et al., 1996) to 58.0% in *Enteromorpha* species (Nisizawa et al., 1987) for green algal members.

The amount of estimated lipids varied from 9.8% to 23.2% on dry weight basis. The minimum value of lipids 9.8% was recorded in Spirogyra rhizoides, while maximum 23.2% lipid content was recorded in *Enteromorpha intermedia*. This range of lipid content was also very much higher as compared to the earlier reported range, minimum of 0.3% in *Enteromorpha species* (Nisizawa et al., 1987) to a maximum 15.7% in *Cladophora fascicularis* (Jayshankar et al., 1990)

Table 1 Estimated amount of constituents

Sr. No.	*Name of the alga*	*Proteins %*	*Total carbohydrates %*	*Lipids %*	*Phenols %*
1	*Hydrodictyon reticulatum* (L.) Lagerheim	2.14	39.5	11.6	0.368
2	*Spirogyra crassa* Kuetzing	3.06	46.0	11.0	1.672
3	*Spirogyra rhizoides* Randhawa	3.55	77.0	9.8	1.77
4	*Dichotomosiphon tuberosus* (A. Br.) Ernst	4.69	46.0	17.2	0.32
5	*Enteromorpha intermedia* Bliding	3.86	67.5	23.2	0.63

for green algae.

The protein content of studied algae ranged from 2.14% to 4.69%. Comparatively lower concentration of proteins in these algae studied might be due to fact that four members were in vegetative growth phase and *Dichotomosiphon tuberosus* with few sex organs. *Hydrodictyon reticulatum* exhibited minimum protein content whereas maximum protein content was revealed in *Dichotomosiphon tuberosus*. The observed range of protein content follows the earlier reported range, minimum of 1.38% in *Valoniopsis pachynema* (Devi et al., 1996) to a maximum of 57.0% in *Chlorella vulgaris* (Parsons et al., 1961).

Naturally growing freshwater algae were first time analyzed for phenol content. Amongst the studied algae, phenols were ranged from minimum of 0.32% in Dichotomosiphon tuberosus to a maximum of 1.77% in *Spirogyra rhizoides*.

ACKNOWLEDGEMENTS

The first author is thankful to Principal, R. B. N. B. College, Shrirampur for his constant encouragements and Principal, P. V. P. College, Pravaranagar for providing all necessary facilities. Thanks are also due to UGC, Western Regional Office, Pune for sanctioning teacher fellowship during 10th plan.

REFERENCES

Dhargalkar, V., Jagtap, J. and Untawale, A. 1980. Biochemical constituents of seaweeds along Maharashtra coast. *Indi. J. Mar Sci.* 9: 297-299.

Devi, G., Sreekumari, S., Saraswati, A. G., Sobha, V. and Nayar, T. 1996. Organic constituents (mannitol, carbohydrate, lipid and protein) of marine algae of Cape Comorin. A preliminary study. *Seaweed. Res. Utilization.* 18 (1 and 2): 67-72.

Farkas, G. and Kiraly, Z. 1962. Role of phenolic compounds in the physiology of disease and disease resistance. *Phytopathol.* 20: 9-15.

Folch, J., Less, M., Stoare-Stanely, G. 1957. A simple method for the isolation and purification of total lipids from animal tissues. *J. Biol. Chem.* 226: 497-507.

Jayshankar, R., Ramlingam, J. and Kaliaperumal, N. 1990. Biochemical composition of some green algae from Mandapam coast. *Seaweed Res. Utilization.* 12 (1 and 2): 37-40.

Jacob, K. and Venkataraman, G. 1962. Amino acid composition of *Chara corollina*. *Phykos.* 1(2): 105-106.

Joshi, H. and Krishnamurthy, V. 1972. The species of *Enteromorpha* from India. *Bot. J. Linn. Soc.* 65: 119-128.

Lowry, O., Rosebrough, N., Farr, A. and Randall, R. 1951. Protein measurement with Folin–phenol reagent. *J. Biol. Chem.* 193: 265-275.

Nisizawa, K., Noda, H., Kikuchi, R. and Watanabe, T. 1987. The main seaweed foods in Japan. *Hydrobiologia* 151(2): 5-29.

Parekh, K., Parekh, H., Modi, H., and Rao, P. 1985. Total lipid, sterol and chlorophyll content of some of Indian seaweeds. *Seaweed Res. Utilization* 6(1): 23-25.

Parsons, T., Stephens, K. and Strickland, J. 1961. On the chemical composition of eleven species of marine phytoplankton. *J. of the Fisheries Research Board of Canada.* 18: 1001-1016.

Randhawa, M. 1959. Zygnemataceae. Pub. I. C. A. R. New Delhi.

Prescott, G. 1951. In: Algae of the Western Great lakes area. Wm. C. Brown Co. Dubeque, Iowa.

Rathore, D. and Ashok, Kumar 1993. Protein, lipid and carbohydrate content of three microalgae. *Phykos.* 32 (1 and 2): 9-12.

Rao, S. and Tipnis, U. 1963. Protein content of marine algae from Gujarat coast. *Curr. Sci.* 33: 16-17.

Sadashivam, S. and Manickam, A. 1996. In: Biochemical methods, 2nd ed., New Age International Publishers, New Delhi.

Sarojini, Y. and Subbarangaiah, G. 1999. Seasonal variation in biochemical composition of some macroalgae along Visakhapatnam, east coast of India. *Phykos.* 38(1 and 2): 71-79.

Venkataraman, Kumar 1993. Biochemical constituents of marine algae from Tuticorin coast. *Indian J. Mar. Sci.* 22:138-141.

8

Effect of Blue-Green Algae as Biofertilizer on Growth and Yield of Chilli (*Capsicum annum* L.) Var. Phule Jyoti

A.R. Abhang, S.D. Pingle, A.M. Bhosale and B.K. Auti
Arts, Science and Commerce College, Rahuri
Post-Graduate Department of Botany
P.V.P. College, Pravaranagar, Loni-413713, Maharashtra

ABSTRACT

The present investigation deals with the study of BGA biofertilizers effects on the growth and yield of chilli (*Capsicum annum* L.) var. Phule Jyoti crop. The experiment was carried out during the kharif season. BGA biofertilizer treatment was given to the chilli plants. About 25 gm of BGA biofertilizer was applied by ring method to each plant with a control. Observations were made at the interval of 15 days after plantation. Different physical parameters like height of plant; number of leaves, number of flowers, number of harvested fruits and weight of harvested fruits were studied during the investigation. It has been observed that the application of BGA supplemented with farmyard manure has significantly increased in vigor and yield of the chilli crops.

Keywords: BGA biofertilizer, chilli plant.

INTRODUCTION

Chilli (*Capsicum annum* L.) is an important vegetable as well as condiment crop widely grown for its green and dry fruits throughout India. Although it is a native of Tropical America (Brazil) and is well adopted to prevailing climatic conditions of our country. Its introduction to India is believed to be through the Portuguese in the seventeenth century.

In India, chilli is grown over an area of 917.6 thousand hectares with an annual production of 779.6 thousand metric tons of dry chillies, while in Maharashtra the total area under this crop is about 132.7 thousand hectares with the production of 68.40 thousand metric tons (Anonymous, 1996).

Chilli is one of the most important vegetable crops used in our daily diet. Fruit have pungency due to volatile alkaloid 'oleresin capsicin' which is widely used for medicinal purposes. Chilli as a vegetable is used for its pungency, spicy taste besides the appealing colour it adds to the food. It is also used in different forms, such as a spice, a condiment and in sauces and pickles.

Considering the importance of chilli in daily diet and its demand for export as a condiment and for medicinal purpose, increase its in production by bringing more area under cultivation and increasing yield per unit area of the cultivated is required.

Nitrogen is the main growth and yield-contributing factor which the chilli requires in more quantity. Though nitrogen is abundantly present in atmosphere, but plants are not able to use it directly for their growth. However, in nature there are some microorganisms, which fix the atmospheric nitrogen in the soil and the plant. Besides conversion of N_2 in the available form, the organisms produce growth of plant.

Taking into consideration, the high cost of chemical fertilizers, the biological nitrogen-fixing agents deserve immediate attention. Amongst the nitrogen fixing microorganisms, blue-green algae are important for non-leguminous crops to supplement nitrogenous fertilizers.

Considering this view and to reduce the use of chemical nitrogen fertilizers, it is proposed to study the effect of blue-green algae as biofertilizer on the growth and yield of chilli in pot experiment and field conditions.

MATERIALS AND METHOD

The present investigation on chilli (*Capsicum annum* L.) var. Phule Jyoti was carried out during 2004-05. Material used and methods followed during the investigation are described in the succeeding paragraphs.

(A) Materials

(i) Seeds of chilli var. Phule Jyoti were obtained from senior vegetable breeder, All India Co-ordinated Vegetable Improvement Project of M.P.K.V. Rahuri.

(ii) BGA mixture consisting of dried mass of *Nostoc*, *Anabaena*, *Calothrix*, *Lyngbya*, *Oscillatoria*, *Phormidium*, *Michrochaetae*, *Westiellopsis* and *Cylindrospermum*.

(B) Method

(i) Pot Experiments – Pot experiments were done to access the influence of algal inoculation on the growth and yield for chilli var. Phule Jyoti. Equal sized earthen pots were used for the experiments. Each pot was filled with 7 kg of soil sample. Three replicates were made for each treatment. First three pots were kept under control and remaining fifteen pots were used as shown below.

Sr. No.	Legend	Fertilizers and Manure Supplied	Doses in gm/pot
1	A	Control	—
2	B	Blue-green algae	25
3	C	Blue-green algae + farm yard manure	25 + 50
4	D	Chemical fertilizer (urea)	0.25
5	E	Farm yard manure (FYM)	50
6	F	Blue-green algae + chemical fertilizer (urea)	25 + 0 25

(C) Growth Performance Parameters

(i) Height – Plant height was recorded from surface of pot to the top of shoot at an internal of 15 days, mean height of the plants was calculated in cm.

(ii) Leaves – The total number of leaves per plant were recorded at an interval of 15 days.

(iii) Number of branches, flowers and fruits – The number of branches, flowers and fruits were counted at an internal of 15 days. Mean value was calculated for each treatment. The number of days required for initiation of branching, flowering and fruiting were also recorded for each treatment.

(iv) Yield – Fresh weight of chilli fruits were noted down after each harvesting and finally total fresh weight was calculated for every harvesting.

RESULTS AND DISCUSSION

An investigation was carried out for chilli to know plant height, number of leaves, number of branches and flowers per plant, number of fruits per plant, number of days required for branches, flowers and fruit initiation, and fresh weight of fruits. The data has been presented in Table 1 and Plate 1.

Table 1 Effect of blue-green algae as biofertilizer on growth and yield of chili (*Capsicum annum* L.) var. Phule Jyoti

Sr. No.	Condition	Plant Height (cm)	No. of Leaves	No. of Branches	Days for Branch Initiation	No. of Flowers per plant	Days for Flower initiation	No. of Fruits per plant	Days for Fruits initiation	Fresh wt. of Fruit per plant (gm)
1.	Control	21.87	36	4	62	6	92	3	112	11.14
2.	BGA	30.12	75	8	42	10	62	5	77	21.60
3.	BGA + FYM	33.27	120	12	30	19	50	11	65	41.35
4.	Fertilizer (Urea)	23.15	64	5	59	7	80	4	98	16.12
5.	Farmyard Manure	28.20	92	9	46	15	65	8	85	29.80
6.	BGA + Fertilizer	26.30	86	6	48	11	68	6	93	25.34

(A) Plant Height

The excellent plant height was responded to the mixture of blue-green algae with farmyard manure (33.27 cm) as compared to the control (Table 1).

The maximum plant height was observed in a mixture of blue-green algae and farmyard manure (33.27 cm), followed by blue-green algae (30.12 cm) FYM (28.20 cm) blue green algae with chemical fertilizer (26.30 cm) and fertilizer (23.15 cm). Thus plant showed optimum result in blue-green algae with farmyard manure. The same pattern of result was observed for the number of leaves and branches per plant.

A-Control, B-BGA, C-BGA + FYM, D-Chemical Fertilizer (Urea), E-FYM

Plate 1 Effect of BGA as a biofertilizer on chilli (*Capsicum annum* L.) Var. Phule Jyoti.

(B) Development of Leaves

As mentioned in Table 1, maximum (120) leaves were recorded in the treatment of blue green algae with farmyard manure. Other treatments showed increase in number of leaves as a response to FYM (92), blue-green algae with chemical fertilizer (86), blue-green algae (75), and fertilizer (64). While Control plant could bear only 36 leaves per plants.

The above-mentioned results might be due to the combined effect of blue-green algae (Nitrogen fixing ability) and FYM (organic contents). Such changes in growth parameters were due to the formation of growth stimulating compounds. (Venkatraman and Neelakanthan, 1967; Rodger et al. 1979; Kaushik and Venkataraman, 1979; Rodger and Watanabe, 1982; Shinde, 1995).

It was noticed that the enhancement in growth parameters were more in the treatment with blue green algae with farmyard manure, than the chemical fertilizer with blue-green algae. It indicates that the mixture of blue-green algae along with chemical fertilizer was less effective than the use of blue-green algae along with farmyard manure.

This was confirmed by Kaszubaik (1976) that the chemical fertilizers reduce the growth of soil bacteria and algae. Our results also showed coinciding results like the earlier reports.

(C) Number of Branches, Flowers and Fruits

The combined influence of blue-green algae and farmyard manure showed increase in the number of branches, flowers and fruits per plant (12,19,11) as compared to control (04,06,03). Other treatments show increase in number of branches, flowers and fruits per plant as response to farm yard manure (09,15,08), blue-green algae and chemical fertilizer (6,11,06), blue-green algae (08,10,05), and chemical fertilizer (05,07,04). These behavioural changes might be due to the liberation of auxins, vitamins and amino acids (Okada and Yamaguchi, 1960; Subrihmanyam and Sahay, 1964; Venkatraman and Neelakanthan, 1967; Shinde, 1995).

According to Sundrarao et al. (1963) extracellular growth regulating substance helped the plants for vigorous growth. Our results also support it.

(D) Number of days required for branch flower and fruit initiation

Early branching, flowering and fruiting were observed in chilli plants, which received algal inoculum and farmyard manure. Early branching and fruiting was also observed with the treatments of chemical fertilizer (urea) with blue green algae and farmyard manure alone.

Chemical fertilizer (urea) alone showed late branching, flowering and fruiting (Table 1). Our results also support the view of Shinde (1995).

(E) Fresh Weight of Fruits

Maximum weight of fresh fruits was observed in blue green algae with farmyard manure (41.25 gm). Then the decline in weight was observed in farmyard manure (29.80 gm), blue green algae with chemical fertilizer (25.34 gm), blue-green algae (21.60 gm) chemical fertilizer (16.12 gm) and control (11.14 gm) (Table 1).

Our results clearly support the view of Dadhich et al., (1969), Venkatraman and Neelakanthan (1967), Shinde (1995) that the addition of blue green algae significantly increases the growth and yield of plants.

Thus the present investigation concludes that by using the low cost technology of use of blue-green algae fertilizer with or without farmyard manure and chemical fertilizer, the growth and yield of chilli plant could be boosted to certain extent.

REFERENCES

Anonymous, 1996. Jeevanu Khate Visheshank, Shetkari Mashik (Marathi Publication.)

Dadhich, K., Varma, A. and Venkatraman, G. 1969. Effect of *Calothrix* inoculation on vegetable crops. *Plant and Soil. 31: 377.*

Kaszubiak, H. 1976. I. The influence of mineral fertilizers on microorganisms in various conditions of the soil environment. II. The influence of mineral fertilizers on some groups of microorganisms at different soil moisture. *Pol. J. Soil.* 9: 41-46.

Kaushik, B. and Vankatraman, G. 1979. Effect of algal inoculation on the yield and vitamin C content of two varieties of Tomato. *Plant and Soil*. 52: 135-137.

Okuda, A. and Yamaguchi, M. 1960. Nitrogen fixing microorganism in paddy soils 6. Vitamin B_{12} activity in nitrogen fixing blue-green algae. *Soil and Plant Ed.* 6(2): 76.

Rodgers, G., Bergman, B., Benricksson, E. and Udris, M. 1979. Utilization of blue-green algae as a bio-fertilizer. *Plant and Soil.* 52 : 90-107.

Roger, P. and Watanabe, J. 1982. IRRI Res. Paper series : 78.

Shinde, M. 1995. Studies on soil characterisation and effect of blue-green algae on selected crops from Pravaranagar area, Maharashtra. Ph.D. Thesis, University of Pune.

Subrahmanyam. R and Sahay, M. 1964. Observation nitrogen fixation by some blue-green algae and remarks on its potentialities in rice culture. *Proc. Indian Acad. Sci. B.* 60 (2) : 145.

Sundara, Rao W., Goyal, S. and Venkatraman, G. 1963. Effect of inoculation of *Aulosira fertilissma* on rice plants. *Curr. Sci.* 32 : 366-367.

Vankatraman, G. and Nellakanthan, S. 1967. Effect of the cellualar constituents of the nitrogen fixing blue-green algae. *Cylindospermum musicola* on the root growth of rice seedlings. *J. Gen. Appl. Microbial.* 13: 53-61.

Venkatraman, G. 1979. Algal inoculation of rice fields. In : Nitrogen and Rice. Pub. IRRI, Philippines, pp. 311-321.

9

Cyanophycean Algae from Soils of Pravara Canal Command Area, Ahmednagar, Maharashtra, India

B.K. Auti, S.D. Pingle and A.R. Abhang
Post-Graduate Department of Botany, P.V.P. College, Pravaranagar,
Loni-413 713, Maharashtra

ABSTRACT

Pravara canal command area is a highly irrigated area of Ahmednagar district. For agriculture three tahsils of the district viz. Shrirampur, Rahata and Sangamner receives irrigation water through Pravara canal of Bhandardara dam. In the present investigation attempts were made to study the diversity in cyanophycean algae of Pravara area. In order to study the algal diversity, 250 soil samples were collected from arable lands of the study area. Altogether 68 cyanophycean (BGA) algal forms belonging to 32 genera were isolated and identified. Among the studied genera *Oscillatoria, Phormidium, Lyngbya, Microchaete* and *Hapalosiphon* were found as most common genera. Out of enlisted species, 40 were heterocystous and 28 non-heterocystous. Maximum (68) species were recorded from soil samples of Shrirampur tahsil having soil pH 8.2, organic carbon 0.45% and electrical conductivity 0.35 m mhos/cm^2. Whereas the Rahata tahsil showed minimum (21) species with pH 8.3, organic carbon 0.42 % and electrical conductivity 0.52 m mhos/cm^2. pH, electrical conductivity and organic carbon of the arable land soil samples proved to be an important factor affecting the growth and relative abundance or distribution of cyanophycean algae.

Keywords: Cyanophycean algae, arable land soils, Pravara canal command area, Ahmednagar district.

INTRODUCTION

Pravara canal is an elixir to the northern region of Ahmednagar. Agricultural economy of the society is based solely upon the farming, which is dependent on Pravara canal. Since last three decades there is continuous reduction in the produce due to excessive use of chemical fertilizers and over-irrigation. Microbial population has been greatly affected by these practices.

Cyanophycean algae are important components of soil micro-flora and help to maintain soil fertility by fixing atmospheric nitrogen and by adding organic matter. However major studies on cyanophycean algae were restricted to only paddy fields (Anand, 1989;Venkataraman, 1975; Santra, 1983; Devi et al., 1999; Nayak et al., 2001). Very scanty and sporadic reports were published on investigation to other field crops from Pravara canal command area. Due to excess use of chemical fertilizers and over-irrigation the soils from study area became alkaline and sodic type (Shinde, 1995).

The present work deals with the study of the algal diversity, particularly BGA (cyanophyceae) from arable land soils of Pravara canal command area (PCCA) of Ahmednagar district. An investigation is related to three tahsils (Sangamner, Rahata and Shrirampur). The soil of the area is vertisol type and sugarcane, bajara, cotton, wheat, tomato and onion are the main agricultural crops.

The dominance and abundance of soil BGA are regulated mostly by soil properties, such as pH, electrical conductivity and organic carbon (Metting, 1981). Therefore, an attempt was made to study the BGA diversity and their distribution pattern from soil of arable lands of Pravara canal command area of Ahmednagar district.

MATERIAL AND METHODS

Pravara canal left bank encompasses sites from three talukas of the district (Sangamner, Rahata and Shrirampur). During 2002-2003 a total of 250 soil samples from various crop fields of 14 stations situated in three tahsils of Ahmednagar district were collected scientifically. Soil from the entire tract of the study area is of vertisol type with average pH range 8.2 to 8.4.

Activating the growth of the algae in laboratory conditions made study of the BGA from the arable lands soil. For identification of BGA forms, 5 gram of the soil from each sample was inoculated in the flasks containing 100 ml of sterilized nitrogen free BG-11 and Fogg's medium (Ripka et al., 1979). These flasks were incubated at 28 ± 2°C under 16/8-light/dark cycles with 2500 Lux of white transluscent light for 15-21 days.

Cyanobacterial growth from enrichment cultures was examined microscopically and identified with the help of available literature (Anand, 1989; Desikachary, 1959; Prescott, 1951). Unialgal cultures were also prepared from enrichment cultures for further investigation.

The relative abundance of a cyanophycean algal form was calculated (Table 1) using the following formula:

Relative abundance = $(Y \div X)$ 100

Where X = Total number of soil samples collected (250).

Y = Number of soil samples harboring the alga.

The soil pH, electrical conductivity and organic carbon were determined by standard methods mentioned in Table 2 (Trivedi et al., 1998).

RESULTS AND DISCUSSION

Table 1 shows a total of 68 cyanophycean forms belonging to 32 genera, recorded from the arable land soils of Pravara canal command area. Out of these 18 was heterocystous and 14 non-heterocystous. From the recorded forms, *Lyngbya polysiphoniae* Fremy was the most common form, while *Oscillatoria animalis* Ag.ex.Gomont, *Westiellopsis prolifica* Janet, *Hapalosiphon welwitschii* W.et G.S. West, *Oscillatoria okeni* Ag.ex Gomont and *Microchaete uberrima* Carter,

Table 1 Relative abundance of BGA from soils of PCCA

Sr. No	Name of blue-green alga	No. of soil samples harboring the alga	Abundance	Sr. No	Nam of blue-green alga	No. of soil samples harboring the alga	Abundance
1	*Microcystis aeruginosa*	3	1.2	35	*Nostoc sporangiaeforme*	5	2.0
2	*Chroococcus macrococcus*	6	2.4	36	*Nostoc ellipsosporum*	10	4.0
3	*Chroococcus turgidus*	6	2.4	37	*Nostoc humifusum*	5	2.0
4	*Chroocoocus minutus*	20	8.0	38	*Nostoc calcicola*	14	5.6
5	*Chroococcus minor*	22	8.8	39	*Nostoc muscorum*	16	6.4
6	*Chroococcus varius*	18	7.2	40	*Nostoc microscopicum*	5	2.0
7	*Chroococcus montanus*	10	4.0	41	*Nostoc hatei*	5	2.0
8	*Chroococcus indicus*	20	8.0	42	*Anabaena spiroides*	39	15.6
9	*Gloeocapsa kuetzingiana*	9	3.6	43	*Anabaena oryzae*	41	16.4
10	*Aphanothece biformis*	11	4.4	44	*Anabaena gelatinicola*	3	1.2
11	*Aphanothece pallida*	8	3.2	45	*Anabaena orientalis*	7	2.8
12	*Gomphosphaeria aponina*	12	4.8	46	*Anabaena variabilis*	40	16.0
13	*Chlorogloea microcystoides*	2	0.8	47	*Anabaena circinalis*	5	2.0
14	*Chroococcidiopsis indica*	6	2.4	48	*Aulosira aenigmatica*	15	6.0
15	*Myxosarcina spectabilis*	7	2.8	49	*Scytonematopsis kashyapi*	5	2.0
16	*Oscillatoria obscura*	15	6.0	50	*Scytonema subtile*	55	22.0
17	*Oscillatoria amoena*	5	2.0	51	*Tolypothrix tenuis*	50	20.0
18	*Oscillatoria okeni*	65	26.0	52	*Tolypothrix distorta*	30	12.0
19	*Oscillatoria animalis*	105	42.0	53	*Microchaete uberrima*	60	24.0
20	*Dasyglogea amorpha*	5	2.0	54	*Microchaete tenera*	40	16.0
21	*Phormidium fragile*	35	14.0	55	*Calothrix javanica*	35	14.0
22	*Phormidium tenue*	40	16.0	56	*Calothrix brevissima*	3	1.2
23	*Phormidium rubroterricola*	30	12.0	57	*Calothrix marchica*	6	2.4
24	*Lyngbya baculum*	3	1.2	58	*Gloeotrichia natans*	7	2.8
25	*Lyngbya polysiphoniae*	205	82.0	59	*Nostochopsis lobatus*	10	4.0
26	*Lyngbya bipunctata*	55	22.0	60	*Mastigocladopsis jogensis*	11	4.4
27	*Lyngbya confervoides*	5	2.0	61	*Mastigocladus laminosus*	10	4.0
28	*Microcoleus lacustris*	30	12.0	62	*Hapalosiphon welwitschii*	70	28.0
29	*Cylindrospermum musicola*	30	12.0	63	*Hapalosiphon hibernicus*	4	1.6
30	*Nostoc punctiforme*	6	2.4	64	*Westiella intricata*	9	3.6
31	*Nostoc entophytum*	4	1.6	65	*Westiellopsis prolifica*	101	40.4
32	*Nostoc paludosum*	14	5.6	66	*Camptylonema indicum*	11	4.4
33	*Nostoc linckia*	10	4.0	67	*Stigonema mesentericum*	6	2.4
34	*Nostoc rivulare*	16	6.4	68	*Stigonema dendroideum*	4	1.6

N. were found fairly distributed. *Nostoc, Anabaena, Phormidium, Tolypothrix, Scytonema and Calothrix* species were also showed common distribution. All 68 species were observed in the soil samples from Shrirampur tahsil, having pH 8.2, electrical conductivity 0.35 m mhos/cm^2 and organic carbon 0.45%. The minimum number of species (21) was recorded from Rahata tahsil, having pH 8.3, electrical conductivity 0.42 m mhos/cm^2 and organic carbon 0.52%.

BGA members have been reported to grow well in pH range 7.5 to 8.5 of the media. The present study also showed that lower pH 8.2 has positive impact on the development and growth of BGA members. Whereas soil samples from Sangamner tahsil (25) area showed average diversity. As the site of over-irrigation passed, there is an increase in the hidden micro flora, particularly, cyanophyceae.

Electrical conductivity and pH had a direct correlation for enhancement of BGA forms. As these parameters decrease algal forms increase. Organic carbon of these sites showed no correlation with algal growth, might be due to excess application of press mud in the fields. Thus to maintain soil micro flora it would be beneficial to switchover to the farming from chemical to organic with proper irrigation.

Table 2 Comparative account of soil parameters

Sr. No.	Name of Tahsil	pH	Electric conductivity m mhos/cm^2	Organic carbon %
1	Sangamner	8.4	0.43	0.44
2	Rahata	8.3	0.42	0.52
3	Shrirampur	8.2	0.35	0.45

(The values in Table 2 are the mean values of 10 observations of each from 14 stations)

ACKNOWLEDGEMENTS

The award of teacher fellowship to BKA from the U.G.C. western regional office, Pune and the facilities extended by principals of P.V.P. College Pravaranagar and R. K. Mahila Mahavidyalaya, Ahmednagar are gratefully acknowledged.

REFERENCES

Anand, N. 1989. Handbook of blue-green algae of rice field of south India, Pub. Bishen Singh Mahendra Pal Singh, Dehradun.

Desikachary, T. 1959. Cyanophyta, Pub. Indian Council of Agricultural Research, New Delhi.

Devi, G. Dorycanta, H. and Singh, N. 1999. Cyanobacterial flora of rice field soils of Manipur. *Phykos.* 38: 13-18.

Metting, B. 1981. The systematics and ecology of soil algae. *Bot. Rev.* 47: 195-312.

Nayak, S., Prasanna, R., Dominic, T. and Singh, P. 2001. Floristic abundance and relative distribution of different cyanobacterial genera in rice field soil at different crop growth stages. *Phykos.* 40: 15-22.

Prescott, G. 1951. Algae of the western great lakes area. Pub. Otto Koeltz Science Publishers, Koeningstein.

Ripka, R., Derulles, J., Waterburry, J., Herdman, M. and Stanier, R. 1979. Genetic assessments, strain histories and properties of pure cultures of Cyanobacteria. J. *Gen. Microbiol.* 111: 1-61.

Santra, S. 1983. Biology of rice field blue-green algae. Daya Publishing House, Delhi, pp. 1-184.

Shinde, M. 1995. Studies on soil characterization and effect of blue-green algae on selected crops from Pravaranagar area, M.S. (Ph.D. Thesis, University of Pune).

Trivedi, R., Goel, P. and Trisal, C. 1998. Practical methods in ecology and environmental science. Pub. Enviro Media, Karad (India), pp. 1-340.

Venkataraman, G. 1975. The role of blue-green algae in tropical rice cultivation. In: Nitrogen fixation by free-living microorganisms. Sd. Stewart W.D.P. Pub. Cambridge Univ. Press, London, pp. 207-218.

10

Chlorophycean Algae of Ahmednagar, Maharashtra, India

B.S. Deshmukh and S.D. Pingle
Post-Graduate Department of Botany, P.V.P. College, Pravaranagar,
Loni-413 713, Maharashtra
India

ABSTRACT

The present investigation was undertaken along five stations of Pravara river in Ahmednagar district. This study was carried for two years survey of algal taxa, which deals with the planktonic, epithic and benthic algae. The total of 155 algal taxa of Chlorophyceae, belonging to 56 genera representing 10 orders and 16 families were investigated. Interesting genera *Ulothrix, Chaetophora, Pediastrum, Coelastrum, Chlorella, Oocystis, Tetraedron, Scenedesmus, Cosmarium, Euastrum and Staurastrum* were collected. *Chaetophora pisiformis* (Roth) C.A. Agardh, *Draparnaldiopsis indica* Bharadwaja, *Golenkinia paucispina* West & West, *Dictyosphaerium ehrenbergianum* Naegeli, *Trochiscia reticularis* (Reinsch) Hansgirg, *Selenastrum bibraianum* Reinsch, *Dichotomosiphon tuberosus* (A. Braun) Ernst, *Enteromorpha intermediata* Bliding, *Ankistrodesmus spiralis* (Turner) Lemm., *Staurastrum ophiura* Lund., *Tetraedron gracile* (Reinsch) Hansgirg were reported for the first time from Maharashtra. The present report can enrich our knowledge of algal flora of the state.

Keywords: Chlorophycean algae, Ahmednagar district.

INTRODUCTION

Freshwater algae have great importance due to their role in the natural environment and serve as the base of food chain in the aquatic ecosystem.

The earlier workers studied the Chlorophycean algal diversity of different aquatic system of Maharashtra (Kamat, 1965,1974; Gunale and Chaugule, 1980; Barhate and Tarar, 1981; Pingle, 1988; Bodas, 1991, Mandhare and Pingle, 1995; Nandan, 1993; Nandan and Kumawat, 2003; Zaware and Pingle, 2003, Pingle and Deshmukh, 2005; More et al., 2005). Looking to the literature cited, an extensive study about the freshwater algal flora from Ahmednagar district, Maharashtra was not found. Hence the work has been undertaken.

The present survey was done along five stations of Pravara river (from its inception to the Ozar pick up weir) in Ahmednagar district. Pravara river is tributary of Godavari river. It rises in the eastern slope of the Sahyadri mountain of Western Ghats. The upper water of river in the amphitheater have been developed in hues reservoir lake, the 'Arthur' behind 'Wilson dam', which impounds 11 thousand million cubic feet of water. For irrigation and drinking purpose, water is released in the river and which is picked up at Ozar pick up weir.

MATERIALS AND METHODS

The algal samples were collected from five sampling stations at monthly interval for a period of two years (2003-05). The collected samples were preserved in 4% formalin. Algae were identified by relevant monographs and recent available literature of Prescott, 1951; Randhawa, 1959; Ramanathan, 1964; Philipose, 1967; Joshi and Krishnamurthy, 1972; Fritsch, 1979; Gonzalves, 1981; Iyengar and Desikachary, 1981; Anand, 1998. Identified taxa were enumerated systematically.

RESULTS AND DISCUSSION

The present investigation reports 155 algal taxa of Chlorophyceae, belonging to 56 genera representing 10 orders and 16 families. Most of them were planktonic and benthic and some were epiphytic. Members of Chlorococcales were found dominant and represented by 25 genera, which is followed by Zygnematales and represented by 14 genera. Order Siphonales and Ulvales are noted only by a single genera and species.

SYSTEMATIC ENUMERATION OF ALGAE

Chlorophyceae

Volvocales

Family–Chlamydomonadaceae

Pandorina morum (Muell.) Bory, *Pandorina cylindricum* Iyengar, *Eudorina elegans* Ehrenberg, *Volvox aureus* Ehrenberg, *Volvox globator* Linn.

Tetrasporales

Family–Asterococcaceae

Sphaerocystis schroeteri Chodat, *Pseudosphaerocystis lacustris* (Lemm.) Novakova, *Chlamydocaspa ampla* (Kuetz.) Fott, *Chlamydocaspa planctonica* (W. et G.S. West) Fott

Ulotrichales

Family–Ulotrichaceae

Ulothrix variabilis Kuetzing, *Ulothrix tenuissima* Kuetz., *Ulothrix cylindricum* Prescott, *Ulothrix zonata* (Weber & Mohr) Kuetz.

Chaetophorales

Family–Chaetophoraceae

Stigeoclonium pachydermum Prescott, *Stigeoclonium tenue* (C.A. Ag.) Kuetz., *Chaetophora elegans* (Roth) C. A. Agardh, *Chaetophora attenuata* Hazen, *Chaetophora pisiformis* (Roth.) C. A. Agardh, *Draparnaldiopsis indica* Bharadwaja

Cladophorales

Family–Cladophoraceae

Cladophora glomerata (L.) Kuetzing, *Pithophora mooreana* Collins, *Rhizoclonium crassipellitum* West & West

Oedogoniales

Family–Oedogoniaceae

Bulbochaete nana var. *longa* Kam., *Oedogonium biformae* (Nordst.) Hirn., *Oedogonium latiusculum* Tiff., *Oedogonium* sps.

Chlorococcales

Family–Chlorococcaceae

Golenkinia radiata (Chod.) Wille, *Golenkinia* paucispina West & West, *Golenkiniopsis minutissima* Iyengar et Balkr.

Family–Characiaceae

Characium acuminatum A. Braun in Kuetzing, *Characium ambiguum* Hermann

Family–Hydrodictyaceae

Hydrodictyon reticulatum (L.) Lagerheim, *Pediastrum simplex* (Meyen) Lemm., *Pediastrum simplex* var. *duodenarium* (Bailey) Rabenhorst, *Pediastrum ovatum* (Ehr.) A. Braun, *Pediastrum duplex* var. *reticulatum* Lagerheim, *Pediastrum duplex* var. *subgranulatum* Raciborski, *Pediastrum duplex* var. *cohaerens* Bohlin, *Pediastrum duplex* var. *gracillimum* W. et G. S. West, *Pediastrum duplex* var. *asperum* (A. Braun) Hansgirg, *Pediastrum boryanum* (Turp.) Meneghini, *Sorastrum spinulosum* Naegeli

Family–Coelastraceae

Coelastrum cambricum Archer, *Coelastrum* sps., *Coelastrum microporum* Naegeli, *Coelastrum reticulatum* (Dang.) Senn, *Coelastrum sphaericum* Naegeli

Family–Oocystaceae

Chlorella ellipsoidea Gerneck, *Chlorella vulgaris* Beyerinck, *Dictyosphaerium, ehrenbergianum* Naegeli, *Dictyosphaerium pulchellum* Wood, *Dictyosphaerium reniforme* Bulnheim, *Trochiscia reticularis* (Reinsch) Hansgirg, *Planktosphaeria gelatinosa* G. M. Smith, *Oocystis irregularis* (Petkof) Printz, *Oocystis borgei* Snow, *Oocystis pusilla* Hansgirg, *Oocystis elliptica* W. West., *Treubaria triappendiculata* Bernard, *Pachycladon umbrinus* G. M. Smith,

Nephrocytium lunatum W.West, *Ankistrodesmus spiralis* (Turner) Lemm., *Ankistrodesmus falcatus* (Corda) Ralfs, *Ankistrodesmus spiralis* var. *fasciculatus* G.M. Smith, *Closteriopsis longissima* Lemm., *Selenastrum bibraianum* Reinsch, *Kircheneriella lunaris* (Kirchner) Moebius, *Kircheneriella lunaris* var. *irregularis* G. M Smith, *Tetraedron regulare* Kuetzing, *Tetraedron pusillum* (Wallich) W. et G. S. West, *Tetraedron enorme* var. *pentaedricum* Prescott, *Tetraedron limneticum* Borge, *Tetraedron gracile* (Reinsch) Hansgirg, *Closteridium obesum* (W. et G.S. West) G. M. Smith, *Cerasterias staurastroides* West & West.

Family–Scenedesmaceae

Scenedesmus bijugatus (Turpin) Kuetzing, *Scenedesmus bijugatus* var. *alternans* (Reinsch) Hansgirg, *Scenedesmus bijugatus* var. *flexuosus* Lemm., *Scenedesmus bijugatus* var. *irregularis* Wille, *Scenedesmus bijugatus* var. *graevenitzii* (Bernard) comb. nov., *Scenedesmus arcuatus* var. *platydisca* G. M. Smith, *Scenedesmus acuminatus* (Lagerheim) Chodat, *Scenedesmus opoliensis* P. Richter, *Scenedesmus dimorphus* (Turpin) Kuetzing, *Scenedesmus indicus* sp. nov., *Scenedesmus denticulatus* Lagerheim, *Scenedesmus muzzanensis* Huber-Pestalozzi, *Scenedesmus platydiscus* (G.M. Smith) Chodat, *Scenedesmus hytrix* Lagerheim, *Scenedesmus quadricauda* (Turpin) Brebssion, *Scenedesmus abundans* (Kirch.) Chodat, *Scenedesmus* sps., *Actinastrum hantzschii* Lagerheim, *Actinastrum hantzschii* var. *elongatum* G. M. Smith, *Crucigenia lauterbornii* Schmidle, *Crucigenia rectangularis* (A. Braun) Gay, *Micractinium pusillum* var. *elegans* G. M. Smith, *Micractinium Pusillum* Fresenius.

Siphonales

Family–Dichotomosiphonaceae

Dichotomosiphon tuberosus (A. Braun) Ernst

Zygnematales

Family–Zygnemataceae

Mougeotia abnormis Kisselew, *Mougeotia genuflexa* (Dillw.) C. A. Agardh, *Mougeotia floridana* Transeau, *Spirogyra aequinoctialis* G. S West, *Spirogyra stictica* (Engl.Bot.) Wille, *Spirogyra mirabilis* (Hass.) Kuetzing, *Spirogyra formosa* (Trans.) Czurda, *Spirogyra crassa* Kuetzing, *Spirogyra rivularis* (Hass.) Rabenhorst, *Spirogyra communis* (Hass) Kuetzing, *Zygnema pectinatum* (Vauch.) C. A. Agardh, *Zygnema sterile* Transeau *in* Transeau, *Zygnema gangeticum* Rao, *Zygnema extenue* Jao., *Zygnemopsis spiralis* (Fritsch) Transeau *in* Transeau, *Zygnemopsis pectinata* (Fritsch) Transeau

Family–Desmidaceae

Closterium cynthica De. Not., *Closterium gracile* Breb., *Cylindrocystis brebissonii* Menegh, *Cosmarium cyclicum* Lund, *Cosmarium cyclicum* var. *articum* Nordst., *Cosmarium contractum* Kirchn, *Cosmarium contractum* var. *incrassatum* Scott & Prescott, *Cosmarium, pseudoprotuberans* var. *tumidum* Borge, *Cosmarium speciosissimum* Schmidle, *Cosmarium granatum* Breb., *Cosmarium hexagonum* Nordstedt, *Cosmarium auriculatum* var. *bogoriense* Bernard, *Cosmarium obtusatum* (Schmidle) Schmidle, *Cosmarium botrytis* Ehr., *Cosmarium baileyi* wolle, *Cosmarium viridis* (Chorda) Josh, *Cosmarium bipunctatum* Borgesen, *Cosmarium pyramidatum* Breb., *Cosmarium scrobiculosum* Borge, *Cosmarium protuberans* var.

semicircularis Salazar C. ver. nov, *Cosmarium dispersum* Johnson, *Cosmarium formosulum* Hoff, *Spondylosium planum* (Wolle) W. & G. S. West, *Euastrum dubium* Nageli, *Euastrum subintegrum* Nordsted, *Euastrum ansatum* var. *didelitiforme* Duc., *Euastrum spinulosum* Delponte, *Staurastrum manfeldtii* Delp., *Staurastrum anatinum* Cooke and Wills, *Staurastrum lunatum* Ralfs, *Staurastrum chaetoceras* (Schrod.) G. M. Smith, *Staurastrum leptocladum* Nordst., *Staurastrum* sps., *Staurastrum ophiura* Lund., *Staurastrum brasiliense* var. *lundellii* W. & G. W. West, *Staurastrum anatinum* forma *paradoxum* Brook, *Xanthidium antilopaeum* (Breb.) Kuetzing, *Anthrodesmus mucronulatus* Nordst., *Micrasterias zeilanica* Fritsch.

Ulvaeles

Family–Ulvaceae

Enteromorpha intermediata Bliding

Interesting and dominant genera *Ulothrix, Chaetophora, Pediastrum, Coelastrum, Chlorella, Oocystis, Tetraedron, Scenedesmus, Coesmarium, Euastrum, Staurastrum* were observed. *Coelastrum* was found in bloom formation while *Spirogyra* and *Zygnema* were contributed in mat formation. The species *Chaetophora pisiformis* (Roth) C. A. Agardh, *Draparnaldiopsis indica* Bharadwaja, *Golenkinia* paucispina West & West, *Dictyosphaerium ehrenbergianum* Naegeli, *Trochiscia reticularis* (Reinsch) Hansgirg, *Treubaria triappendiculata* Bernard, *Scenedesmus indicus* sp. nov., *Selenastrum bibraianum* Reinsch, *Dichotomosiphon tuberosus* (A. Braun) Ernst, *Enteromorpha intermediata* Bliding, *Ankistrodesmus spiralis* (Turner) Lemm., *Staurastrum ophiura* Lund., *Tetraedron gracile* (Reinsch). Hansgirg were reported for the first time from Maharashtra. This report can enrich our knowledge of algal flora of the state.

ACKNOWLEDGEMENTS

Authors thank the principals of Arts, Commerce and D. R. Science College, Akole and P. V. P. College, Pravaranagar for providing valuable co-operation during research work. Also thankful to the UGC, New Delhi for the awarding of Teacher Fellowship under 10th plan period.

REFERENCES

Anand, N. 1998. Indian fresh water micro algae. Bishen Singh Mahendra Pal Singh, Deharadun.

Barhate, V. and Tarar, J. 1981. The algal flora of Tapi river, Bhusawal, Maharashtra. *Phykos.* 20: 75-78.

Bodas, K. 1991. Hydrobiological and Taxonomical studies of some Lotic and Lentic waters in and around Nashik. *Biol. Ind.* 2(1&2): 13-21.

Fritsch, F. 1979. The structure and reproduction of the algae. Vikas Pub. House Pvt. Ltd., New Delhi.

Gonzalves, E. 1981. Oedogoniales. I.C.A.R., New Delhi.

Gunale, V. and Chaugule, B. 1980. Algae of sludge samples from Pavana, Mula and Mutha rivers. *M. V. M. Patrika.* 15(2): 75-78.

Joshi, H. and Krishnamurthy, V. 1972. The species of *Enteromorpha* from India. *Bot. J. Linn. Soc.* 65: 119-128.

Kamat, N. 1965. Ecological note on the algae of Kolhapur. *J. Biol. Sci.* 8: 47-54.

Kamat, N. 1974. Algae of Marathwada (MS). *Phykos.* 13: 22-32.

Mandhare, S and Pingle, S. 1995. Occurrence and tolerance of algae in distillery waste water. *J. Aqua. Bio.* 10 (1&2): 31-33.

More, Y., Mali, M. and Nandan, S. 2005. Algal diversity of Panzara river (Maharashtra). *Eco. Env. & Cons.* 11(2): 319-323.

Nandan, N. 1993. Algal flora of fish pond in Dhule, Maharashtra. *Indian Bot. Rept.* 12 (1+2): 61-63.

Nandan, S. and Kumawat, M. 2003. Limnological study of algae of Aner river of Dhule (MS), *India. Eco. Env. & Cons.* 11(2): 9-11.

Pingle, S. 1988. Algae of Khadakwasla Lake. *Indian Bot. Rept.* 7(1&2): 91-92.

Pingle, S. and Deshmukh, B. 2005. Some fresh water algae from Amphitheatre of Wilson Dam. *Indian Hydrobiology*, 7(Suppl.): 97-100.

Philipose, M. 1967. Chlorococcales. I.C.A.R., New Delhi.

Prescott, G. 1951. Algae of the Western Great Lakes Area, Cranbook Inst. of Sci., Bloomfield Hills, Mich.

Ramanathan, K. 1964. Ulotriachales. I.C.A.R., New Delhi.

Randhawa, M. 1959. Zygnemaceae. I.C.A.R., New Delhi.

Iyengar, M. and Desikachary, T. 1981. Volvocales. I.C.A.R., New Delhi.

Zaware, B. and Pingle, S. 2003. Fresh water algae from Pashan lake. *Ind. Hydrobiol.* 6(1&2): 9-11.

11

Studies on Hydrobiology of Godawari River at Nasik, Maharashtra, India

R.R. Sanap, A.K. Mohite[1], S.D. Pingle[2] and V.R. Gunale
Department of Environmental Science, University of Pune,
[1]R.B.N.B. College, Shrirampur. Dist. Ahmednagar
[2]P.G. Department of Botany, P.V.P. College, Pravaranagar. Dist. Ahmednagar, Maharashtra, India

ABSTRACT

The water quality of Godawari river was studied for a period of one year. (June 2004 –May 2005). Monthly collections of water samples from five different sampling stations were carried out and analyzed for physicochemical parameters and phytoplankton studies. Present study revealed that station-wise seasonal variation was noticed in majority of parameters and algal groups. Raised values of TDS, alkalinity, chlorides, hardness, BOD and COD showed considerable pollution load on river water. During the present investigation in all, 85 genera and 148 species of phytoplankton were encountered. Chlorophyceae was dominant group consisting of 54 genera and 84 species followed by Bacillariophyceae consisting 16 genera and 34 species. Cyanophyceae ranked third and comprised 11 genera and 18 species while Euglenophyceae comprised 3 genera and 11 species. Dinophyceae was sparsely reported having only one genus and species. Phytoplankton population was dominant during summer while it was declined during monsoon season. Some physicochemical parameters showed ranges beyond permissible limits suggested by WHO and ISI. Planktonic groups also showed pollution tolerant genera indicating pollution of riverine ecosystem which is unfit for drinking purpose.

Keywords: Godawari river, water quality, phytoplankton, pollution, BOD, COD.

INTRODUCTION

Godawari river is the largest river flowing through South India. It emerges out at Trymbakeshwar in Western Ghats just 30 km upstream of Nasik city. Flowing through Maharashtra, AP, it joins the Bay of Bengal. Nasik is one of the largest cities situated at the bank of river at starting stretch. In the river water, huge quantity of domestic wastes, municipal sewage, faecal matter is being dumped daily in addition to industrial effluents, agricultural run-off and thermal power effluents resulting the organic pollution of river water. Many workers have studied the hydrobiology of many rivers. Pandey et al. (1995), Vyankateswarlu and Jayanti (1968), Ragothaman and Jaswal (1995). However, hydrobiological studies of Godawari river at starting stretch remain untouched. Therefore present study was undertaken to access the water quality of Godawari river.

MATERIALS AND METHODS

Five sampling stations were selected along the river covering a stretch of 50 km. Station I was situated upstream of Nasik city. Station II was the Ramkund at Panchvati, III was near village Eklahare, IV was point at confluence of Darna and Godavari river, while V was Nandur Madhameshwar pick up well. Monthly collections of water samples were made for 1 year (June 2004-May 2005). For analysis of physicochemical parameters, water samples were collected in plastic containers while for phytoplankton studies, plankton collections were made by using plankton net (200 mesh/ linear inch) and preserved in 4% formaldehyde and Lugol's solution. Parameters like pH, light penetration and temperature were detected at sampling stations while remaining parameters were analyzed after reaching the laboratory. For physicochemical parameters standard methods described in APHA (1985), Trivedi and Goel (1986), were followed. Phytoplankton identification was done with the help of Desikachary (1959), Philipose (1967), Prescott (1962), and Sarode and Kamat (1984).

RESULTS AND DISCUSSION

Seasonal data of physicochemical parameters is depicted in Table 1. It showed fluctuation of parameters throughout the year. Parameters like temperature, pH (except during summer at certain stations), nitrates, phosphates, free CO_2 and D.O. showed minimum values which were within the permissible limits suggested by WHO and ISI. While parameters like light penetration, TDS, chlorides, alkalinity, hardness, BOD and COD showed raised values, which were beyond the permissible limits suggested by WHO and ISI.

Light penetration is low during monsoon season due to flooding condition and soil erosion. While E.C. at some stations showed the values beyond permissible limits suggested by WHO i.e. 1m mho/cm^2. It might be due to impact of effluence and chloride contents. The pH values during summer season is low and might be due to discharge of industrial effluents. During winter and summer chlorides values were also more than permissible limits, which might be due to release of municipal sewage and domestic waste in river water. Venkateswarlu and Jayanti (1968) were of opinion that high values of chlorides in summer may be due to quantity of water and discharge of municipal sewage.

Alkalinity values were frequently increased up to 320 mg/lit by addition of municipal sewage of Nasik city and industrial wastes, which were more than permissible limits of WHO i.e. 100 mg/lit. Detergents, soaps, sodium hydroxide, potassium hydroxides were main contributors to increase the alkalinity. According to Sanker et al. (2002), the high alkalinity values during summer might be due to concentration of salts in water and dumping of domestic wastes. BOD values were also increased during summer and riches up to 88 mg/lit which lies more than WHO standards i.e. 6 mg/lit. BOD refers the amount of O_2 used by microorganisms in aerobic oxidation of organic matter. So BOD increases with amount of dead organic matter and stagnant water

Table 1 Seasonal variation in physicochemical parameters of Godawari river (June 2004-May 2005)

Sr. No.	Parameter	Ranges of Parameter			WHO standard
		Monsoon	Winter	Summer	
1.	Light penetration (in cm)	15 - 80	13 - 168	16 - 128	——
2.	Temp. (°C)	16 - 23	16 - 30	18 - 32	40°C
3.	E.C. m mho/cm^2	0.24 - 1.13	0.26 - 1.39	0.31 - 1.06	1.00 m mho/cm^2
4.	T.D.S	154 - 570	166 - 816	218 - 640	500 mg/lit.
5.	pH	7.1 - 8.2	7 - 8.4	6 - 8.1	7- 8.5
6.	Chlorides	85 - 190	47 - 215	50 - 221	200 mg/lit
7.	Nitrates	0.98 - 1.98	0.01 - 2.04	0.96 - 1.98	45 mg/lit
8.	Phosphate	1.1 - 2.33	0.98 - 1.9	0.56 - 2.05	——
9.	Alkalinity	38.1 - 320	58 - 252	45 - 230	100 mg/lit
10.	Hardness	64 - 218	69 - 200	48 - 151	100 mg/lit
11.	BOD	1.28 - 54.3	7.2 - 39	44 - 88	6 mg/lit
12.	COD	61 - 448	138 - 302	122 - 398	10 mg/lit
13.	Free CO_2	1.1 - 4.4	2.2 - 11	0 - 24.2	30 mg/lit
14.	D.O.	0.9 - 1.8	1.1 - 8.1	1.4 - 6.8	5 mg/lit

*All values expressed in mg/lit except light penetration, temperature and EC.

condition. COD showed high values at all stations during investigation period and ranged from 64 to 448 mg/lit which were more than permissible limits suggested by WHO i.e. 10 mg/lit. Organic matter, thermal power effluents and anthropogenic activities were main contributing factors for high COD. Throughout the investigation period mean total dissolved solid values were high and it may have resulted due to the presence of a larger quantity of colloidal soil particles.

PERIODICITY OF PHYTOPLANKTON

The present study revealed that there was a seasonal variation in number of genera and species of different phytoplankton group and showed high degree of abundance. Table 2 shows monthly and seasonal variation in phytoplankton.

Total crop of phytoplankton varied from season to season. The greatest crop of phytoplankton of Godawari river was recorded during winter and summer seasons and lowest during rainy season. Water flow and flooding condition during monsoon affects the abundance. Lakshminarayana (1965) made similar observations from the Ganges.

Chlorophyceae was dominant group throughout the year. It's more no. was recorded and might be due to eutrophic condition, raised values of BOD and CO_2 during winter and summer season. At all stations most common pollution tolerant genera were recorded as suggested by Palmer (1969). These are *Ankistrodesmus, Chlorella, Scenedesmus, Closterium, Chlamydomonas, Pediastrum, Actinastrum, Stigeoclonium, Coelastrum, Ulothrix, Cladophora and Crucigenia.*

Bacillariophyceae was second largest dominant group during winter season. It was represented by 16 genera and 34 species. Dominancy of diatoms during winter was similar to the observations reported by Lakshminarayana (1965).

Table 2 Monthly and seasonal variation in phytoplankton of Godawari river (June 2004 - May 2005) monsoon season

Month and Year	Chloro-phyceae		Cyano-phyceae		Bacillario-phyceae		Eugleno-phyceae		Dino-phyceae		Total	
	G	S	G	S	G	S	G	S	G	S	G	S
June 2004	14	17	4	4	8	11	1	1	-	-	27	33
July	18	20	5	5	10	14	1	1	-	-	34	40
Aug	30	39	5	7	11	12	2	4	-	-	48	62
Sept	17	26	6	6	7	9	2	5	-	-	32	46
Total	79	102	20	22	36	46	6	11	-	-	141	181
					Winter Season							
Month and Year	Chloro-phyceae		Cyano-phyceae		Bacillario-phyceae		Eugleno-phyceae		Dino-phyceae		Total	
	G	S	G	S	G	S	G	S	G	S	G	S
Oct	28	39	6	6	12	23	2	4	1	1	49	73
Nov	21	33	7	8	17	25	2	5	1	1	48	72
Dec	31	46	8	8	16	27	2	4	1	1	58	86
Jan 2005	30	52	7	9	15	27	3	6	1	1	56	95
Total	110	170	28	31	60	102	9	19	4	4	211	326
					Summer Season							
Month and Year	Chloro-phyceae		Cyano-phyceae		Bacillario-phyceae		Eugleno-phyceae		Dino-phyceae		Total	
	G	S	G	S	G	S	G	S	G	S	G	S
Feb 2005	39	57	9`	11	12	15	3	8	1	1	64	92
March	32	42	10	13	10	21	3	10	1	1	56	87
April	37	57	10	13	12	18	3	9	1	1	63	98
May	45	62	10	15	10	16	3	7	1	1	69	101
Total	153	218	39	52	44	70	12	34	4	4	252	378

During winter due to cold conditions and low temperature diatoms flourishes (Ragothaman and Jaiswal, 1995). Dominant genera recorded were *Cyclotella, Fragillaria, Nitzchia, Synedra, Navicula, Gomphonema, Cymbella, Pinnularia, Melosira, Mastogloia,* etc. Domestic waste, municipal sewage, low water level and raised values of alkalinity, nitrate and organic matter favours the growth of diatoms.

Cyanophyceae was ranked third and represented by 11 genera and 18 species. During summer season due to eutrophic condition, stagnant water, availability of more sunlight and CO_2 for photosynthesis this group flourishes. The most common pollution tolerant genera recorded were *Spirulina, Oscillatoria, Anabaena, Lyngbya, Anacystis,* etc.

Euglenophyceae was the fourth group comprising 3 genera and 11 species. These genera were *Phacus, Euglena* and *Haematococous.* During late winter and summer, their number was increased while during monsoon it was sparsely recorded. High values of DO during winter were responsible for maximum number of Euglenoids. (Pandey et al., 1995). Low pH also favours the growth of Euglenoids, (Rishi and Kacharoo, 1984).

Dinophyceae ranked last as represented by single genus *Ceratium* and was found during summer when water was stagnant and more polluted. Fluctuation in water flow of Godawari river due to rains and floods with change in light penetration, affects the growth of phytoplankton. So, during monsoon, phytoplankton population was low.

CONCLUSION

In the present investigation of Godawari river physicochemical parameters fluctuate during various seasons of year (June 2004-Feb 2005) River water shows raised values of TDS, chlorides, alkalinity, hardness, BOD, COD and low pH at certain stations during summer indicates the pollution of river. Physical factors like water flow, light penetration and temperature affects the growth of phytoplankton. Alkaline water and rapid utilization of CO_2 with intense sunlight during summer favours the growth of phytoplankton. During winter the diatoms were flourished due to domestic waste, alkalinity, higher pH and organic matter. During investigation period at all stations different pollution tolerant genera were recorded. Due to deposition of municipal sewage, domestic waste, faecal matter at Nasik city, this riverine ecosystem got more polluted particularly during winter and summer seasons. To protect this ecosystem there should be proper management and planning of dumping the municipal sewage for health hygiene and sustainable environment.

ACKNOWLEDGEMENTS

The first two authors are grateful to UGC for providing FIP fellowship. Authors are also thankful to Head, Department of Environmental Science, University of Pune, and Principal of S. S. G. M. College Kopargaon for encouragement.

REFERENCES

APHA, AWWA, WEDF. 1985. Standard methods for the examination of water and waste water, 16th edition, American Public Health Association,Washington, USA.

Desikachary, T. 1959. Cyanophyta. Indian Council of Agricultural Research, New Delhi, pp. 1-688.

Lakshminarayana, J. 1965. The seasonal growth and succession of plankton algae in the river Ganges. *J. Hydrobiologia.* 25: 138-165.

Palmer, C. 1969. Composite rating of algae of tolerating organic pollution. *British Phycology Bulletin.*5 : 78-92.

Pandey, B., Jha, A., Das, P., Pankaj, P. and Mishra, A. 1995. On the seasonal abundance of phytoplankton in relation to certain ecological conditions in the stretch of Kosi river. *Poll. Res.* 14(3): 283-293.

Philipose, M. 1967. Chlorococcales, I.C.A. R., New Delhi, pp.1-365.

Prescott, G. 1962. Algae of the Western Great Lakes Area. Wm. C. Brown Co. Dubeque, Iowa. pp. 1-935

Ragothamem G. and Jaiswal R (1995) Studies on the hydrobiology of Tapi river from Jalgaon region (Maharashtra) with reference to phytoplankton. *Poll Res.* 14(2): 181-194.

Rishi, V. and Kachroo, P. 1984. Euglenophyceae of Dudh Ganga stream, Kashmir, *Phykos.* 23 (1 and 2): 65-70.

Sanker, P., Jayaraman, P. and Ganga, Devi, T. 2002. Studies on the hydrography of a lotic ecosystem. 'Killar' at Thiruvananthapuram, Kerala, India. *Poll. Res.* 21(2): 113-121.

Sarode, P. and Kamat, N. 1984. Fresh water diatoms of Maharashtra, Sai Krupa Prakashan, Aurangabad.

Trivedi, R. and Goel, P. 1986. Chemical and biological methods for water pollution studies. Environmental Publications, Karad, Maharashtra, India. Pp. 248.

Venkateswarlu, T. and Jayanti, T. 1968. Hydrological studies on the river Sabarmati to evaluated water quality. J *Hydrobiologia.* 31 : 442-448.

12

Impact of Heavy Metal(s) on Accessory Pigments of *Anabaena oryzae* (Fritsch)

***Ch. Sreedhar Rao, B. Raju, M. Venkateshwarulu and B. Digamber Rao**
*Department of Botany, Kakatiya University, Warangal, Andhra Pradesh

ABSTRACT

Cyanobacteria exhibit an extraordinary resistance to many environmental factors including metal pollution. *Anabaena oryzae* 4trichomes collected from the rice fields were used to study the basic mechanisms underlying the toxicity of heavy metal(s) tolerance. The heavy metals namely manganese (II), cobalt (III), copper (II), zinc (II) and mercury (II) were used to study the chlorophyll, carotenoid and phycobiloprotein contents of the cyanobacterium. Observations were made on every 4th day for the short-term duration of 28 days. The species growth response under different concentrations (0.001, 0.01, 0.1 & 1.0%) was determined and we reported heavy metal tolerance by using different parameters. The accessory pigments content decreased drastically, revealing the toxic effects that intracellular heavy metals imposes on cellular metabolism and extracellularly, the metal ions may bind to mucilaginous material found outside the Gram-negative cell wall. The mercury, zinc, copper treated trichomes at higher concentrations showed lethality to the cells and manganese and cobalt showed less impact on the cells under investigation. The toxicity of the studied metals were analyzed and are in the order of Hg > Cu > Zn > Co > Mn.

Keywords: Heavy metal(s), toxicity, *Anabaena oryzae,* accessory pigments.

INTRODUCTION

Metals occur naturally, and several of them are essential components of global ecosystems. Heavy metals, depending on their oxidation states can be highly reactive and as a consequence, toxic to most organisms. They are produced by expanding a variety of anthropogenic sources suggesting an increasingly important role for this form of pollution. Environmental pollution by metals became extensive as mining and industrial activities increased severely. Mine production includes Cu, Cd, Pb, and Hg in considerable amounts (Kennish, 1996). These pollutants, ultimately derived from growing number of diverse anthropogenic sources industrial effluents and wastes, urban run-off, sewage treatment, plants, boating activities, agricultural fungicide run-off, domestic garbage dumps, etc. have progressively affected more and more ecosystems

(Macfarlane and Burchett 2001, Wickfors and Ukeles 1982, Rebhun and Amotz 1984, Cotte-Krief et al. 2000, Bu-Olayan et al. 2001, Esser and Volpe 2002).

Cyanobacteria are important providers of nitrogen fertilizer used in cultivation of rice and beans due to their ability to fix atmospheric nitrogen. Nutrient availability (in particular, nitrogen) dramatically increased the ability of algae to accumulate heavy metals (Wang and Dei, 2001 a, b) suggesting that agricultural run-off into the fresh water and coastal areas will greatly increase the entry of heavy metals into the food chain. The physiological and toxicological effects imparted by metal ions are considered to be biological continuum, since algae are considered to be basic in food chain as primary producers.

The heavy metals can cause membrane depolarization and acidification of cytoplasm (Cumming and Gregory, 1990, Conner and Schimid, 2003), and infact membrane injury is one important effect of metal ions that may lead to cellular disruption of homeostasis. Thus cellular adaptation such as exudation of chelating compounds and active efflux of metal ions by primary ATPase pumps can provide some degree of metal tolerance (Cumming and Gregory, 1990; Rosen, 1996). Though living organisms for various metabolic processes need heavy metals (Whitton and Say, 1975) their excess concentration showed lethal effect. The accumulated metals are chemically detoxified and physically sequestered by the algae rendering them inactive. The toxic effects of metals, their uptake and accumulation and detoxification mechanisms operating in the tolerant algae vary from species to species. It is essential to study the toxicity of metals towards algae and the influence of various environmental factors on their toxicity, uptake and accumulation. Various detoxification mechanisms in algae and changes in their metabolism on exposure to the heavy metals are to be understood in order to develop strategies for the treatment of industrial effluents, recycling of wastewater and recovery of essential metals using algae. The present study was attempted to obtain an insight into some of the aspects of metal toxicity in the trichomes of *Anabaena oryzae*.

MATERIALS AND METHODS

Anabaena oryzae was isolated from paddy fields of Warangal district, Andhra Pradesh. One single trichome from each colony was isolated (Venkataraman, 1979) and transferred to fresh solid medium and subjected to repeated sub-culturing on fresh solid media before transfer to sterilized liquid BG-11 medium. The culture illumination was provided by fluorescent tube lamps giving a light intensity of 16:8 hours of light: dark cycles at 28±2°C. Filaments from actively growing healthy cultures were transferred to fresh medium one day prior to the treatment with heavy metals. In the experiment an aliquot of 30 ml of the exponentially growing algal trichomes are transferred to sterilized Erlenmeyer flasks supplemented with 300 ml of nutrient medium containing separately 0.001%, 0.01%, 0.1%, & 1.0% of manganese, cobalt, zinc, copper and mercury. The treated filaments were observed for short-term duration from the 4th day and then continued for the period of 28 days (4th, 8th, 12th, 16th, 20th, 24th & 28th) to screen the morpho-physiological variations. The total carotenoid in the *NN*-Dimethyl formamide (DMF) extract was estimated by the method of Chamovitz et al. (1993) and the total chlorophyll content is measured from the dimethyl formamide extract by the method of Moran, (1982) and phycobiloprotein were estimated by the method of Bennet and Bogorad, (1971). Optical density was recorded by using the UV-SL-171 spectrophotometer at 461 nm for carotenoids, 664 nm for chlorophyll and 615, 650 and 565 nm for phycocyanin, allolophycocyanin, and phycoerythrin respectively.

Results

The results present in Table 1, Table 2 & Table 3 give a clear picture about the biochemical, and that morphological changes were dependent on the concentration and duration of exposure of the cells to the heavy metals. *Anabaena oryzae* exposed to 0.001%, 0.01%, 0.1% and 1% concentrations of manganese, cobalt, zinc, copper and mercury revealed different abnormalities in the trichomes cells, which has been induced by metal ions. Observations were made initially on the 4th day after the treatment and then continued on every four days for a period of 28 days. The trichomes exposed to higher concentrations of metals exhibited severe bleaching, chlorosis, reduction in cell size with shrunken chloroplast, enlargement of cells, fragmentation, loss of cellular contents, cell lysis, diad cells, tetrad cells and clumping of cells.

Table 1 Effect of heavy metals on carotenoid content in mg/ml over different durations

Metals		4th Day	8th Day	12th Day	16th Day	20th Day	24th Day	28th Day
	Control	1.934	3.392	1.982	0.51	0.486	0.27	0.235
	0.001%	2.966	4.377	1.795	1.521	0.897	0.225	0.161
Manganese	0.01%	2.385	3.043	2.362	1.49	0.613	0.302	0.1
	0.1%	2.945	2.995	1.569	0.789	0.394	0.092	0.023
	1.0%	2.96	3.223	1.179	0.368	0.083	0.021	0.12
	0.001%	3.225	5.359	2.81	1.494	0.792	0.193	0.043
Cobalt	0.01%	3.676	4.604	1.983	0.428	0.107	0.085	0.035
	0.1%	3.133	4.644	1.983	0.45	0.09	0.065	0.029
	1.0%	3.953	4.125	1.234	0.306	0.069	0.018	0.004
	0.001%	2.943	4.312	1.672	0.444	0.223	0.055	0.04
Zinc	0.01%	2.778	3.178	1.428	0.324	0.19	0.045	0.024
	0.1%	2.058	3.082	0.316	0.292	0.073	0.036	0.018
	1.0%	2.416	2.189	0.654	0.229	0.055	0.015	0.008
	0.001%	2.072	6.652	0.446	0.412	0.206	0.051	0.025
Copper	0.01%	2.298	5.277	0.412	0.357	0.178	0.045	0.018
	0.1%	1.86	4.631	0.433	0.298	0.15	0.03	0.012
	1.0%	2.15	2.277	−0.092	−0.264	−0.312	−0.049	−0.034
	0.001%	1.327	2.409	0.363	−0.082	−0.021	−0.044	−0.063
Mercury	0.01%	1.069	1.698	0.355	−0.249	−0.298	−0.312	−0.356
	0.1%	0.682	1.2	−0.423	−0.486	−0.51	−0.538	−0.596
	1.0%	0.739	1.44	−0.546	−0.596	−0.64	−0.686	−0.712

Manganese Chloride

Manganese plays a vital role in synthesis of glycolate and galactocyl glyceride and for activity of photosystem II. Manganese has an important role in enzymatic reactions in Krebs cycle. Results from Table 1 show that manganese at 0.001% on the 8th day showed maximum carotenoid content

(4.377 mg/ml) over the control. A result from Table 2 indicates that manganese at 0.001% showed maximum chlorophyll content (14.935 mg/ml) over the control and also on other metals studied. However at 1.0% trichomes showed chlorosis inhibiting the algal growth. The result is coincides with Csatorday et al. (1984) which might be due to the inhibition of chlorophyll biosynthesis induced by the metal. Table 3 reveals that in 0.001% concentration phycocyanin content (0.124 mg/ml) was maximum, on the 8th day, over the control, and of all the metals studied in the experiment.

Cobalt Nitrate

Cobalt is one of the essential trace elements for the growth of nitrogen fixing and non-nitrogen fixing cyanobacteria. Cobalt specificity affects protein-synthesizing machinery. In the present observation results from Table 1 show that cobalt at 0.001% concentration on the 8th day showed maximum carotenoid content (5.359 mg/ml) over the control and of all the concentrations studied. Results from Table 2 show that at 0.001% concentration on the 8th day showed maximum chlorophyll content (14.804 mg/ml) of all the concentrations (0.01%, 0.1%, & 1.0%) and was over the control.

Table 2 Effect of heavy metals on total chlorophyll content in mg/ml over different durations

Metals		4thDay	8thday	12thday	16thday	20thday	24thday	28thday
	Control	7.032	12.432	6.317	1.406	0.366	0.092	0.045
	0.001%	9.965	14.935	8.49	1.311	0.305	0.077	0.034
Manganese	0.01%	8.178	13.767	7.581	1.277	0.286	0.074	0.026
	0.1%	7.834	13.29	5.325	1.215	0.244	0.061	0.012
	1.0%	6.678	13.016	5.544	1.187	0.223	0.051	0.004
Cobalt	0.001%	12.372	14.804	8.344	1.644	0.411	0.105	0.036
	0.01%	11.562	12.408	7.795	1.43	0.36	0.09	0.026
	0.1%	10.596	12.194	7.152	1.394	0.322	0.08	0.016
	1.0%	13.398	12.623	0.905	0.727	0.132	0.03	0.008
	0.001%	10.918	14.768	8.57	1.37	0.34	0.085	0.04
Zinc	0.01%	7.867	7.914	5.423	0.786	0.152	0.036	0.028
	0.1%	7.509	6.83	3.218	0.57	0.194	0.046	0.022
	1.0%	7.152	5.364	3.003	0.678	0.169	0.042	0.013
	0.001%	12.909	4.934	1.292	1.192	0.293	0.073	0.037
Copper	0.01%	12.582	3.278	1.406	1.703	0.426	0.106	0.021
	0.1%	10.871	2.503	2.133	1.252	0.31	0.092	–0.023
	1.0%	4.8	3.266	0.405	0.286	–0.072	–0.021	–0.161
	0.001%	5.244	4.076	1.072	–0.643	–0.161	–0.044	–0.011
Mercury	0.01%	4.839	2.741	–0.81	–0.715	–0.134	–0.032	–0.008
	0.1%	4.231	2.384	–0.774	–0.774	–0.241	–0.318	–0.116
	1.0%	3.671	1.573	–3	–0.858	–1.317	–2.436	–3.535

Zinc Sulphate

Zinc is one of the important micronutrients for the growth and metabolism of algae and plays a key role in maintaining the integrity of ribosomes. At high concentrations zinc inhibited the growth of algae. Results from Table 1 revealed at 0.01%, 0.1% and 1.0% concentration on the 4th day (2.778 mg/ml, 2.058 mg/ml, 2.416 mg/ml) and at 0.001% concentration on the 8th day (4.312 mg/ml) showed more carotenoid content over the control. The trichomes were observed with shrunken cells with enlarged heterocysts at 0.01% and 0.1% concentration treated cells appeared with maximum chlorosis. The results obtained were in accordance with other workers, Guikema and Sherman, (1983); Rosko and Rachlin, (1977); Gupta and Verma, (1989). Results from Table-2 at 0.001% on the 4th day (12.372 mg/ml), 8th day (14.768 mg/ml), 12th day (8.57 mg/ml), and at the concentrations of 0.01%, 0.1% and 1.0% on the 4th day showed chlorophyll content over the control. The phycobiloprotein content from Table 3 shows that at 0.001% concentration allphycocyanin content on the 8th day, 24th day, and 28th day and at 0.01% phycocyanin content on the 16th day, 20th day and phycoerythrin content on the 20th day showed more than control.

Table 3 Effect of heavy metals on phycobiloprotein content in mg/ml over different durations

Metals			4 th Day	8 th Day	12 th Day	16 th Day	20 th Day	24 th Day	28 th Day
		PC	0.13	0.11	0.095	0.085	0.076	0.07	0.065
	Control	APC	0.125	0.106	0.11	0.095	0.085	0.06	0.055
		PE	0.135	0.112	0.106	0.08	0.08	0.075	0.06
		PC	0.11	0.124	0.115	0.11	0.075	0.06	0.04
	0.001%	APC	0.115	0.11	0.105	0.095	0.08	0.055	0.035
		PE	0.095	0.12	0.1	0.08	0.07	0.06	0.045
		PC	0.095	0.11	0.09	0.08	0.07	0.065	0.035
Manganese	0.01%	APC	0.11	0.105	0.085	0.07	0.065	0.065	0.03
		PE	0.085	0.115	0.08	0.075	0.06	0.07	0.04
		PC	0.09	0.105	0.085	0.07	0.065	0.03	0.025
	0.1%	APC	0.105	0.1	0.075	0.065	0.06	0.055	0.025
		PE	0.075	0.095	0.075	0.07	0.055	0.045	0.02
		PC	0.085	0.1	0.06	0.05	0.045	0.025	0.015
	1.0%	APC	0.09	0.095	0.055	0.035	0.025	0.015	0.02
		PE	0.09	0.085	0.05	0.04	0.035	0.02	0.01
		PC	0.115	0.105	0.097	0.085	0.07	0.065	0.06
	0.001%	APC	0.105	0.115	0.095	0.075	0.065	0.07	0.055
		PE	0.125	0.12	0.11	0.09	0.085	0.075	0.06
		PC	0.11	0.105	0.095	0.08	0.065	0.06	0.055
Cobalt	0.01%	APC	0.11	0.1	0.09	0.07	0.06	0.065	0.05
		PE	0.12	0.11	0.1	0.085	0.07	0.065	0.05
		PC	0.105	0.1	0.085	0.075	0.06	0.055	0.045
	0.1%	APC	0.11	0.095	0.085	0.065	0.055	0.05	0.04
		PE	0.115	0.09	0.08	0.07	0.06	0.05	0.04
		PC	0.095	0.085	0.075	0.055	0.05	0.035	0.025
	1.0%	APC	0.08	0.07	0.065	0.05	0.045	0.03	0.028

Contd.

		PE	0.11	0.085	0.075	0.068	0.058	0.048	0.025
		PC	0.115	0.105	0.095	0.085	0.075	0.06	0.055
	0.001%	APC	0.12	0.115	0.1	0.09	0.08	0.065	0.06
		PE	0.125	0.11	0.09	0.08	0.07	0.06	0.05
		PC	0.11	0.1	0.095	0.09	0.087	0.07	0.055
Zinc	0.01%	APC	0.11	0.095	0.09	0.085	0.08	0.06	0.05
		PE	0.105	0.09	0.085	0.08	0.085	0.065	0.05
		PC	0.105	0.095	0.085	0.08	0.075	0.065	0.05
	0.1%	APC	0.125	0.09	0.09	0.075	0.065	0.055	0.045
		PE	0.105	0.09	0.085	0.07	0.06	0.05	0.04
		PC	0.094	0.08	0.075	0.06	0.055	0.05	0.045
	1.0%	APC	0.09	0.085	0.08	0.065	0.055	0.045	0.04
		PE	0.095	0.085	0.075	0.06	0.05	0.045	0.04
		PC	0.11	0.105	0.095	0.08	0.065	0.055	0.045
	0.001%	APC	0.115	0.11	0.095	0.085	0.075	0.065	0.05
		PE	0.108	0.105	0.09	0.085	0.07	0.06	0.05
		PC	0.105	0.095	0.075	0.065	0.045	0.035	0.025
Copper	0.01%	APC	0.09	0.085	0.08	0.06	0.04	0.03	0.02
		PE	0.095	0.09	0.085	0.055	0.035	0.03	0.025
		PC	0.095	0.065	0.035	−0.025	−0.04	−0.025	−0.01
	0.1%	APC	0.09	0.08	0.03	−0.03	−0.045	−0.02	−0.015
		PE	0.085	0.07	0.04	−0.02	−0.045	−0.015	−0.005
	PC	0.09	0.05	−0.025	−0.035	−0.045	−0.065	−1.05	
	1.0%	APC	0.085	0.045	−0.03	−0.045	−0.035	−0.075	−1.07
		PE	0.095	0.055	−0.035	−0.05	−0.065	−0.07	−1.95
		PC	0.115	0.095	−0.035	−0.045	−0.06	−0.08	−0.06
	0.001%	APC	0.12	0.09	−0.03	−0.055	−0.045	−0.075	−0.065
		PE	0.105	0.085	−0.025	−0.04	−0.05	−0.06	−0.065
		PC	0.11	0.075	−0.045	−0.06	−0.08	0.06	−0.035
Mercury	0.01%	APC	0.115	0.08	−0.055	−0.065	−0.075	−0.065	−0.04
		PE	0.1	0.085	−0.04	−0.5	−0.06	−0.04	−0.045
		PC	0.095	0.045	−0.035	−0.045	−0.07	−0.09	−1.025
	0.1%	APC	0.085	0.04	−0.03	−0.04	−0.06	−0.085	−0.095
		PE	0.09	0.04	−0.03	−0.04	−0.065	−0.08	−1.03
		PC	0.05	−0.02	−0.06	−0.08	−0.095	−1.35	−2.025
	1.0%	APC	0.06	−0.025	−0.055	−0.085	−0.09	−1.25	−2.05
		PE	0.055	−0.035	−0.065	−0.075	−0.085	−1.3	−2.075

Copper Sulphate

Copper acts as an essential component for plastocyanin (Katoh and Takamiya, 1961), photosystem² (Bishop, 1971) and enzyme amine oxidase (Palenik and Morel, 1991). Results given in Table 1 shows that Copper at 0.001% on the 8th day showed maximum carotenoid content (6.652 mg/ml) of all the metals studied and was over the control. From Table 2 copper at 0.001%, 0.01%, and 0.1% day showed more chlorophyll content compared to control and it was found in other concentrations (0.01%, 0.1% & 1.0%) at different durations the contents were decreased. At

1% concentration cells bleaching was observed which might be due to the replacement of free radical production or increasing activity of chlorophyllase or replacement of magnesium ions (an important requirement for chlorophyll and phycobiloprotein synthesis). However phycobiloprotein content was observed less than control for all the concentrations in the investigation.

Mercuric Chloride

Mercury inhibits photosystem I reaction center (Terashima and Ono, 2002). The toxic potential of mercury depends upon its multiple binding capacity with electron transport chain (Rai et al., 1991). Results from Table 1, Table 2, and Table 3 show that mercury at all levels inhibited the cell growth under investigation and accessory pigments contents declined when compared to control.

DISCUSSION

According to Welch (1980) zinc, cadmium, lead, cobalt, mercury and nickel are well-known toxic metals that occur in a variety of wastes and cause either acute or chronic effects on an organism. A common characteristic of heavy metals is that they interact with and bind themselves to biological macromolecules, particularly, enzymes. This is a determinant factor in their toxicity (Eichorn, 1974). Uptake and accumulation of heavy metals has been investigated to understand various aspects of selective permeability and execution mechanisms of cells (Whitton, 1970; Sorentio, 1979; Venkataraman et al., 1992). During the last decade, research has been centered on metal toxicity in order to evolve strategies for the possible removal of potential toxicants from effluents. Thus studies on the uptake, accumulation and consequent toxicity of metals in the algae assume greater significance. Our results showed that uptake of the metals are directly proportional to their concentration in the ambient medium. The species metal uptake was dependent upon cell size and medium in which it is grown. The progressive increase in heavy metal(s) concentration (0.001%, 0.01%, 0.1%, 1%) for *Anabaena oryzae* caused reduction in pigment content in a concentration dependent fashion, the order of metal toxicity was Hg<Cu<Zn<Co<Mn. Carotenoids appeared to be more resistant for all the heavy metal concentrations compared to total chlorophyll and phycobiloprotein. The obtained results would further contribute to the basic database in ascertaining the toxic potential of the metals and also in determining the sensitive algal indices for biomonitoring of metal pollution.

ACKNOWLEDGEMENTS

We are thankful to UGC for financial assistance and Head, Department of Botany, Kakatiya University for providing facilities.

REFERENCES

Bennet, A. and Bogoard, L. 1971. Properties of subunits and aggregates of blue algal biloproteins. *Biochemistry*. 10: 3625-3634.

Bishop, N. I. 1971. Photosynthesis: The electron transport system of green plants. *Ann. Rev. of Biochem.* 40: 197-200.

Bu-Olayan, A. H., Al-Hassan, R., Thomas, B.V. and Subramanyam, M. N. V. 2001. Impact of trace metals and nutrient levels on phytoplankton from the Kuwait coast. *Environ. Int.* 26: 199-203.

Chamovitz, D., Sandmann, G. and Hirschberg, J. 1993. Molecular and biochemical characterization of herbicide resistant mutants of cyanobacteria reveals that phytoene desaturation is a rate limiting step in carotenoid biosynthesis. *J. Biol. Chem.* 268: 17348-17353.

Conner, S. D. and Schimid, S. L. 2003. Regulated portals of entry in to the cell. *Nature.* 422: 37-44.

Cotte-Krief, M. C., Guiea, C., Thomas, A. J. and Martin, J. M. 2000. Sources of Cd, Cu, Ni and Zn in portugese coastal waters. *Mar. Chem.* 71: 199-214

Csatorday, K., Gombos, Z and Szalontai, B. 1984. Manganese and Cobalt toxicity in chlorophyll biosynthesis. *Proc. Nat. Acad. Sci. USA.* 81: 476-478

Cumming, J. R. and Gregory, J. T. 1990. Mechanisms of metal tolerance in plants: Physiological adaptations for exclusion of metal ions from cytoplasm. *In* Alscher, R.G. & Cumming, J. R. [Eds.] *Stress Responses in Plants: Adaptation and Acclimatation Mechanisms.* Wiley-Liss, New York, pp. 338-55.

Eichorn, G. L. 1974. Active sites of biological macro-molecules and their interaction with heavy metals. Ecological toxicology research A D Mc Inture and C F Mills (Eds) In: *Pleunum press New York.*

Esser, B. K. and Volpe, A. 2002. At sea high-resolution trace element mapping: San Diego bay and its plume in adjacent coastal ocean. *Environ. Sci. Technol.* 36: 2826-32.

Guikema, J. A. and Sherman L. A. 1983. Organization and function of chlorophyll in membranes of cyanobacteria during iron starvation. *Plant Physiol* 73: 250-256.

Gupta, R. S. and Verma, U. 1989. Growth responses of prokaryotic algae towards heavy metals. In Adv. Appl. Phycol. EE. Today and Tomorrow Publ., New Delhi, pp. 69-72.

Katoh, S. and Takamiya, A. 1961. A new leaf copper protein plastocyanin, a natural hill oxidant. *Nature (London).* 189: 665-666.

Kennish, M. J. 1996. *Practical Handbook of Estuarine and Marine Pollution.* CRC Press, New York.

Macfarlane, G. R. and Burchett, M. D. 2001. Photosynthetic pigments and peroxidase activity as indicators of heavy metal stress in the grey mangrove, *Avicennia marina* (Forsk). Vierh. *Mar. Pollut. Bull.* 42: 233-40.

Moran, R. 1982. Formulae for determination of chlorophyllous pigments extracted with *N, N*- Dimethylformamide. *Plant Physiol.* 69: 1376-1381.

Palenik, B. and Morel, F. M. M. 1991. Amine oxidases of marine phytoplankton. *Appl. Microbiol.* 57: 2440-2443.

Rai, L. C., Mallick, N., Singh, N., J. B., and Kumar, H. D. 1991. Physiological and biochemical characteristics of a copper tolerant and wild type strain of *Anabaena doliolum* under copper stress. *J. Plant Physiol.* 138: 68-74.

Rebhun, S. and Amotz, A. B. 1984. The distribution of cadmium between the marine algae *chlorella stigmatophora* and seawater medium. *Water Res.* 18: 173-8.

Rosen, B. 1996. Bacterial resistance to heavy metals and metalloids. *J. Biol. Chem.* 1: 273-7

Rosko, J. J. and Rachlin J. W. 1977. The effect of cadmium, copper, mercury, zinc and lead on cell division, growth and chlorophyll-a content of the Chlorophyta *Chlorella vulgaris*. *Bull. Torr. Bot.* Club 104: 226-233.

Sorentino, C. 1979. The effect of heavy metals on phytoplankton. A review. *Phykos.* 18: 149-161.

Terashima and Ono (2002). Effect of $HgCl_2$ on CO_2 dependence of leaf photosynthesis. *Plant. Cell. Physiology.* 43: 70-78.

Venkataraman, G. S. 1979. Algal inoculation of rice fields. In: Nitrogen and Rice, International Rice Research Institute, Philippiness, Los Banos, pp. 311-321.

Venkataraman, L. V., Survanalatha, G. and Manoj. G. 1992. Uptake, accumulation and toxicity of heavy metals in algae. *Phykos*. 31: 173-195

Wang, W. X. and Dei, R. C. H. 2001a. Metal uptake in coastal diatom influenced by major nutrients (N, P and Si). *Water Res*. 35: 315-21.

Wang, W. X. and Dei, R. C. H. 2001b. Effect of major nutrient additions on metal uptake in phytoplankton. *Environ. Poll.* 111: 233-40.

Whitton, B. A. 1970. Toxicity of zinc, copper, and lead to Chlorophyta from flowing waters. *Arch. Microbiol.* 72: 353-360

Whitton, B. A. and Say P. J. 1975. Heavy metals. In River Ecology (ed.) B.A. Whitton Oxford: Blackwell Scientific Pub. pp. 286-311.

Wickfors, G. H. and Ukeles, R. 1982. Growth and adaptation of estuarine unicellular algae in media with excess Copper, Cadmium or Zinc. *Mar. Ecol. Progr. Ser.* 7: 191-206.

13

Cyanobacterial Flora of Rice Fields of North Telangana Region, Andhra Pradesh, India

***B. Digamber Rao, Ch. Sreedhar Rao and D. Srinivas**
*Department of Botany, Kakatiya University, Warangal, Andhra Pradesh

ABSTRACT

A total of 241 rice field soil samples were collected from four districts of North Telangana region of Andhra Pradesh and samples were analyzed for blue-green algal content. Altogether 84 algal forms belonging to 32 genera were recorded. *Aphanothece littoralis* was abundant followed by *Anabaena variabilis*, *Nostoc punctiforme*, *Oscillatoria major* and *Scytonema simplex* respectively. The dominant species observed throughout the period were *Merismopedia convoluta*, *Phormidium tenue*, *Anabaena variabilis*, *Anabaena muscorum*, *Nostoc commune*, *Cylindrospermum majus* and *Scytonema javanicum*. About 52% of total recorded blue-green algae were found to be heterocystous. Non-heterocystous forms showed limited distribution and diversity, with 13 species belonging to 4 genera, *Spirulina*, *Lyngbya*, *Oscillatoria* and *Phormidium* respectively. The range of pH observed was 5.9-7.3.

Keywords: Blue-green algae, rice fields, North Telangana, abundance.

INTRODUCTION

Studies on blue-green algae distribution in the rice fields of Andhra Pradesh state are scanty. The occurrence and succession of blue-green algae, which grow on the soil surface, is regulated by the soil properties. The distribution of nitrogen fixing cyanobacteria in rice fields varies from 4% to 80% (Watanabe, 1959; Singh, 1961; Aiyer, 1965; Matarassi and Balloni, 1965; Pandey, 1965). Venkataraman (1975) reported that 33% soils harboured nitrogen fixing forms in the rice fields of India. Many workers have reported occurrence of cyanobacterial flora from rice fields of the country (Tiwary, 1972; Kolte and Goyal, 1985; Goyal 1993; Saikia and Bordoloi, 1994; Singh et al., 1997; Sahu et al., 1996; Dorycanta, 1997; Amita Devi et al., 1999). Reports are available on blue-green algal flora of the rice fields of certain areas of Uttar Pradesh (Shukla, 1971), Bihar (Verma et al., 1990), West Bengal (Chatterjee, 1983), Tamilnadu (Anand and Revati, 1987), Kerala (Anand and Hopper, 1986; Madhusoodanan and Dominic, 1995), Maharashtra (Tiwari, 1979; Chaporkar and Gangawane, 1984; Patil and Satav, 1986), North eastern region (Saikia and

Bardoloi, 1994), Punjab (Grover and Pandhol, 1975), Madhya Pradesh (Tiwari, 1979; Mishra and Purohit, 1979), and Jammu and Kashmir (Goyal et al., 1984). Though rice is cultivated in the region of North Telangana, no concerted attempt has so far been made on the occurrence and distribution of blue-green algae in the area. Dikshit and Pandey (1986) surveyed blue-green algal flora from paddy fields of Andhra Pradesh. However, the component of nitrogen fixing blue-green algae in rice fields of North Telangana is not known. Therefore, the present study was made to identify different nitrogen fixing blue-green algae from rice fields of North Telangana region.

MATERIALS AND METHODS

During 2004-2005 a total 241 rice field soil samples were collected from five districts of North Telangana region (Adilabad, Karimnagar, Khammam, Adilabad and Warangal). Fogg's medium (Fogg, 1949) with or without nitrogen source was used for raising enrichment cultures. 5 gm of soil sample was inoculated into 250 ml conical flask containing 100 ml of sterile medium and incubated under controlled conditions of continuous fluorescent light (2-3 K Lux) and temperature (28 ± 1°C) for 20 days. The pH of the medium was adjusted to 7.5. Identification of cyanobacteria was done with the help of published literature (Desikachary, 1959; Anand, 1989 and Santra, 1993). Water holding capacity and pH of the soil samples were determined (Kaushik, 1987). The relative abundance of blue-green algal type was also studied.

Relative abundance = (Y/X) 100

where X = total number of soil samples collected

Y = number of soil samples from which a particular cyanobacteria type isolated.

Rice field blue-green algae from North Telangana region of Andhra pradesh.

Cyanobacteria	Abundance(%)
Chroococcales	
Chroococaceae	
Microcystis aeruginosa Kütz	(10) 5.3
M. flosaquae Wittr. Kirchner	(6) 3.9
M. elongata Desik	(7) 2.8
Microcystis sp. Kütz	(5) 3.1
Chroococcos minutus (Kütz) Nag.	(4) 2.4
C. pallidus Nag.	(7) 1.9
C. indicus Zeller.	(5) 2.2
Gloeocapsa calcerea Tilden.	(4) 1.34
G. compacta Kütz.	(2) 2.4
G. montana Kütz.	(3) 3.1
Gleothece rupestris (Lyngb) Born.	(2) 1.7
G. rupestris var. maxima West, w.	(6) 2.4
Aphnocapsa biformis A. Br.	(8) 3.6

A. littoralis Hansg.	(21) 7.8
A. montana Cramer	(4) 1.75
Aphanothece bullosa (Menegh) Rabenh.	(7) 1.70
A. pallida (Kütz) Rabenh	(1) 0.52
A. saxicola Nag.	(4) 1.2
A. stagnina (Spreng.) A. Br.	(3) 1.4
Merismopedia convoluta Breb.	(2) 0.081
Chaemosiphonales	
Dermocarpaceae	
Dermocarpa sp. Crouun	(2) 0.4
Stichosiphon sansibaricus (Hieron) D.D.	(2) 1.1
Pleurocapsales	
Pleurocapsaceae	-nil-
Nostocales	
Nostocaceae	
Anabaena oryzae Fritsch.	(11) 5.1
A. fertilissima Rao.	(8) 4.6
A. variabilis Kutz.	(18) 6.9
A. ambiqua Rao.	(6) 4.2
Anabaena sp. Bory.	(5) 1.7
A. muscorum (Born) Flah.	(3) 2.1
A. naviculoides Fritsch.	(2) 1.3
A. laxa	(3) 1.6
Aulosira prolifica Bharadwaja	(7) 3.4
A. fertilissima Ghose.	(5) 2.7
Nostoc punctiforme (Kütz) Hariot.	(14) 6.2
N. muscorum Ag.	(11) 5.4
N. carneum Ag.	(6) 3.7
Nostoc sp. Vaucher.	(5) 2.4
N. calcicola Breb.	(4) 2.1
N. commune Vaucher.	(5) 3.7
N. linckia (Roth) Born.	(11) 5.6
Nodularia spumigena (Mertens) Flah.	(4) 2.3
Cylindrospermum musicola Kütz ex Born.et. Flah.	(11) 4.8
Cylindrospermum majus Kütz ex Born. et Flah.	(8) 3.7
Oscillatoriaceae	
Spirulina major (Kütz) Gomont.	(2) 1.0

S. princeps W.W.	(3) 1.4
Lyngbya birgei Smith.	(6) 2.8
L. laxespiralis Skuja.	(5) 3.1
L. ceylanica Wille.	(4) 2.0
L. majuscula (Harvey) Gomont.	(2) 1.7
Lyngbya sp. Ag.	(9) 4.2
Oscillatoria acuta Bruhl.	(10) 5.1
O. annae (Goor)	(7) 4.0
O. major West.	(14) 6.0
Phoridium ambiguum (Gomont)	(10) 5.5
P. tenue (Meageh)	(6) 3.2
P. rotunda Santra	(5) 2.7
Rivulariaceae	
Calothrix braunii Bom et. Flah.	(7) 3.4
C. ghosei Bharadwaja.	(4) 2.0
C. javanica De Wilde	(1) 0.18
C. marchica Lemm.	(6) 2.4
C. parietina Thuret ex. Born.	(8) 3.8
C. thermalis (Schmidle) Hansg.	(7) 4.1
Gloeotrichia intermedia (Lemm) Geitler.	(6) 3.8
G. pisum Born et. Flah.	(7) 4.1
G. natans Born et. Flah.	(5) 2.8
Dichothrix sp. NT	(6) 3.8
Scytonemataceae	
Scytonema simplex Bharadwaja.	(10) 5.8
S. javanicum (kutz) Born. et. Flah.	(4) 2.1
S. mirabile (Dillw.) Born.	(5) 2.7
S. cincinnatum (Thuret) Born. et. Flah.	(7) 4.5
Tolypothrix tenuis (kutz) Schmidt.	(6) 3.7
T. nodosa Bharadwaja.	(4) 1.08
Tolypothrix sp. Kutz	(2) 1.5
Plectonema notatum	(6) 3.2
P. nostocorum Born. ex Gomont.	(5) 2.8
Microchaetaceae	
Microchaete tenera Born. et Flah.	(2) 0.89
M. uberrima Carter.	(1) 1.1
Stigonematales	

Stigonemataceae

Haplosiphon luteolus West W. and G. S. (2) 1.1

Westiellopsis sp. (4) 2.4

W. prolifica Janet (2) 0.89

Westellia intricate Baorzi. (6) 2.0

Stigonema ocellatum Born. et. Flah. (3) 1.7

Camptylonema indicum Schmidle. (4) 1.2

Mastigocladiaceae

Mastigocladus laminosus Cohn. (2) 1.5

Mastigocladopsidaceae

Mastigocladopsis sp. Iyengar and Desik (3) 2.1

Cyanobacteria observed in rice the fields of North Telangana

No.	Order and Family	Genera	Species
1.	Chroococcales		
	(a) Chroococcaceae	07	20
2.	Chaemosiphonales		
	(a) Dermocarpaceae	02	02
3.	Pleurocapsales		
	(a) Pleurocapsaceae	-nil-	-nil-
4.	Nostocales		
	(a) Nostocaceae	05	20
	(b) Oscillatoriaceae	04	13
	(c) Rivulariaceae	03	10
	(d) Scytonemataceae	03	09
	(e) Michrochaetaceae	01	02
5.	Stigonematales		
	(a) Stigonemataceae	05	06
	(b) Mastigocladaceae	01	01
	(c) Mastigocladopsidaceae	01	02
	Total	**32**	**84**

A total of 84 taxa of blue-green algae belonging to 32 genera were identified from the rice fields of North Telangana region. These members represented unicellular, colonial, non-heterocystous heterocystous and heterotrichous forms. They belonged to four orders namely Chroococcales, Chemaesiphonales, Nostocales and Stigonematales of myxophyceae. Among them above 52% of them was heterocystous forms showed clear dominance over non-heterocystous forms the genus *Anabaena* showed wide distribution with 8 species. This was followed by genera like *Nostoc, Calothrix* and *Scytonema* with 7, 6, 4 species respectively. The other genera like *Tolypothrix, Gloeotrichia* represented by 3 species each, *Aulosira, Cylindrospermum, Plectonema, Microchaete, Westiellopsis* had 2 species each and *Nodularia, Dichothrix, Westiellia, Stigonema*, and *Camptylonema, Mastigocladus, Hapalosiphon* and *Mastigocladopsis* have shown one species each. Non-heterocystous filamentous forms showed limited distribution and diversity, with 13 species belonging to 4 genera such as, *Spirulina, Lyngbya, Oscillatoria* and *Phormidium* respectively. Unicellular and colonial forms were represented by 7 genera namely, *Microcysties*, *Chroococcus*, *Gloeocapsa*, *Gloeothece, Aphanocapsa, Aphanothece* and *Merismopedia.* The genus *Dermocarpa* was identified with two species under the order of Chamaesiphonales. No species has been reported from the order of Pleurocapsales. The dominant species noticed throughout the period was *Merismopedia convoluta*, *Phormidium tenue*, *Anabaena variabilis*, *Anabaena muscorum*, *Nostoc commune*, *Cylindrospermum majus*, and *Scytonema javanicum*. Soil pH varied from 5.9-7.3. The most abundant blue-green algae *Aphanocapsa littoralis* was observed in 7.8% of the total soil samples analysed. It was followed by *Anabaena variabilis* (6.9%), *Nostoc punctiforme* (6.2%), *Oscillatoria major* (6.0%) and *Scytonema simplex* (5.8%). Such variation was also reported by many workers (Chaporkar and Gangawane, 1992; Anand and Santakumar, 1995; Dorycanta, 1997 and Ahmed et al., 1990.

The number of algal populations was found to be highest during tillering period (30-40 days) probably because of the progressively diminishing fertilizer quantity, increased moisture and shade provided by gradually spreading crop canopy. There are reports on the increase in the pH as a result of waterlogging (Roger and Reyand, 1976).

The relative abundance of cyanobacteria may be due to this rise in soil pH.

ACKNOWLEDGEMENTS

We are thankful to the UGC, New Delhi for providing the financial assistance.

REFERENCES

Ahmed, S. U., Kalita, M. C., Manab Deka and Hazarika, Binay Medhi, S. 1999. Distributional pattern of blue-green algae in rice field soils of Nagaon sub-division. *Phykos.* 38(1&2): 101-107.

Aiyer, R. S. 1965. Comparative algological studies in rice fields in Kerala state. *Agri. Res. J. Kerala.* 3(1): 100-104.

Amita Devi, G., Dorycanta, H and Singh, N. I. 1999. Cyanobacterial flora of rice-field soils of Manipur. *Phykos.* 38(142): 13-18.

Anand, N. and Revati, G. 1987. Blue-green algae from rice fields of Tamilnadu. *Phykos.* 26: 17.

Anand, N. and Shanthakumar Hopper, R. S. 1986. Blue-green algae (Cyanobacteria) from several rice fields of Kerala state, India. *Hydrobiologia.* 144: 223-232.

Anand, N. 1989. Handbook of blue-green algae of rice fields of South India. Pub. Bishen Singh Mahendra Pal Singh. Dehradun.

Anand, N. and Shanthakumar Hopper R. S. 1995. Distribution of blue-green algae in rice fields of Kerala state. India. *Phykos*. 34(1&2): 55-64.

Chaporkar, C. B. and Gangawane, L. V. 1984. Blue-green algae of some cultivated soils of Marathwada, Maharashtra-2. *Phykos*. 23: 55-58.

Chaporkar, C. B. and Gangawane, L. V. 1992. Nitrogen fixing blue-green algae of some cultivated soils of Marathwada region. In Biofertilizer Technology Transfer Ed. Gangawane. V.L. pp. 165-168

Chatterjee, M. and Chatterjee, S. P. 1983. Nitrogen fixing cyanobacteria from paddy field soils of Burdawan district. *Phykos*. 22: 64-66.

Desikachary, T. V. 1959. *Cyanophyta.* Pub. Indian Council of Agricultural Research, New Delhi.

Dikshit, G. and Pandey, D. C. 1986. Blue-green algal flora of paddy fields of Andhra Pradesh. University of Allahabad studies. 18: 85-96.

Dorycanta, H. 1997. Nitrogen fixing blue-green algae from rice fields of Mizoram and Nagaland for bio-fertilizer technology Ph.D. thesis. Manipur University. Imphal.

Fogg, G. E. 1949. Growth and heterocyst production in *Anabaena cylindrica* Lemm. In relation to Carbon and Nitrogen metabolism. *Ann. Bot.,* 13: 241-259.

Goyal, S. K. 1993. Algal biofertilizer for vital soil and free nitrogen. *Proc. Indian. Natn. Sc. Acad.* B-59: 295.

Goyal, S. K., Sharma, B. M. and Gupta, R. S. 1984. Algal flora of rice field soils of Jammu and Kashmir state. *Phykos*. 23: 59-62.

Grover, I. S. and Pandhol, R. K. 1975. Algal flora of paddy fields of Ludhiana and its adjacent areas. *Phykos*. 14: 89-97.

Kaushik, B. D. 1987. Laboratory methods for blue-green algae. Associated Publ. Company, New Delhi.

Kolte, S. O. and Goyal, S. K. 1985. Distribution pattern of blue-green algae in rice field soils of Vidarbha region of Maharashtra state *Phykos,* 24: 156-162.

Madhusoodanan, D.V. and Domnic, T. K. 1995. Variations in the diversity of blue-green algae in rice field of Kerala. *Phykos*. 34: 65-69.

Matarassi, R. and Balloni, W. 1965. *Quelques observation surla presence de microorganism's autotrophs fixations d'azote dans les reirires*. Ann. Inst. Pastens. *Ann. Inst. Paster, Paris*. 3: suppl. 218-334.

Mishra, G. P. and Purohit, J. 1979. A preliminary list of soil algae from Sagar. *Phykos*. 18: 63-67.

Pandey, D. C. 1965. A study of algae from paddy soils of Ballia and Ghazipur districts of Uttar Pradesh. India. Cultural and ecological considerations. *Nova Hedwigia*, 9: 299-334.

Patil, P. L. and Satav, S. D. 1986. A study on nitrogen fixing blue-green algae from rice fields of Western Maharashtra. *Phykos*. 25: 113-116.

Roger, P. and Reynaud, D. 1976. *Dynamique de la population algale all cours d'un cycle de culture dans une riziere sahelienne* (Dynamics of algal populations during a culture cycle in a Sahel rice field) *Rev. Ecol. Biol.* 13: 545-560.

Sahu, J. K., Nayak, H. and Adhikary, S. P. 1996. Blue-green algae of rice fields of Orissa state. Distributional pattern in agroclimatic zones. *Phykos*, 35(1&2): 93-110.

Sahu, J. K., Nayak, H. and Adhikary, S. P. 1997. Blue-green algae of rice fields of Orissa state. Distributional pattern in different agroclimatic zones. *Phykos*. 35: 93-110.

Saikia, P. and Bordoloi, R. P. M. 1994. Blue-green algal flora from the rice fields of Assam. *Phykos*. 33(1&2): 53-57.

Santra, S. C. 1993. *Biology of rice fields blue-green algae*. Daya Publishing House, Delhi.

Shukla, A. C. 1971. Systematic description of algae from Panki rice fields. India. *Rev. Algol*. **3:** 257-270.

Singh, N. I., Singh, N. S., Devi, G. A. and Singh, S. M. 1997. Cyanobacterial flora of rice field soils of Tripura. *Phykos*. 36: 121-126.

Singh, R. N. 1961. Role of blue-green algae in nitrogen economy of Indian Agriculture Pub. Indian council for Agricultural Research, New Delhi.

Tiwari, G. L. 1972. A study of blue-green algae from paddy fields of India. *Hydrobiologia*, 39: 335-350.

Tiwari, G. L. 1979. A study of the blue-green algae from paddy field soils of India. Taxonomic considerstions of Nostocales and Stigonematales. *Nova Hedwigia*. 63: 133-159.

Venkataraman, G. S. 1975. The role of blue-green algae in tropical rice cultivation. In *Nitrogen fixation by free living microorganisms*. Ed. Stewart W. D. P Cambridge Univ. Press, London, pp. 207-218.

Verma, B. N., Sinha, R., Prasad, P. K., and Das. H. N. 1990. Blue-green algae from rice fields of North Bihar. *Phykos*. 29: 55-56.

Watanabe, A. 1959. Distribution of nitrogen-fixing blue-green algae in various areas of South-West Asia. *J. Gen. Appl. Microbiol*. 5: 21-29

14

Seaweeds as a Source of Bioethanol

Savindra Kumar and Dinabandhu Sahoo
Marine Biotechnology Laboratory, Department of Botany, University of Delhi
Delhi-110007, India

ABSTRACT

Alcohol especially ethanol is used as an alternative fuel to gasoline in some parts of the world and is being widely used in automobiles mainly in Brazil, US and some European countries. Ethanol is easy to manufacture and process from many common plants. Sugars can be directly converted to ethanol, while starches and cellulose first have to be hydrolyzed to fermentable sugars and then they are converted to ethanol. Various studies have shown that seaweeds can be a good source of bioethanol. Seaweeds have been broadly classified into 3 groups Green, Brown and Red. Out of which brown seaweeds lack lignin and contain high amount of carbohydrates such as alginate, mannitol, laminaran and cellulose. Thus, brown seaweeds should be an easier material for biological degradation than land plants for the production of fuel ethanol or 3rd generation fuel. Alginate is being used in various industrial processes and is a very good source of revenue. So the alginate industry waste which contain high amount of mannitol and laminaran may be used a source for raw material for ethanol production.

INTRODUCTION

The term biofuel includes solid biomass (wood, wood waste, straw, manure, sugar cane, maize and many other by-products from a variety of agricultural processes), liquid fuels and various biogases. Today biofuels are providing 10-20% of total energy of the world. Biofuels are gaining interest by the society as there is tremendous increase in the price of crude oil. There is need to look for alternative resources for energy as the fossil fuels are expected to be depleted in next 70-80 years (Duncan, 2005-2006). Nowadays, various forms of biofuel are available such as bioalcohols, biodiesel, bioethers, biogas and syngas. Ethanol fuel is the principal biofuel produced by microorganisms which is now one of the most important non-fossil fuels in some parts of the world. Ethanol is a high-octane fuel and can be used in various combinations with gasoline like E85 (85% ethanol). With the hike in oil prices in 1970s the first major fuel ethanol programme started in 1975 in Brazil (National Alcohol Programme or ProAlcool) followed by USA and Canada. The past few years have seen a fast increase in the number of vehicles in the world and our dependency on fossil fuels become problematic, because of the depletion of limited resources. Climate change, food and fuel security are creating a requirement for alternative, renewable energy. Biomass, the stored solar energy in the form of chemical energy is an example of renewable energy. About 2 million years ago *Homo erectus* first used biomass in the form of wood and sticks as a fuel and since then the burning of wood has been our most important source of energy. Even today about 15% of the global energy production is based on biomass in developing countries (Hall and House, 1995).

Ethanol, unlike gasoline or petrol, is an oxygenated fuel that contains 35% oxygen, which reduces particulate and nitrogen oxides emissions. The most important feature of fuel ethanol is that the CO_2 released by ethanol combustion has been fixed by growing plants (Bastianoni and Marchettini, 1996; Wheal et al., 1999). When added to motor fuel bioethanol facilitates in the reduction amount of cancer-causing compounds such as benzene, toluene, xylene, and ethyl benzene. Emissions from ethanol production may vary slightly depending on the process, design and feedstock.

The global economy literally runs on energy. Virtually every sector of our economy is affected from the rapidly expanding ethanol industries (Gopinathan and Sudhakaran, 2009). Ethanol production offers significant economic opportunities for virtually all countries due to the diversity of feedstock from which ethanol can be produce. Skyrocketing prices, environmental pollution and global warming, and the movement to alternate fuels, the world ethanol market is projected to reach 27.7 billion gallons per annum by the year 2012 (Martin et al., 2010).

METHODS OF ETHANOL PRODUCTION

Ethanol can be produced synthetically from petroleum or by microbial conversion of biomass through fermentation (Badger, 2002). Fermentation technology is the oldest of all biotechnological process. The term is derived from the Latin verb *fevere* (to boil). Fermentation is the conversion of a carbohydrate into acid or alcohol or it is a process of deriving energy from the oxidation of organic compounds, using an endogenous electron acceptor. Fermentation include upstream processing of sugar solution and downstream processing of fermented solution. Upstream processing generally carried out in two steps: (1) preparation of fermentable sugar solution and (2) the fermentation of sugar solution, by appropriate microbial strain into ethanol under controlled and axenic condition. Downstream processing includes separation of ethanol from fermented solution and purification of ethanol usually by distillation. In ethanol production the carbohydrates are converted into ethanol by microbes that on average bring the concentration of ethanol to 8% in the broth with 92% water. Large amounts of energy are required to remove the 8% ethanol from the 92% water (Pimentel and Patzek, 2005). Fermentation does not necessarily have to be carried out in an anaerobic environment for instance yeast cells greatly prefer fermentation as long as sugars are readily available for consumption even in the presence of abundant oxygen.

FEEDSTOCKS FOR ETHANOL PRODUCTION

Sugars are the most common substrate for fermentation. Many fungi, bacteria, and yeast can be used for fermentation but *Saccharomyces cerevisiae* is frequently used to ferment glucose into ethanol. These microorganisms typically use the 6-carbon sugars. Therefore, biomass containing high levels of glucose or its precursors are the easiest to convert into ethanol. One example of a sugar feedstock is sugarcane which contains 12-17% of sugar on wet basis (Wheal et al., 1999). Theoretically, 100 grams of glucose will produce 51.4 g of ethanol and 48.8 g of carbon dioxide. However, in practice, the microorganisms use some of the glucose for growth and the actual yield is less than 100%.

Starch is also a potential ethanol feedstock. Starch molecules are made up of long chains of glucose molecules. Thus, starchy materials like cereal grains, potato, sweet potato, and cassava can also be fermented. Corn is the primary feedstock for most ethanol production in US. Starchy materials require a reaction of starch with water (hydrolysis) to break down the starch into fermentable sugars (saccharification). Typically, hydrolysis is performed by mixing the starch with water to form slurry which is then stirred and heated to rupture the cell walls. Specific enzymes that will break the chemical bonds are added at various times during the heating cycle (Badger, 2002). Bioethanol from sugar and starch comes under 1st generation biofuel.

Cellulose is also used for the production of ethanol in some countries during fuel shortage. The cellulosic feedstocks are widespread and abundant such as paper, cardboard, wood, and other fibrous plant material. For example, forests comprise about 80% of the world's biomass. Being abundant and outside the human food chain makes cellulosic materials relatively less expensive feedstocks in comparison to sugar and starch for ethanol production (Badger, 2002). Many improvements have been achieved by researcher regarding pretreatment, enzymatic hydrolysis and fermentation of feedstock for ethanol (Galbe and Zacchi, 2002; Groenestijn et al., 2007). There are three basic types of methods for the production of ethanol from cellulose (EFC) processes—(1) acid hydrolysis, (2) enzymatic hydrolysis, and (3) thermochemical— out of which acid hydrolysis is most common one. Although any acid can be used sulfuric acid is the most commonly used one since it is the least expensive when compared to other acids. Ethanol-from-cellulose (EFC) holds great potential due to the widespread availability, abundance, and relatively low cost of cellulosic materials. However, several EFC processes are technically feasible, cost-effective processes still have been difficult to achieve.

NEEDS FOR AN ALTERNATIVE SOURCE FOR ETHANOL PRODUCTION

According to World Energy Outlook (2008), current energy supplies are unsustainable from environmental, economic, and societal standpoints. In addition, it is projected that world energy demands will continue to expand by 45% from 2008 to 2030, an average rate of increase in 1.6%/ yr. Corn (starch) and sugar cane (sugar) are the major feed-stocks for the bioethanol industry, and the continued use of these crops will drive the food versus fuels debate even more as demand for ethanol increases because these are components of the human food chain (Walker, 2009). Large-scale production of corn and sugar damage the environment by the use of harmful pesticides. Not only this but also it uses two other valuable resources: arable land and enormous quantities of water (Martin et al., 2010). Another alternative is cellulosic materials which comprises of lignin, hemicellulose, and cellulose and are thus sometimes called lignocellulosic materials (Roland et al., 1989; Carpita and Gibeut, 1993). Cellulose molecules consist of long chains of glucose molecules as do starch molecules, but have a different structural configuration. Cellulosic ethanol also has many hurdles such as high cost of production including the cost of pretreatment and enzymes production. The structural characteristics plus the presence of lignin makes cellulosic materials more difficult to hydrolyze. Hemicellulose also comprises of long chains of sugar molecules; but contains, pentose (5-carbon sugars) in addition to glucose (a 6-carbon or hexose sugar). Ethanol production from lignocellulose requires microorganism that produce ethanol with a high yield from all sugar present (Olsson and Hägerdal, 1996). Recently, GMO's (genetically modified organisms) have been developed which can ferment 5-carbon sugars into ethanol with relatively high efficiency. One example is a genetically engineered microorganism (US patent 5,000,000) developed by the University of Florida that has the ability to ferment both 5- and 6-carbon sugars (Badger, 2002). The ethanol yield and productivity obtained during fermentation of lignocellulosic hydrolysates decreases due to formation of inhibiting compounds, such as weak acids, furans and phenolic during hydrolysis (Palmqvist and Hägerdal, 2000). The scarcity of fossil fuel in near future and climate change compels the whole world to think over the alternatives fuel and other energy sources.

WILL SEAWEEDS BE AN ALTERNATIVE SOURCE?

Brazil and North America are still only two regions that produce large quantities of ethanol for fuel (Table 1). The efficiency of ethanol production has steadily increased and valuable co-products are produced, but only tax credits make fuel ethanol a fuel commercially viable. These problems encountered with the above source of ethanol, draw our attention to look for new feedstock. The challenge is to find a feed-stock which is abundant and carbohydrate-rich. That feedstock use less or no agricultural inputs (pesticides, fertiliser, land, water) and not to be part of the human or animal food chain. Macro-marine algae called seaweeds may be an answer because

Table 1 World fuel ethanol production by country (million gallons)

Country	2007	2008	2009
USA	6,499	9,000	10750
Brazil	5,019	6,472	6578
Europe	570	734	1040
China	486	502	542
Canada	211	238	291
Thailand	79	90	435
Colombia	75	79	83
India	53	66	82
Australia	26	26	57
Other	82	128	247
WORLD	**13,101**	**17,335**	**19 535**

Data Source: Renewable Fuels Association, Ethanol Industry Outlook 2008 and 2009, p. 16 and 29. Available at http://www.ethanolrfa.org/pages/statistics/#E

seaweeds do not require agricultural land and can grow much faster than land plants. In brown seaweeds, alginate is the main structural compound (Kloareg and Quatrano, 1988), while mannitol and laminaran are common storage materials. Thus, the absence of lignin, low content of ash and high content of carbohydrates in brown algae make them a simpler material for biological degradation than land plants (Horn, 2000) because they are the only type of primary biomass that needs no pretreatment prior to digestion (Gunaseelan, 1997). The cell wall of brown algae consists of cellulose, alginate, fucoidan and protein (Kloareg et al., 1986) so their degradation requires a microbial community with the ability for mixed substrate utilization. The use of marine biomass energy was investigated in US and Japan as an alternative energy in 1970s after the oil crisis, but the studies were discontinued when oil prices stabilised (Yokoyama et al., 2007). In the present scenario seaweeds may be a key link between energy, local environment and climate change.

BROWN SEAWEEDS CONTENTS AND THEIR DEGRADATION

Alginate: Alginate a polysaccharide embedded in the cell wall of brown seaweeds where it may account for more than 80% of the organic matter. Alginate cements cells together, giving both mechanical strength and flexibility to the algal tissue. Alginates are salts of alginic acid, a linear copolymer of β-1,4-D-mannuronic acid (M) and α-1,4-L-guluronic acid (G) (Chapman, 1950; Sahoo, 2000; Usov et al., 2001 and Zubia et al., 2008). The two uronic acids are organised in blocks of polymannuronate (M-blocks) and polyguluronate (G-block), as well as heteropolymeric sequences (MG-block) (Horn and Østgaard, 2001). Alginates rich in guluronate form gels with a high mechanical rigidity, and a good stability towards competing Na-ions. In contrast, mannuronate-rich alginates form softer and more elastic gels (Horn, 2000). Alginate can be depolymerised chemically (by acid and alkali hydrolysis and by oxidative reductive depolymerisation) or enzymatically (Moen, et al., 1997a). Enzymatic degradation of alginate is catalysed by alginases or alginate lyases (Horn and Østgaard, 2001). All alginases cleave a hexose-1,4-α- or β-uronic acid sequence by β-elimination and create an unsaturated uronic acid at the new nonreducing end (Sutherland, 1995). Most lyases are endo-acting enzymes and the major product is unsaturated triuronide (Gacesa, 1992). The ultimate product is the monosaccharide 4-deoxy-L-erythro-5- hexoseulose uronic acid, in equilibrium with its open chain form 2-keto-3-deoxy glucoaldehyde (Preiss and Ashwell, 1962). Alginate lyases classified in two groups as mannuronate or guluronate lyase according to their substrate specificity (Østgaard, 1993) Alginases are of widespread occurrence (Sutherland, 1995), obtained from microorganisms, brown algae, marine molluscs and echinoderms (Boyen et al., 1990; Brown and Preston, 1991; Favorov and Vaskovsky, 1971; Kennedy et al., 1992 and Larsen et al., 1993). Linker and Evans (1984) isolated the alginases from *Pseudomonas* species. Moen et al. (1997b) showed that soluble Na-alginates were consumed 6-8 times faster than Ca-alginate gels. G-rich parts of the alginate may be less accessible for lyases due to the calcium junction zones. On 21 Oct 2008 Weiner, from University of Maryland got US patent (US743903482) for alginases, systems containing alginases and methods of cloning, purifying and/or utilizing alginases. Microbial cleavage of alginate leads to a range of oligosaccharides of different sizes with an unsaturated uronic acid at the non-reducing end (Haugen et al., 1990).

Storage carbohydrates (Mannitol and Laminaran): Mannitol is a sugar alcohol corresponding to mannose and a readily utilisable carbohydrate in contrast to the polysaccharides which first must be cleaved. Depending upon the organism involved, mannitol is taken into the cell either by a facilitated diffusion mechanism or by an energy-dependent phosphoenolpyruvate phosphotransferase system (Horn, 2000). In the latter case, mannitol enters in the cell as mannitol-1-phosphate, and is further converted to fructose-6-phosphate by a NAD(P)+ dependent dehydrogenase (Horn, et al., 2000b). Mannitol transferred into the cell by diffusion, on the other hand, is converted to fructose via another NAD(P)+ dependent dehydrogenase, and further to fructose-6-phosphate via a kinase (Forro, 1987). Fructose-6-phosphate may then enter into the glycolysis. The net reaction for the glycolysis is:

$$1 \text{ mannitol} + 2 \text{ ADP} + 3 \text{ NAD}^+ \rightarrow 2 \text{ pyruvate} + 2 \text{ ATP} + 3 \text{ NADH}$$

Thus, compared to glucose, one extra NADH is produced, so regeneration of all the NAD+ then requires either oxygen or transhydrogenase to convert NADH to NADPH (Horn, et al., 2000b). Yeast lack transhydrogenase, and anaerobic growth of *Saccharomyces* on pure mannitol is not possible (Quain and Boulton, 1987). Prokaryotes, like *Bacteroides* on the other hand, usually possess transhydrogenase (Horn, 2000) and should be able to ferment mannitol under truly anaerobic conditions (Forro, 1987). Presence of other substrates such as glucose does not shut off mannitol metabolism. A *Cytophaga* strain, when grown on glucose or alginate, produce acetate and propionate but when grown on mannitol its produced by ethanol (Horn, 2000). The same was seen for a *Lactobacillus* strain when substrate was changed from glucose to mannitol. Both strains apparently lack hydrogenase activity (Forro, 1987). The metabolic shift reflects the ability of the organism to eliminate the extra reducing equivalents found in mannitol.

Laminaran is a β-(1→3)-D-glucan containing about 25 glucosyl residues (Horn, 2000). A majority of laminaran molecules terminates with a non-reducing 1-linked D mannitol residue and are designated M-chains, whereas the a small proportion terminate with a reducing 3-linked glucose residue designated G-chains (Horn, et al., 2000a). Generally, the side chains consist of a single glucose. The variations in the degree of branching do affect the solubility of the polysaccharide in water. Laminaran, containing only β-(1→3)-linked residues, is water-insoluble, while branched laminaran tend to be water soluble (Read et al., 1996). β-(1→3)-glucanases are relatively widespread, and many microorganisms can hydrolyse laminaran to its glucose monomer.

Yeast, *Saccharomyces cerevisiae* and the bacterium *Zymomonas mobilis*, two most important microorganisms for ethanol production (Dumsday et al., 1997) but have a very narrow substrate range. Yeast lacks transhydrogenase (Van Dijken and Scheffers 1986), so it cannot grow on mannitol anaerobically (Quain and Boulton, 1987). *Zymobacter palmae* a facultative, anaerobic catalase-positive oxidase-negative, non sporeforming and peritrichously flagellated, Gram-negative proposed as a new ethanol-fermenting bacterium isolated from palm sap in Qkinawa, Japan. It ferments hexoses, c~-linked di- and tri-saccarides, and sugar alcohols (fructose, galactose, glucose, mannose, maltose, melibiose, saccharose, raffinose, mannitol and sorbitol). (Okamato et al., 1993). As discussed earlier in this review mannitol fermentation demonstrated a clear dependency on air. Ethanol yield was highly sensitive to the oxygen supply since excess aeration led to production of organic acids and thereby decrease in the yield. The best procedure for optimization of ethanol yield is start up at a low oxygen transfer rate (OTR), and then gradually increases the OTR. The mixture of both laminaran and mannitol found in a seaweed extract might have a positive effect on the ethanol production (Horn, 2000). *Z. palmae* metabolised glucose under strictly anaerobic conditions but ethanol from mannitol in seaweed extract only possible if some supply of oxygen was provided (Horn et al. 2000b). Initial experiments with seaweed extract showed that *Pichia angophorae* is able to utilise both mannitol and laminaran for ethanol production. The consumption rate of mannitol increased with the growing biomass but laminaran consumption rate remained constant for a while, and then fell strongly at the end of fermentation. Some production of acetate and propionate also observed in laminaran fermentation by *P. angophorae* because this strain is unable to hydrolyse polymeric branching points due to a lack of β-(1→6)-glucanase. *P. angophorae* is a more suitable organism for ethanol production from seaweed extract. It can utilise both substrates mannitol as well as laminarn simultaneously, and is not inhibited before the substrate is being consumed. Three strains of *Cytophaga*, isolated from a *Macrocystis pyrifera* anaerobic degrading culture, were able to utilise alginate, mannitol and laminaran. The major metabolites produced by *Cytophaga* when grown on alginate were acetate and propionate (Forro, 1987).

OTHER ORGANIC COMPOUNDS: While the intercellular matrix is dominated by alginate, the cell wall of brown algae also contains cellulose, fucoidan and protein. Fucoidan is found in most brown algae, but is most abundant in species that grow in the intertidal zone. Anaerobic degradation of fucoidan has not been reported (Forro, 1987). Generally, the protein fraction of brown seaweeds is low (3-15 % of the dry weight) compared with that of green and red seaweeds (10-47 % of the dry weight). Presence of polyphenols and salt (Ghosh et al., 1981) reduce the biodegradability of seaweeds. For most seaweeds, aspartic and glutamic acids constitute together a large part of the amino acid fraction (Fleurence, 1999). Degradation of cellulose is catalysed by cellulases, and occurs both under aerobic and anaerobic conditions. However, in brown algae cellulose is found in the cell wall in close association with other structural components which may limit the enzymatic access to cellulose. A combined enzymatic attack of alginate lyases, proteases and cellulases may be necessary to degrade the algal cell wall, as seen in the case of protoplast isolation (Butler et al., 1989). Thus, the cell walls seem to be more recalcitrant to microbial degradation than the intercellular matrix.

Ethanol production is not straight growth associated, giving same ethanol yields both during growth and stationary phases. This mismatch in the balance could be due to evaporation of ethanol, which is known to be a problem in ethanol fermentations.

CURRENT GLOBAL ACTIVITIES

In the last few years many activities have been accelerating very fast in this filed. Little work has been published on the fermentation of macroalgae to ethanol to date apart from the work of Horn (2000) and Adams et al. (2009). There is huge scope for ethanol production from seaweeds because the 2/3rd of the earth is covered by the water. Using algae for ethanol production is in such an early stage that not much can be concluded yet about its strengths and weaknesses and is therefore not investigated further in this report. Some of the major steps in this direction are as follows:

- Seaweed Bioethanol Production in Japan, entitled the "Ocean Sunrise Project", aims to produce seaweed bioethanol by farming and harvesting *Sargassum horneri,* utilizing 4.47 million km^2 (sixth largest in the world) of unused areas of the Exclusive Economic Zone (EEZ) and maritime belts of Japan. The Project aims to combat global warming by contributing an alternative energy to fossil fuel.
- Algenol, headquartered in Bonita Springs, FL, is a company, developing a process to produce ethanol directly from the algae. The company collected 10,000 strains of algae and used molecular biology to enhance certain traits.
- The philippine government's Department of Science and Technology in collaboration with Korean Institute developed seaweeds ethanol production technology for industries.

CONCLUSIONS AND FUTURE CHALLENGES

Twenty-first century comes with two main problems of clean environment and sustainable source of energy. Biofuels offer a potential source of renewable energy and possible large new markets for agricultural producers. But few current biofuels programs are economically viable, and most have social and environmental costs: upward pressure on food prices, intensified competition for land and water, and possibly, deforestation (Walker, 2009; Adams, et al., 2009). For more than three decades, critics have tried to cast ethanol as a "food versus fuel" but we don't have to make a choice. We can do both. We must do both—and we do.

The future of biofuels depends on the accelerated diffusion of new technologies, with an appropriate and market-friendly regulatory environment. New use of seaweeds can be developed without risk to existing ones. However, utilizations of seaweed carbohydrates for ethanol production is probably only of economic interest when integrated with a balanced and total utilization of the seaweed material. The largest organic fraction of brown seaweeds is alginate, and brown seaweed is today exploited on industrial scale for alginate production. In the alginate extraction process, mannitol and laminaran is washed out and disposed into the sea, representing an organic load for the recipient. In such a case alginate industry waste with mannitol and laminaran may be considered to be non-cost material for ethanol production. Ethanol can be produced from the mannitol and laminaran using different strain and conditions. The complex composition of seaweeds makes it a difficult substrate to ferment into ethanol by one or a few strains of microbes. An optimization of the extraction process is necessary to obtain higher ethanol concentrations. An obvious practical problem with ethanol production is that the microbial culture may have to be protected against contamination of other microbes. Future research should be concentrated on continuous or semi-continuous fermentations. Seaweeds appear to be the only source of ethanol that capable of meeting the global demand for transport fuel. So in the present study an attempt has been made for the evaluation of ethanol production from seaweeds.

REFERENCES

Adams, J. M., Gallagher, J. M. and Donnison, I. S. 2009. Fermentation study on *Saccharina latissima* for bioethanol production considering variable pretreatments. *J. Appl. Phycol.* 21:569-574.

Badger, P. C. 2002. Ethanol From Cellulose: A General Review. Reprinted from: Trends in new crops and new uses. J. Janick and A. Whipkey (Eds.). ASHS Press, Alexandria, VA.

Bastianoni, S. and Marchettini, N. 1996. Ethanol production from biomass: Analysis of process efficiency and sustainability. *Biomass. Bioenergy.* 11(5): 411-418.

Boyen, C., Kloareg, B., Polne-Fuller, M. and Gibor, A. 1990. Preparation of alginate lyases from marine molluscs for protoplast isolation in brown algae. *Phycologia* 29:173-181.

Brown, B. J. and Preston, J. F. 1991. L-Guluron alginate lyase from a marine bacterium associated with *Sargassum*. *Carbohydr. Res.* 211:91-102.

Butler, D. M., Østgaard, K., Boyen, C., Evans, L. V., Jensen, A. and Kloareg, B. 1989. Isolation conditions for high yields of protoplasts from *Laminaria saccharina* and *L. digitata*. (Phaeophyceae). *J. Exp. Bot.* 40: 1237-1246

Carpita, N. C. and Gibeaut, D. M. 1993. Structural models of cell walls in primary plants: Consistency of molecular structure with the physical properties of the walls during growth. *Plant J.* 3:1-30.

Chapman, V. J. 1950. Seaweeds and their uses. London, Methuen & Co.

Dumsday, G. J., Jones, K., Stanley, G. A., and Pamment, N. B. 1997. Recombinant organisms for ethanol production from hemicellulosic hydrolyzates: A review of recent progress. *Australian Biotechnol.* 7:285-295.

Duncan, R.C. 2005-2006. The olduvai theory energy, population, and industrial civilization. The social contract. *Winter.* 1-12.

Ethanol Across America Spring, 2006. Economic impact of ethanol production.

Ethanol Across America Summer, 2009. Environmental impacts of ethanol production.

Favorov, V. V. and Vaskovsky, V. E. 1971. Alginase from marine invertebrates. *Comp. Biochem. Physiol.* 38B: 689-696.

Fischer, G. and Schrattenholzer, L. 2001. Global bioenergy potentials through 2050. *Biomass. Bioenergy.* 20 : 151-159.

Fleurence, J. 1999. Seaweed proteins: Biochemical, nutritional aspects and potential uses. *Trends Food Sci. Technol.* 10:25-28.

Forro, J. 1987. Microbial degradation of marine biomass. In Bird K. T. and Benson P. H. (Eds). Seaweed cultivation for renewable resources. Elsevier, Amsterdam, 305-325.

Gacesa, P. 1992. Minireview: Enzymatic degradation of alginates. *Int. J. Biochem.* 24: 545- 552.

Galbe, M. and Zacchi, G. 2002. A review of the production of ethanol from softwood. *Appl. Microbiol. Biotechnol.* 59: 618-628.

Ghosh, S., Klass, D. L. and Chynoweth, D. P. 1981. Bioconversion of *Macrocystis pyrifera* to methane. *J. Chem. Tech. Biotechnol.* 31:791-807.

Gopinathan, M. C. and Sudhakaran, R. 2009. Biofuels: opportunities and challenges in India. *In Vitro Cell. Dev. Biol. Plant.* 45:350-371.

Groenestijn, J. W. V., Hazewinkel, J. H. O. and Bakker, R. R. 2007. Pretreatment of lignocellulose with biological acid recycling (the biosulfurol process). *TNO* :1-6.

Gunaseelan, V. N. 1997. Anaerobic digestion of biomass for methane production: A review. *Biomass. Bioenergy.* 13:83-114.

Hall, D. O. and House, J. 1995. Biomass: An environmentally acceptable fuel for the future. *J. Power. Energy.* 209:203-213.

Haugen, F., Kortner, F. and Larsen, B. 1990. Kinetics and specificity of alginate lyases: Part I, a case study. *Carbohydr. Res.* 198:101-109.

Horn, S. J. 2000. Bioenergy from brown seaweeds. Thesis Ph.D. Department of Biotechnology, Norwegian University of Science and Technology, Trondheim, Norway.

Horn, S. J. and Østgaard, K. 2001. Alginate lyase activity and acidogenesis during fermentation of *Laminaria hyperborea*. *J. Appl. Phycol.* 13:143-152.

Horn, S. J., Aasen, I. M. and Østgaard, K. 2000a. Ethanol production from seaweed extract. *J. Ind. Microbiol. Biotechnol.* 25:249-254.

Horn, S. J., Aasen, I. M. and Østgaard. K. 2000b. Production of ethanol from mannitol by *Zymobacter palmae*. *J. Ind. Microbiol. Biotechnol.* 24:51-57.

Kennedy, L., McDowel, K. and Sutherland, I. W. 1992. Alginases from *Azotobacter* species. *J. Gen. Microbiol.* 138: 2465-2471.

Kloareg, B. and Quatrano, R. S. 1988. Structure of cell walls of marine algae and ecophysiology functions of the matrix polysaccharides. Ocenogr. *Mar. Biol. Ann. Rev.* 26:259-315.

Kloareg, B., Demarty, M. and Mabeau, S. 1986. Polyanionic characteristics of purified sulphated homofucans from brown algae. *Int. J. Biol. Macromol.* 8:380-386.

Larsen, B., Hoøen, K. and Østgaard, K. 1993. Kinetics and specificity of alginate lyases. *Hydrobiologia*. 260/261:557-561.

Linker, A. and Evans, L. R. 1984. Isolation and characterization of an alginase from mucoid strains of *Pseudomonas aeruginosa*. *J. Bacteriol.* 159 (3): 958-964.

Martin, P., Mair, C. and Kraan, S. 2010. Seaweeds for second generation bioethanol: Can we compete with corn? XX International Seaweed Symposium, México.

Michel, C., Lahaye, M., Bonnet, C., Mabeau, S. and Barry, J. L. 1996. *In vitro* fermentation by human faecal bacteria of total and purified dietary fibres from brown seaweeds. *Br. J. Nutr.* 75: 263-280.

Moen, E., Horn, S. J. and Østgaard, K. 1997a. Alginate degradation during anaerobic digestion of *Laminaria hyperborea* stipes. *J. Appl. Phycol.* 9:157-166.

Moen, E., Horn, S. J. and Østgaard, K. 1997b. Biological degradation of *Ascophyllum nodosum*. *J. Appl. Phycol.* 9:347-357.

Okamoto, T., Taguchi, H., Nakamura, K., Ikenaga, H., Kuraishi, H. and Yamasato, K. 1993. *Zymobacter palmae* gen. nov., sp. nov., a new ethanol-fermenting peritrichous bacterium isolated from palm sap. *Arch. Microbol.* 160: 333-337

Olsson, L. and Hägerdal, B.H. 1996. Fermentation of lignocellulosic hydrolysates for ethanol production. *Enz. Microbiol. Tech.*18: 312-331.

Østgaard, K. 1993. Determination of alginate composition by a simple enzymatic assay. *Hydrobiologia*. 260/261:513-520.

Palmqvist, E. and Hägerdal, B. H. 2000. Fermentation of lignocellulosic hydrolysates. I: inhibition and detoxification. *Bioresource Technol.* 74: 17-34.

Pimentel, D. and Patzek, T. W. 2005. Ethanol production using corn, switchgrass, and wood; biodiesel production using soybean and sunflower. *Nat. Res.* 14(1): 65-76

Preiss, J. and Ashwell, G. 1962. Alginic acid metabolism in bacteria. Enzymatic formation of unsaturated oligosaccharides and 4-deoxy-L-erythro-5-hexoseulose uronic acid. *J. Biol. Chem.* 237:309-316.

Quain, D. E. and Boulton, C. A. 1987. Growth and metabolism of mannitol by strains of *S. cerevisiae*. *J. Gen. Microbiol.* 133:1675-1684.

Read, S. M., Currie, G. and Bacic, A. 1996. Analysis of the structural heterogeneity of laminarin by electrospray-ionisation-mass spectrometry. *Carbohydr. Res.*, 281:187-201.

Roland, J. C., Reis. D., Vian, B. and Roy, S. 1989. The helicoidal plant cell wall as a performing cellulose-based composite. *Biol. Cell.* 67:209-220.

Sahoo, Dinabandhu 2000. *Farming the Ocean: Seaweeds Cultivation and Utilization*. Aravali Publication Corporation, New Delhi.

Sutherland, I. W. 1995. Polysaccharide lyases. *FEMS Microbiol. Rev.* 16:323-347.

Usov, A. I., Smirnova, G. P. and Klochkova, N. G. 2001. Polysaccharides of Algae: 55 Polysaccharide Composition of Several Brown Algae from Kamchatka. *Russian J. Bioorg. Ch.* 27(6):395-399

Van Dijken, J. P. and Scheffers, W. A. 1986. Redox balances in metabolism of sugar by yeasts. *FEMS Microbiol. Rev.* 32, 199-224.

Walker, D. A. 2009. Biofuels, facts, fantasy and feasibility. *J Appl. Phycol.* 21: 509-517.

Wheal, A. E., Basso, L. C., Alves, D. M. G. and Amorim, H. V. 1999. Fuel ethanol after 25 years. Focus 17: 482-487.

World Energy Outlook Fact Sheet: Global Energy Trends. OECD/IEA, Paris; 2008. http://www.worldenergyoutlook.org/docs/weo2008/fact_sheets_08.pdf.

Yokoyama, S., Jonouchi, K. and Imou, K. 2007. Energy production from marine biomass: Fuel cell power generation driven by methane produce from seaweed. *Proc. World. Acad. Sci. Engg. Technol.* 22:320-323.

Zubia, M., Payri, C. and Deslandes, E. 2008. Alginate, mannitol, phenolic compounds and biological activities of two range-extending brown algae, *Sargassum mangarevense* and *Turbinaria ornata* (Phaeophyta: Fucales), from Tahiti (French Polynesia) *J. Appl. Phycol.* 20:1033-1043.

15

Ecology and PCR based Molecular Characterization of Cyanobacteria with Special Emphasis on Oscillatoriales and Nostocales

Gunapati Oinam, Indrama Th. and O.N. Tiwari
Microbial Bioprospecting Laboratory
Microbial Resources Division
Institute of Bioresources and Sustainable Development
Takyelpat, Imphal-795001, Manipur, India

ABSTRACT

Cyanobacteria have an ancient history dating almost 3.5 billion years and diversified extensively to become one of the most successful and ecologically significant organisms on the earth, with respect to longevity of lineage and impact on earth's early environment. Despite the availability of various monographs based on morphological and ecological variants, the identification and classification of cyanobacteria remain a difficult and confusing task leading to uncertain identifications. Therefore, molecular approaches based on PCR techniques and DNA fingerprinting has been adopted for a clear and unambiguous idea of the taxonomy. Molecular markers such as RAPD, RFLP and AFLP are being used for the PCR techniques which have proven more authenticity for complete identification of the organisms.

INTRODUCTION

Cyanobacteria are an ancient group of prokaryotic microorganisms exhibiting the general characteristics of Gram-negative bacteria. They are unique among the prokaryotes in possessing the capacity of oxygenic photosynthesis. In addition, some cyanobacteria also have the capacity for fixation of atmospheric nitrogen within the same organism. These qualities make cyanobacteria the most successful and widespread group among the prokaryotes found in diverse terrestrial and aquatic environments. Cyanobacteria are a morphologically diverse group ranging from unicellular to colonial and filamentous forms. Taxonomically, cyanobacteria are grouped into unicellular forms that divide by binary or multiple fission and filamentous forms that are non-heterocystous or differentiate heterocysts in non-branching or branching filaments. Unicells cyanobacteria may divide in one, two or three planes. Some unicells dividing in one plane show asymmetric division; in *Chamaesiphon* the smaller cells glide down to the base of the larger cell and repeats of this process eventually give rise to a colony of many cells. The resulting colonies can often be seen as brownish spots on submerged rocks. Branching of filaments may be false or true. False branching is when division in a filament with a sheath leads to the formation of two separate sheathed filaments and one of these grows separately from the other. Maintenance of the colonial structure in many of these forms is aided by the presence of exopolysaccharides such as mucilage and/or a firm sheath. The presence or absence of a heterocyst is an important feature separating genera. However, gas vacuoles, structures that aid buoyancy, are found in species of many different genera.

A review by Castenholz (1992) and the sister volume (Bryant, 1994) to the present one make clear how important is an understanding of cyanobacterial molecular biology not just as an aid to taxonomy, but for interpreting ecological phenomena in general. Suboptimal light and nutrient conditions result in a number of responses that strongly influence the physiology of the cell. The responses can be striking or subtle and subsequent changes take place rapidly or very slowly. Cyanobacteria tend to show resistance to multiple environmental stresses, and it is probable that the response pathways to the different stimuli overlap.

ORIGIN AND EARLY EVOLUTION OF CYANOBACTERIA

The Evolution of the Cyanophytes

Although the evidence for the time and manner of origin of cyanophytes is still meager, and most of it subject to varying interpretation, it seems likely that the cyanophytes arose between 2 and 3 $\times 10^9$ years ago (Schopf, 1970) probably from a photosynthetic bacterium. After the discovery of gliding photosynthetic bacterium *Chloroflexis* (Pierson and Castenholz, 1971) is a reasonable candidate as a forerunner of the cyanophytes. It is virtually certain that the cyanophytes were probably responsible for a marked increase in the O_2 concentration of the atmosphere during the Precambrian. Another reason why a cyanophyte is the best candidate for the first O_2-evolving photosynthetic organism is that at least some cyanophytes are able to grow an aerobically, a property not normally associated with eukaryotic algae (Stewart and Pearson, 1970). If the bulk of the atmospheric O_2 is indeed of biogenic origin, the organisms first involved in O_2 formation would of course have developed in an anaerobic environment and would have had to be able to live anaerobically.

Stromatolites and the Evolution of the Cyanophytes

Stromatolites are rocks consisting of many layers, most commonly developing from the mats of cyanophytes. The rocks may be either siliceous or calcareous, although the later are most common. The layers probably build as a result of sediment trapping and carbonate precipitation by unicellular and filamentous cyanophytes. In shallow marine habitats, calcareous sediments deposit continuously on the surface of the mat, and the cyanophytes move up through the sediments during the day (Monty, 1967; Gebelein, 1969). In some cases the layers are diurnal, resulting from phototrophic movement of the cyanophytes.

ECOLOGY OF CYANOPHYTES

To what kind of habitats are the cyanophytes as a group adapted, in the Precambrian they may have been found in a very wide variety of habitats but today their distribution is more restricted, since in many habitats they must compete with eukaryotes. There is only one known type of habitats on the earth today where cyanophytes are the exclusive O_2-evolving photosynthetic organisms, and that is in thermal springs (Brock, 1970). However, even in the thermal springs cyanophytes are not always successful. In acid springs with pH less than 4.0, cyanophytes are never found, irrespective of temperature. In neutral and alkaline thermal springs, cyanophytes are the exclusive O_2-evolving photosynthetic organisms at temperatures above 55-60°C, and are far in the dominance at temperatures down to about 40°C. The upper temperature limit for cyanophytes is about 72-73°C and at temperatures above this only non-photosynthetic bacteria are present.

In other habitats, even in those in which cyanophytes are widely held to be successful, they are never exclusive and often not dominant. Cyanophytes are the only group of O_2-evolving photosynthetic organism, which fixes nitrogen. The fixation of nitrogen is found under aerobic conditions in heterocystous cyanophytes and under anaerobic condition also in some non-heterocystous forms. The ability of a variety of cyanophytes to fix nitrogen in natural aquatic systems is well established (Stewart, 1969). The presence in cyanophytes of phycobilins as accessory photosynthetic pigments may confer a considerable ecological advantage to them under conditions of low light intensity. Perhaps one of the most important factors controlling cyanophytes development may be the ability of this group to grow under conditions of low O_2 concentration. It might be in such a niche that the first eukaryotic alga could have developed. In such an alga, with its photosynthetic apparatus concentrated in a chloroplast, the cytoplasm surrounding the chloroplast could have provided an effective pH buffer which would have served to keep H^+ from the very acid sensitive chlorophyll molecule.

MECHANISM OF MOVEMENTS IN CYANOBACTERIA

Motility in cyanobacteria is gliding motility. Gliding in contrast to swimming is movement in contact with a solid or semi-solid substrate, but without a visible change in the shape of the organism. A filament may have considerable passive flexibility, but there does not seem to be any possibility of steering. The relatively slow progress of gliding is accompanied in some species by a rotation of the trichome along its axis which is either right handed or left handed. In liquid microscope mounts with or without a cover slip, free ends of trichome are seen to jerk back and forth, wave to and fro, or oscillate by tracing a cone. It appears that all of these oscillations are passive responses to the rotation of the trichome when any part of it is able to bear against a substrate and glide.

Gliding Behaviour

Gliding is movement in contact with a solid or semi-solid surface without flagella like propulsive organs. In most cases it is a smooth, non-jerking movement resembling somewhat the progress of a snail. In many cases the movement is continuous in one direction for prolonged periods; in others frequent or regular reversals occur. Gliding of trichome in almost all circumstances probably occurs within a thin sheath casing or film of mucus which adheres to the substrate and is shed and left behind as a 'mucous' trail. When measured microscopically, the velocity of movement without reversals varies with species and environmental conditions. There is no consistent correlation between trichome diameter of different species and the velocity of gliding.

The Mechanism of Motility

One of the earliest suggested mechanisms for gliding motility (Doetsch and Hageage, 1968) has been supported and modified by Halfen and Castenholz (1971) that the *Oscillatoria* is that numerous microfibrils strung uninterrupted in helices around the exterior portion of the trichome are the propulsive organelles which move the chain of cell against an elastic sheath or other suitable substrate by a rapid succession of unidirectional waves with submicroscopic amplitude. The origin of wave propagation may shift from one pole to the other, reversing the direction of wave movement and of gliding. Unicellular species or those with trichome of more disjointed cell could have a similar system but continuous for only a single cell length. Species that do not rotate as they glide should have longitudinal microfibrils which are not helically strung.

Orientated Movements

Phototaxis is the only tactic response of cyanobacteria that has received much attention. Phototopotaxis is the orientation and movement towards (+) or away from (–) the incident light. Photophobotaxis is the reversal of direction of movement following a sudden change from high to low light intensity (+) or from low to high light (–).

GAS VACUOLES IN CYANOBACTERIA

Fine Structure of Gas Vesicles

The fine structure of gas vacuoles has been investigated in a total of eleven species from the genera *Microcystis*, *Oscillatoria*, *Trichodesmium*, *Anabaena*, *Aphanizomenon*, *Nostoc* and *Gloeotrichia* and in each case they have been found to comprise gas vesicles of the same basis morphology originally described by Bowen and Jensen (1965). The vesicles have the form of hollow, cylindrical tubes with conical caps at each end. The tubes are possibly of uniform diameter, usually given as being about 70 nm the reported extreme being 65 nm for *Oscillatoria rubescens* and 75-85 nm for *Trichodesmium erythraeum*. Gas cylinder aptly described these structures in cyanobacteria but is inapplicable to homologous structure of different shape found in various bacteria (Cohen-Bazire et al., 1969; Walsby, 1972). 'Gas vacuole subunits' might be confused with the proteinaceous subunits of which the gas vesicles membrane is formed. However, it is felt that it is often useful to be able to distinguish the composite structures, seen under the light microscope, form the elements which comprise them. In many species the gas vacuole are irregular in form and distributed throughout the cell but in some algae they have a characteristic shape and position.

Chemical Composition of Gas Vesicles

Chemical analyses have now been made on gas vesicles isolated from cyanobacteria and the results have confirmed the idea, suggested by electron microscopy, that the gas vesicles membrane is fundamentally different from typical unit membrane. To avoid collapsing the gas vesicles, the cell were lysed by method which did not involve exposing the preparations to pressure, the filamentous alga being disrupted by osmotic shrinkage in hypertonic sucrose solution and the unicellular form by osmotic shock after weakening the cell wall by penicillin treatment. The intact vesicles were then purified by a process of repeated, centrifugally accelerated flotation, followed by combination of membrane filtration, molecular sieving and liquid polymer partitioning (Buckland and Walsby, 1971).

Functions of Gas Vesicles

Of the three functions that gas vacuoles might fulfill that of storing gas is no longer tenable. The gas vesicle membranes are far too permeable to retain any particular gas. It has been pointed out that the two other functions, providing light shielding and buoyancy do not depend on the presence of gas in the vesicles, but rather on these structures being kept free of liquid or solids.

Light Shielding

The peculiar optical quantities of gas vacuoles have suggested (Waaland et al., 1971) that they might provide light shielding for cells exposed to high light at water surfaces. The path length of light passing through a gas vacuole may differ from that passing through the neighbouring cytoplasm by more than ¼ wavelength (the normal retardation of the objective phase plate) and result in 'false reversal' so that the vacuole looks brighter instead of darker than the background. With high resolution optics, gas vacuoles appear almost invisible under bright field illumination but with low numerical apertures they are characterized by diffraction fringes.

Spectrophotometric analysis confirm that gas vesicles show strong light scattering properties, and that over most of the visible spectrum the intensity of the scattered light shows the relationship $i \alpha \lambda^{-4}$ which is characteristic of Tyndall scattering. By refraction, scattering and internal reflection gas vacuoles may so redirect the path of light from the cell as to provide shielding of underlying, light sensitive structures such as photosynthetic pigments and the genetic material (which may be damaged by ultra violet light).

Light Intensity in Cyanobacteria

Within certain limits, an inverse correlation between light intensity and photo pigment content is general among photosynthetic organisms. As might be expected, the ratio of phycocyanin to chlorophyll can vary widely in response to simultaneous change in light intensity, temperature and carbon dioxide partial pressure. In chromatically adapting strains, growth in green light stimulates the synthesis of the red coloured phycoerythrin (absorbing in the green) whereas when growth occurs in red light it is the blue protein, phycocyanin (absorbing in the red), which is the dominant biliprotein. After a brief 'red' suspension immediately, phycocyanin alone is synthesized whereas after a brief 'green' irradiation, phycoerythrin and phycocyanin are both synthesized. These effects are repeatedly reversible; the last 6 irradiation determines the pattern of biliprotein synthesis in the dark. The chromatic effects are clearly photochemical, being independent of the temperature used during the irradiations (Fujita and Hattori, 1962).

MODERN APPROACHES AND CONCEPTS ON OSCILLATORIALES

Included all filamentous cyanobacteria that undergo binary fission in a single plane and that produce "vegetative" cells only. Heterocysts and akinetes do not occur. Trichome diameters may range from about 0.4 µm to well over 100 µm. The terminal cell of some species may be distinctly shaped but is apparently still capable of photosynthesis. Sometimes the terminal cell is tapered and with a cap or calyptra, but in a few forms the taper may include several sub-terminal cells as well. Generally, within a strain the trichome division varies little (<10%) in contrast to some other orders of cyanobacteria. As a rule, all cells retain the ability to divide; in practice, certain portions of a trichome may be more active meristematically than terminal regions. The terminal cell in some may never divide.

An apparent sheath may be present, but even species without an easily visible sheath leave behind at least a very thin, gossamer sheath when moving by gliding. When short fragments of a few cells separate from the remainder of the trichome near the free, open end of a sheath, these free trichomes may glide out, eventually forming new sheaths. Although trichomes without apparent sheaths also fragment, a separable, migrating, hormogonial phase is difficult to distinguish, since all lengths of trichome are generally motile. Fragmentation of trichomes occurs in some forms where a cell loses much of its contents and dies. In some cases, there appears to be an orderly sacrificial death of these cells (necridial cells) which determines the sites of trichome breakage (Lamont, 1969; Ciferri, 1983).

The range in mole % G + C of the DNA is large (40-67). The range of genome sizes is also great (2.14-5.19 $\times$ 10^9 daltons), but all sizes are generally less than those in the Nostocales or Stigonematales (Herdman et al., 1979b). The range in DNA base composition and in genome size indicates a probable artificiality in the grouping oscillatoriales, a conclusion further supported by the degree of disparity in some sequences of 800-900 continuous nucleotides of 16S rRNA from 11 strains of 7 genera of this group (Giovannoni et al., 1988). The triviality of some generic distinctions used here should also be emphasized. Often only one characteristic is used, a characteristic that may be the result of a slight difference in genetic code. In the "Geitlerian" system, generic distinction in the oscillatoriales is based primarily on the diversity of sheaths or their absence. Although still used as a characteristic in the present system, knowledge of physiology, biochemistry, and nucleotide base sequence homologies will eventually determine degree of relatedness. Members of oscillatoriales occur in an enormous diversity of habitats: freshwater and marine, both as plankton, mats, and periphyton. Terrestrial crusts, mats, and turfs are also common. Hot spring mats of some oscillatorian cyanobacteria develop upto temperature of about 62°C (Castenholz, 1969).

MODERN APPROACHES AND CONCEPTS ON NOSTOCALES

Members of all nostocales are distinguished from all other cyanobacteria by being filamentous organisms dividing exclusively by binary fission in one plane only (some with the possibility of false branching) and having the potential to produce heterocysts. False branching may occur frequently in some genera, but this does not involve division in more than one plane. Nostocales are defined into three well-circumscribed families namely; Nostocaceae, Sytonemataceae and Rivulariaceae.

Nostocaceae: Trichomes are composed of vegetative cells of uniform diameter (the trichomes do not taper), and false branching does not occur or, if the organism is sheathed, occurs only rarely. In culture, this family is known primarily for strains that fit under the generic names: *Anabaena*, *Aphanizomenon*, *Nodularia*, *Cylindrospermum* and *Nostoc*.

Scytonemataceae: Trichomes, uniseriate, usually untapered and sheathed. False branching, single or double (germinate), frequent or infrequent. Meristematic regions instead of diffuse cell divisions usually develop near the apex of the trichome while cell degeneration progresses from the base. Heterocysts are predominantly intercalary and are often associated with false branches. Akinetes are absent or rare. Non-heterocyst containing hormogonia may be formed and released. Hormogonia, after coming to rest, differentiate a single terminal heterocyst. Organisms are usually attached, aquatic (freshwater or marine) or terrestrial.

Rivulariaceae: In culture, the family Rivulariaceae is best known by the strains most easily classified in the genus *Calothrix*. The key characteristic is trichome polarity as seen by taper at one end. However, the polar (basal) heterocyst under appropriate environment may be an even more reliable criterion of true relatedness. Trichome is uniseriate, false branching is rare to common; trichome is tapered, although not under all culture conditions. Heterocysts are terminal and single pored; at widened end under appropriate conditions of low combined nitrogen (intercalary heterocysts may also be present). Cell divisions are usually intercalary and localized. Tapered end of the trichome commonly extends into long, pale multicellular hair. Hormogonia, short lengths of cells smaller in diameter than the larger cells of the trichome, are formed in culture in response to phosphate repletion, following depletion. After a period of gliding, hormogonia come to rest, produce a basal heterocyst and begin tapered growth. Trichomes are usually attached to substrates individually or in groups forming hemispherical or sub-spherical "cushions". Colonial planktonic forms (*Gloeotrichia*) with radically arranged trichomes forming spheres also contain gas vesicles, as do hormogonia of many sessile members of this group.

CONVENTIONAL/TRADITIONAL METHODS FOR SYSTEMATICS

Traditionally, the classification of cyanobacteria has been based on morphological characters such as trichome width, cell size, division planes, shape and arrangement, pigmentation and the presence of characters such as gas vacuoles and a sheath (Baker, 1991, 1992; Bourrelly, 1970; Komarek and Anagnostidis, 1986, 1989). The importance of structural and developmental characters for classification of both field and cultured material makes it essential for researchers to isolate and purify the specimen of interest under optimal cultural conditions and provide it with a taxonomic identity. Many cyanobacteria are conspicuous in nature and collection of samples is therefore greatly facilitated. To date, several hundred cyanobacteria have been isolated and purified using standard microbiological techniques. The first and foremost step taken is collection of samples from various natural sources such as water and soil.

Direct Observation: This is the most suitable for samples that contain good number of organisms. Rice field floodwater carrying suspended soil particles are filtered on membrane filters and the population is smeared on the slide and observed under microscope. For proper enumeration of small cells of cyanobacteria, microscopy as well as epifluorescence microscopy is also used to distinguish between cyanobacteria and small sized bacteria. This method is based on the chlorophyll a fluorescence of algae that distinguishes them from bacteria. Combination of excitation and suppression filters may be used to differentiate between phycocyanin and phycoerythrin (Hawes and Davey, 1989). Since cultured samples give better performance, there is a need to improve it for field application.

BIOCHEMICAL AND PHYSIOLOGICAL CHARACTERIZATION

Photosynthesis, N_2 fixation, NH_4 excretion, pigment profile, carbohydrates, total soluble proteins and N-assimilatory enzymes like nitrogenase, nitrate reductase and glutamine synthetase activity plays a key role for characterization.

Measurement of pigments: This is useful in quantifying surface growth dominated by cyanobacteria (Davey, 1988; Whitton and Roger, 1989) but less suited for sparse populations spread over the upper part of the soil column. The organisms can be frozen also to avoid degradation of pigments. The use of television image analysis with epifluorescence microscopy for cyanobacteria and other microorganisms has also been described (Wynn- Williams, 1990). Satellite sensor images offer a completely different approach to quantifying surface crusts. This method is similar to those used for studying semiarid vegetation index (Karnieli et al., 1996).

MOLECULAR APPROACHES FOR CLASSIFICATION

The considerable expertise required to identify species by certain criteria based on morphological characters such as trichome width, cell size, division planes, shape and arrangement, pigmentation and the presence of characters such as gas vacuoles and a sheath, subjective judgement by operators can lead to errors, resulting in incorrect assignment of isolates. Komarek and Anagnostidis (1989) estimated that more than 50% of the strains in culture collections are misidentified. Moreover, some diagnostic features, such as gas vacuoles or akinetes, can show variations with different environmental or growth conditions and even be lost during cultivation (Rudi et al., 1997; Lyra et al., 2001). Such limitations of phenotypic characters have highlighted the requirement for more reliable methods and promoted molecular approaches in cyanobacterial taxonomy, including DNA base composition (Kaneko et al., 2001), DNA hybridizations (Kondo et al., 2000), gene sequencing (Nubel et al., 1997) and PCR fingerprinting (Rasmussen and Svenning, 1998; Versalovic et al., 1991). Cyanobacterial specific methods not requiring axenic cultures are of utmost importance since such cultures are difficult to obtain (Choi et al., 2008). PCR based techniques proved to be more reliable than the conventional ones in various aspects of identification of the cyanobacterial population and to uncover cryptic variations of strains or closely related species.

The polymerase chain reaction (PCR) is a technique widely used in molecular biology. It derives its name from one of its key components, a DNA polymerase used to amplify a piece of DNA by in vitro enzymatic replication. As PCR progresses, the DNA generated is used as a template for replication. With PCR it is possible to amplify a single or few copies of a piece of DNA across several orders of magnitude, generating millions or more copies of the DNA piece. PCR can be extensively modified to perform a wide array of genetic manipulations.

The purpose of PCR is to make a huge number of copies of a gene. This is necessary to have enough starting template for sequencing. There are three major steps in a PCR, which are repeated for 30 or 40 cycles. This is done on an automated cycler, which can heat and cool the tubes with the reaction mixture in a very short time. During the denaturation, the double strand melts open to single stranded DNA, all enzymatic reactions stop. The primers are jiggling around, caused by the brownian motion. Ionic bonds are constantly formed and broken between the single stranded primer and the single stranded template. The more stable bonds last a little bit longer (primers that fit exactly) and on that little piece of double stranded DNA (template and primer), the polymerase can attach and starts copying the template. Once there are a few bases built in, the ionic bond is so strong between the template and the primer, that it does not break anymore. Elongation is the ideal working temperature for the polymerase. The primers, where there are a few bases built in, already have a stronger ionic attraction to the template than the forces breaking these attractions. Primers that are on positions with no exact match get loose again (because of the higher temperature) and don't give an extension of the fragment. The bases (complementary to the template) are coupled to the primer on the 3' side (the polymerase adds dNTP's from 5' to 3', reading the template from 3' to 5' side, bases are added complementary to the template).

PCR Amplification and Sequencing of 16S rRNA and rpoC1 Genes

Since universal primers for direct sequencing of 16S rRNA genes are usually designed to be used with axenic cultures, and available primers for rpoC1 amplification and sequencing are highly degenerate, specific primers were selected or designed in order to obtain clean sequences for both genes without the need for a cloning step.

Restriction Fragment Length Polymorphisms (RFLPs)

The restriction fragment length polymorphisms (RFLPs) of particular PCR products can provide signature profiles specific to the genus, species, or even strain. Genetic characterization of cyanobacterial strains has been undertaken using RFLPs of the 16S rRNA gene (16S-ARDRA) (Lyra et al., 1997) and of the intergenic transcribed spacer region (ITS-ARDRA) (Lu et al., 1997; West and Adams, 1997). Furthermore, amplification of the 16S–23S rRNA ITS, which has been shown to be variable in length (Rocap et al., 2002; Iteman et al., 2002; Neilan, 2002, Laloui et al.,2002) and number (West and Adams, 1997) in cyanobacteria, can also be used as an identification tool. Neilan et al., (1995) have also found heterogeneity in the cluster containing mostly heterocystous planktonic strains of *Anabaena* and *Aphanizomenon* genera by using PCR/RFLP of the phycocyanin locus with intergenic spacers. RFLP has revealed the genetic proximity of the *Aphanizomenon* with *Anabaena* and *Nodularia* with *Nostoc*. Also, it was found that the neurotoxic starins of *Anabaena* were identical while the hepatotoxic strains formed a heterogenous group.

Random Amplified Polymorphic DNA (RAPD)

Random amplified polymorphic DNA (RAPD) markers are DNA fragments from PCR amplification of random segments of genomic DNA with single primer of arbitrary nucleotide sequence. Unlike traditional PCR analysis, RAPD does not require any specific knowledge of the DNA sequence of the target organism: the identical 10-mer primers will or will not amplify a segment of DNA, depending on positions that are complementary to the primers' sequence. Furthermore, by RAPD, polymorphisms can be easily analyzed by small amounts of template DNA RAPD was used to generate unique and identifying profiles for members of cyanobacterial genera *Anabaena* and *Microcystis*. This method is based on the combination of the two 10-meroligonucleotides in a single PCR and provided specific and repeatable DNA fingerprints which made it possible to discriminate among all toxigenic cyanobacteria studied, to the taxonomic ranks of genera, species and strains. Analysis of DNA typing results obtained by this method clearly discriminates between genera *Anabaena* and *Microcystis*. Strains of *Microcystis* with identical allozyme patterns *(M. novacekii, M. viridis, M. wesenbergii)* provided similar RAPD profiles (Nishihara et al., 1997); whereas no similarities in the profiles were observed for the strains with different allozyme genotype (*M. aeruginosa*). Several amplified DNA fragments which were expected to be markers for a particular taxon with identical allozyme genotype were also observed on RAPD patterns. The good accordance between the RAPD and allozyme divergency, indicate the high reliability of RAPD analysis for the easy and rapid discrimination of cyanobacteria.

Limitations of RAPD

Nearly all RAPD markers are dominant, *i.e.* it is not possible to distinguish whether a DNA segment is amplified from a locus that is heterozygous (1 copy) or homozygous (2 copies). Co-dominant RAPD markers, observed as different sized DNA segments amplified from the same locus, are detected only rarely. PCR is an enzymatic reaction, therefore the quality and concentration of template DNA, concentrations of PCR components, and the PCR cycling conditions may greatly influence the outcome. Thus, the RAPD technique is notoriously laboratory dependent and needs carefully developed laboratory protocols to be reproducible. Mismatches between the primer and the template may result in the total absence of PCR product as well as in a merely decreased amount of the product. Thus, the RAPD results can be difficult to interpret.

Amplified Fragment Length Polymorphism (AFLP)

AFLP-PCR is a highly sensitive method for detecting *polymorphisms* in *DNA*. In detail, the procedure of this technique is divided into three steps: *i*) Digestion of total cellular DNA with one or more *restriction enzymes* and ligation of restriction half site specific adaptors to all restriction fragments and *ii*) selective amplification of some of these fragments with two PCR primers that have corresponding adaptor and restriction site specific sequences; *iii*) electrophoretic separation of amplicons on a gel matrix, followed by visualisation of the band pattern.

The AFLP technology has the capability to detect various polymorphisms in different genomic regions simultaneously. It is also highly sensitive and reproducible. As a result, AFLP has become widely used for the identification of genetic variation in strains or closely related species of plants, fungi, animals, bacteria and cyanobacteria. There are many advantages to AFLP when compared to other marker technologies including randomly amplified polymorphic DNA (*RAPD*), restriction fragment length polymorphism (RFLP), and microsatellites. AFLP not only has higher reproducibility, resolution, and sensitivity at the whole genome level compared to other techniques, but it also has the capability to amplify between 50 and 100 fragments at one time. In addition, no prior sequence information is needed for amplification (Meudth and Clarke, 2007). As a result, AFLP has become extremely beneficial in the study of taxa including bacteria, fungi, and plants, where much is still unknown about the genomic makeup of various organisms. AFLP analysis is based on selective amplification of DNA restriction fragments. It is technically similar to restriction fragment length polymorphism analysis, except that only a subset of the fragments are displayed and the number of fragments generated can be controlled by primer extensions. The advantage of AFLP over other techniques is that multiple bands are derived from all over the genome. This prevents over-interpretation or is interpretation due to point mutations or single locus recombination, which may affect other genotypic characteristics. The main disadvantage of AFLP markers is that alleles are not easily recognized (Majer et al., 1998). PCR has proven to be successful in detecting genetic variation amongst plant pathogenic fungi, as well as bacteria (Janssen et al., 1996). The utility, repeatability and efficiency of the AFLP technique are leading to broader application of this technique in the analysis of cyanobacterial populations.

DNA Fingerprinting based on STRR and LTRR Sequences

Repetitive sequences constitute an important part of the prokaryotic genome (Van Belkum et al., 1998). Despite their unknown function, and lack of knowledge about how they are maintained and dispersed, the presence, widespread distribution and high conservation of these sequences make them methodologically important for DNA fingerprinting and allow their use as an alternative for the identification of species or strains and in diversity studies among related prokaryotes and for identification (fingerprinting) of microorganisms in general. However it has been suggested that they may regulate transcription termination or be the target of DNA binding proteins responsible for chromosomal maintenance in the cell. In the particular case of cyanobacteria, a family of repetitive sequences, the short tandemly repeated repetitive sequences (STRRs), has been described (Mazel et al., 1990). These heptanucleotide sequences have been identified in several cyanobacterial genera and species, so far mostly in heterocystous cyanobacteria (Wilson et al., 2000; Teaumroong et al., 2002; Lyra et al., 2005) but also in some non-heterocystous ones (Rasmussen and Svenning, 1998). Furthermore, a 37 bp long tandemly repeated repetitive sequence (LTRR) has also been identified in some cyanobacterial species (Prasanna et al., 2006). Analysis of STRRs and LTRRs has been described as a powerful tool for taxonomic studies. Moreover, the specificity of these sequences has made STRRs useful even for non-axenic cyanobacterial cultures (Nilsson et al., 2000). As it can be difficult and time-consuming to establish axenic cultures of cyanobacteria, the developed PCR method with STRR

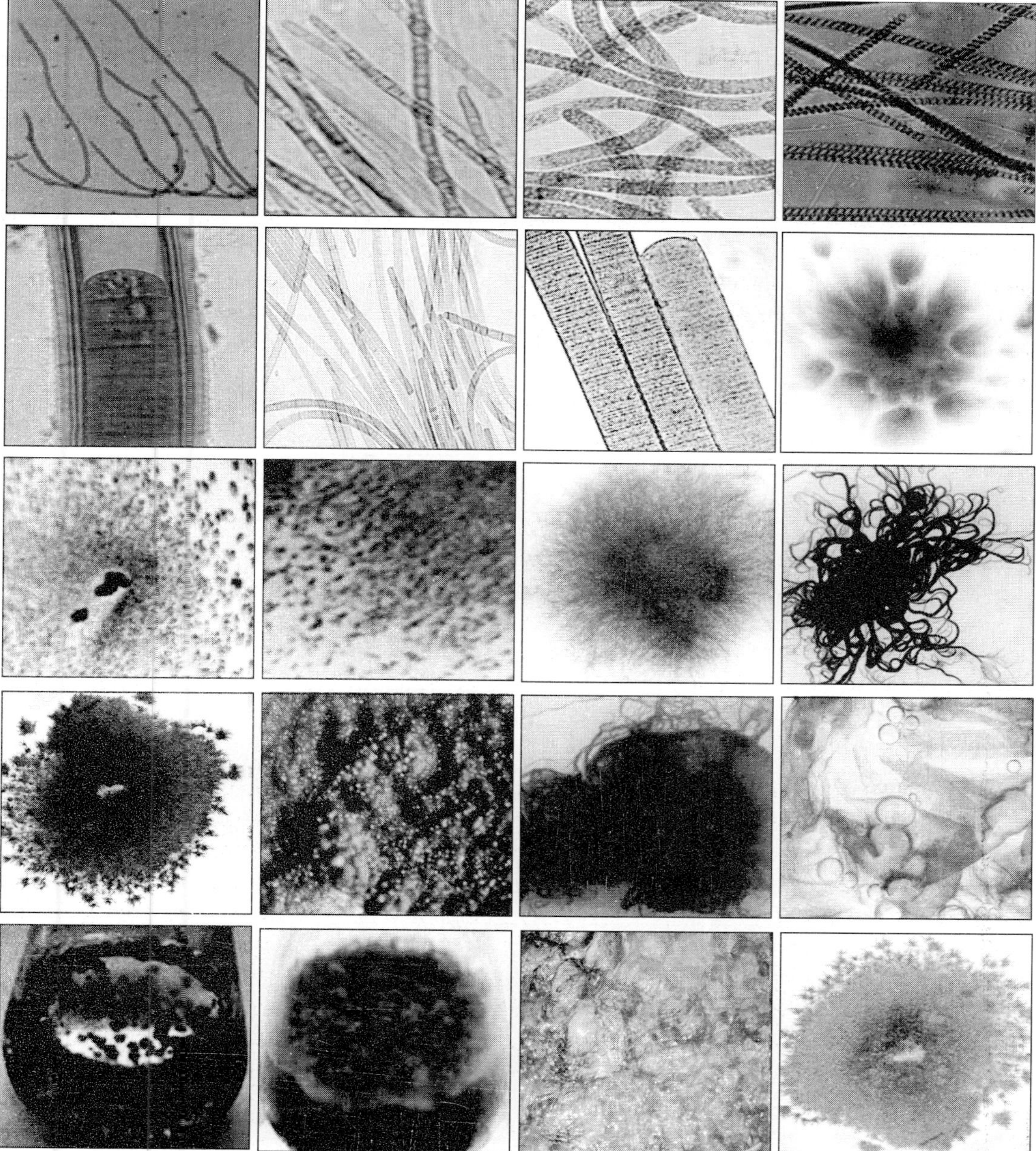

Fig. 1 Cultural studies and microphotograph of non-heterocystous filamentous cyanobacteria of germplasm of IBSD, Imphal.

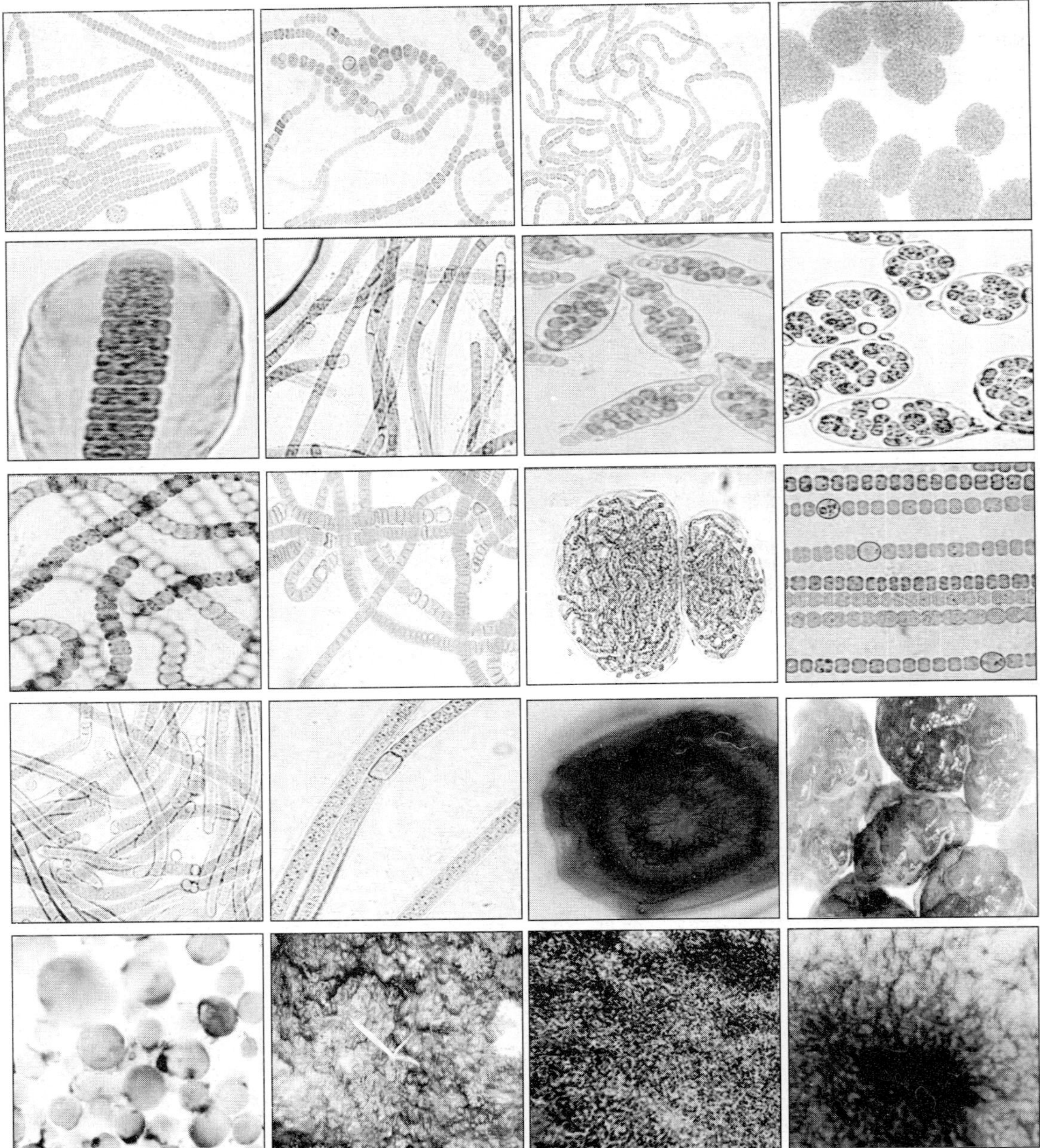

Fig. 2 Cultural studies and microphotograph of non-heterocystous filamentous cyanobacteria of germplasm of IBSD, Imphal.

or LTRR primers provides a useful method for studying the diversity of cyanobacteria in the natural environment, whether free living or symbiotic.

A universal marker for DNA fingerprinting is the oligonucleotide csM13. It has already been tested in a small number of cyanobacteria (Valerio et al., 2005), and has a demonstrated ability even to discriminate strains of the same species. Techniques based on the enterobacterial repetitive intergeneric consensus (ERIC) have also been used for identification and discrimination purposes in some cyanobacteria (Lyra et al., 2001; Bruno et al., 2006). However, the method based on STRR and LTRR sequences is accurate in distinguishing and classifying even closely related strains of cyanobacteria. The described method is a valuable and rapid alternative to other methods used for classification and diversity studies of cyanobacteria.

REFERENCES

Baker, P., 1991. Identification of common noxious cyanobacteria Part I – Nostocales: Urban water research association of Australia. Research Report no. 29.

Baker, P., 1992. Identification of common noxious cyanobacteria Part II– Chroococales and Oscillatoriales: Urban water research association of Australia. Research Report no. 46.

Bowen, C. C. and Jensen, T. E. 1965. Blue-green algae: Fine structure of the gas vacuoles. *Science* 147: 1460-1462.

Bourrelly, P. 1970. Les Algues d'Eau Douce: Les Algues Bleus et al., Rouges, Les Eugle´niens, Peridiniens, et Cryptomonadines. Paris: N. Boube´e & Cie.

Brock, T. D. 1970. High temperature systems. *A Rev. Ecol. System.* 1: 191-220.

Bruno, L., Billi, D., Albertano P, and Urzi, C. 2006. Genetic characterization of epilithic cyanobacteria and their associated bacteria. *Geomicrobiol. J.* 23: 293-299.

Bryant, D. A. 1994. The Molecular Biology of Cyanobacteria, Kluwer Academic Publishers, Dordrecht, The Netherlands.

Buckland, B. and Walsby, A. E. 1971. A study of the strength and stability of gas vesicles isolated from a blue-green alga. *Arch. Microbiol.* 79: 327–337.

Castenholz, R. W. 1969. Thermophylic blue-green algae and the thermal environment *Bact. Rev.* 31(4): 476-504.

Castenholz, R. W. 1992. Species usage, concept, and evolution in the cyanobacteria. *J. Phycol.* 28: 737-745.

Choi, G. G., Bae, M. S., Ahn, C. Y. and Oh, H. M. 2008. Induction of axenic culture of *Arthrospira (Spirulina) platensis* based on antibiotic sensitivity of contaminating bacteria. *Biotechnol. Lett.* 30: 87-92.

Ciferri, O. 1983. *Spirulina*, the edible microorganism. *Microbiol. Rev.* 47: 551-578.

Cohen-Bazire, G., Kunisawa. R. and Pfening, N. 1969. Comparative study of the structure of gas vacuoles. *J Bact.* 100: 1049-1061.

Davey, M.C. 1988 Ecology of terrestrial algae of the fell field ecosystems of signy island. South Orkney Islands. *Br. Antarctica Surv. Bull.* 69-74.

Doetsch, R. N., Hageage, G. J. 1968. Motility in prokaryotic organisms: Problems, points of view, and perspectives. *Biol. Rev.* 43: 317-362.

Fujita, Y. and Hattori, A. 1962. Photochemical interconversion between precursors of phycobilin chromoproteids in *Tolypothrix tenuis. Plant Cell Physiol.* 3: 209-220.

Gebelein, C. D. 1969. Distribution, morphology, and accretion rate of recent subtidal algal stromatolites, Bermuda. *J. Sebium. Petrol.* 39: 49-69.

Giovannoni, S. J., Turner, S. G., Olsen, J., Barns, S., Lane, D. J. and Pace, N. R. 1988. Evolutionary relationships among cyanobacteria and green chloroplasts. *J. Bacteriol. August.* 170(8): 3584-3592.

Halfen, L. N. and Castenholz, R. W. 1971. Gliding motility in the blue-green alga, *Oscillatoria princeps. J. Phycol.* 7: 133-145.

Hawes, I. and Davey, M. C., 1989. Use of florochrome auramine for determination of cell viability in filamentous and thalloid algae. *Phycologia.* 28: 518-523.

Herdman, M., Janvier, M., Waterbury, J. B., Rippka, R. and Stanier, R. Y. 1979. Deoxyribonucleic acid base composition of cyanobacteria. *J. Gen. Microbiol.* 111: 63-71.

Iteman, I., Rippka, R., Tandeau, de Marsac, N. and Herdman, M. 2002. rDNA analyses of planktonic heterocystous cyanobacteria, including members of the genera *Anabaenopsis* and *Cyanospira. Microbiol.* 148: 481-496.

Janssen, P., Coopman, R., Huys, G., Swings, J. and Bleeker, H., Vos, P., Zabeau, M., and Kersters, K. 1996. Evaluation of the DNA fingerprinting methods: AFLP as a new tool in bacterial taxonomy. *Microbiol.* 142: 1881-1893.

Karnieli, A., Shachak, M., Tsoar, H., Zaady, E., Kaufman, Y., Denin, A. and Porter, W. 1996. The effect of macrophytes on the spectral reflectance of vegetation in semiarid regions. *Remote Sensing Environment.* 57: 88-96.

Kaneko, T., Nakamura, Y., Wolk, C. P., Kuritz, T., Sasamoto, Watanabe A., Iriguchi, A., Ishikama, K., Kawashima, K. and other authors. 2001. Complete genomic sequence of the filamentous nitrogen fixing cyanobacterium *Anabaena* sp. strain PCC 7120. *DNA Res.* 8: 205-213.

Komarek, J. and Anagnostidis, K. 1986. Modern approach to the classification system of Cyanophytes. 2. Chroococcales. *Algol. Stud.* 43:157-226.

Komarek, J. and Anagnostidis, K. 1989. Modern approach to the classification system of Cyanophytes. 4. Nostocales. *Algol. Stud.* 56: 247-345.

Kondo, R., Yoshida, T., Yusi, Y. and Hiroishi, S. 2000. DNA–DNA reassociation among a bloom-forming cyanobacterial genus, *Microcystis. Int. J. Syst. Evol. Microbiol.* 50: 767-770.

Laloui, W., Palinska, K. A., Rippka, R., Partensky, F., Tandeau, de Marsac, N., Herdman, M and Iteman, I. 2002. Genotyping of axenic and non-axenic isolates of the genus *Prochlorococcus* and the OMF-'*Synechococcus*' clade by size, sequence analysis or FLP of the internal transcribed spacer of the ribosomal operon. *Microbiol.* 148: 453-465.

Lamout, H. C. 1969. Sacrificial cell death and trichome breakage in an Oscillatoriacean blue-green alga the role of murein. *Arch. Mikrobiol.* 69: 237-259.

Lu, W., Evans, H., McColl S. M. and Saunders V. A. 1997. Identification of cyanobacteria by polymorphisms of PCR-amplified ribosomal DNA spacer region. *FEMS Microbiol. Lett.* 153 141-149.

Lyra, C., Hantula, J., Vainio, E., Rapala, J., Rouhiainen, L. and Sivonen, K. 1997. Characterization of cyanobacteria by SDS-PAGE of whole-cell proteins and PCR/RFLP of the 16S rRNA gene. *Arch. Microbiol.* 168: 176-184.

Lyra, C., Suomalainen, S., Gugger, M., Vezie, C. and Sundman, P., Paulin, L., Sivonen, K. 2001. Molecular characterization of planktic cyanobacteria of *Anabaena*, *Aphanizomenon*, *Microcystis* and *Planktothrix* genera. *Int. J. Syst. Evol. Microbiol.* 51: 513-526.

Lyra, C., Laamanen, M., Lehtima, J. M., Surakka, A. and Sivonen, K. 2005. Benthic cyanobacteria of the genus *Nodularia* are non-toxic, without gas vacuoles, able to glide and genetically more diverse than planktonic *Nodularia. Int. J. Syst. Evol. Microbiol.* 55: 555-568.

Majer, D., Lewis, B. G. and Mithen, R., 1998. Genetic variation among field isolates of *Pyrenopeziza brassicae. Plant Pathol.* 47: 22-28.

Mazel, D., Houmard, J., Castets, A. M. and Tandeau, de Marsac, N. 1990. Highly repetitive DNA sequences in cyanobacterial genomes. *J. Bacteriol.* 172: 2755-2761.

Meudt, H. M., Clarke and A. C., March 2007. "Almost forgotten or latest practice? AFLP applications, analyses and advances". *Trends Plant Sci.* 12 (3): 106-17.

Monty CIV. 1967. Distribution and structure of recent stromatolitic algal mats, eastern Andros Island, Bahamas. *Annls. Soc. Geol. Belg.* 90: 55-99.

Neilan, B. A., Jacobs, D. and Goodman, A. E. 1995. Genetic diversity and phylogeny of toxic cyanobacteria determined by DNA polymorphisms with the phycocyanin locus. *Appl. Enviro. Microbiol.* 61: 3875-3883.

Neilan, B. A. 2002. The molecular evolution and DNA profiling of toxic cyanobacteria. *Mol Biol* 4: 1-11.

Nilsson, M., Bergman, B. and Rasmussen, U. 2000. Cyanobacterial diversity in geographically related and distant host plants of the genus *Gunnera. Arch. Microbiol.* 173: 97-102.

Nishihara, H., Miwa, H., Watanabe, M., Nagashima, M., Yagi, O. and Takamura, Y. 1997. Random Amplified Polymorphic DNA (RAPD) analyses for discriminating genotypes of *Microcystis* cyanobacteria. *Biosci. Biotech. Biochem.* 61: 1067-1072.

Nubel, U., Garcial-Pichel, F. and Muyzer, G. 1997. PCR primers to amplify 16S rRNA genes from cyanobacteria. *Appl. Environ. Microbiol.* 63: 3327-3332.

Pierson, B. K., Castenholz, R. W. 1971. Bacteriochlorophylls in gliding, filamentous prokaryotes from hot springs. *Nature New. Biol.* 233: 25-27.

Prasanna, R., Kumar, R., Sood, A., Prasanna, B. M. and Singh, P. K. 2006. Morphological, physiochemical and molecular characterization of *Anabaena* strains. *Microbiol Res.* 161: 187-202.

Rasmussen, U. and Svenning, M. M. 1998. Fingerprinting of cyanobacteria based on PCR with primers derived from short and long tandemly repeated repetitive sequences. *Appl. Environ. Microbiol.* 64: 265-272.

Rocap, G., Distel, D. L., Waterbury, J. B. and Chisholm, S. W. 2002. Resolution of *Prochlorococcus* and *Synechococcus* ecotypes by using 16S–23S ribosomal DNA internal transcribed spacer sequences. *Appl. Environ. Microbiol.* 68: 1180-1191.

Rudi, K., Skulberg, O. M., Larsen, F. and Jakobsen, K. S. 1997. Strain characterization and classification of oxyphotobacteria in clone cultures on the basis of 16S rRNA sequences from the variable regions V6, V7, and V8. *Appl. Environ. Microbiol.* 63: 2593–2599.

Schopf, J. W. 1970. Precambrian microorganisms and evolutionary events prior to the origin of vascular plants. *Biol. Rev.* 45: 319-352.

Stewart, W. D. P. 1969. Biological and ecological aspects of nitrogen fixation by free living microorganism. *Proc. R. Soc.* 172: 367-388.

Stewart, W. D. P. and Pearson, H. W. 1970. Effects of aerobic and anaerobic conditions on growth and metabolism of blue-green algae. *Proc. R. Soc.* 175: 293-311.

Teaumroong, N., Innok, S., Chunleuchanon, S. and Boonkerd, N. 2002. Diversity of nitrogen fixing cyanobacteria under various ecosystems of Thailand. I. Morphology, physiology and genetic diversity. *World J. Microbiol. Biotechnol.* 18: 673-682.

Valerio, E., Pereira, P., Saker, M. L. and Franca, S., Tenreiro, R. 2005. Molecular characterization of *Cylindrospermopsis raciborskii* strains isolated from Portuguese freshwaters. *Harmful Algae.* 4: 1044-1052

Van Belkum, A., Scherer, S., van Alphen, L. and Verbrugh, H. 1998. Short-sequence DNA repeats in prokaryotic genomes. *Microbiol. Mol. Biol. Rev.* 62: 275-293.

Versalovic, J., Koeuth, T. and Lupski, J. R. 1991. Distribution of repetitive DNA sequences in eubacteria and application to fingerprinting of bacterial genomes. *Nucleic Acids. Res.* 19: 6823–6831.

Waaland, J. R., Waaland, S. D. and Branton, D. 1971. Gas vacuoles. Light shielding in blue-green algae. *J. Cell. Boil.* 48:212-215.

Walsby, A. E. 1972. Structure and function of gas vacuoles. *Bact. Rev.* 36: 1-32.

West, N. J. and Adams, D. G. 1997. Phenotypic and genotypic comparison of symbiotic and free living cyanobacteria from a single field site. *Appl. Environ. Microbiol.* 63: 4479-4484.

Whitton, B. A. and Roger, P. A. 1989. Use of blue-green algae and *Azolla* in rice culture. In: Campbell R and Macdonald R. M. (Eds) *Microbial inoculation of crop plants. IRS Press at Oxford* 25: 89-100.

Wilson, K. M., Schembri, M. A., Baker, P. D. and Saint, C. P. 2000. Molecular characterization of the toxic cyanobacterium *Cylindrospermopsis raciborskii* and design of a species specific PCR. *Appl. Environ. Microbiol.* 66: 332-338

Wynn-Williams, D. D. 1990. Microbial colonization processes in Antarctic fell field soils – an experimental overview. In Proc. *National Institute of Polar Research Symposium on Polar Biology.* 3: 164-178.

16

Free Amino Acids, Total Carbon, Total Nitrogen, and Phycobiliproteins Content in *Porphyra* sp. after a Short Term Nutrient Uptake

A. Pedersen,[1] G. Kraemer[2], S. Hariskov[3] and C. Yarish[4].

[1]Norwegian Institute for Water Research, P.O. Box 173 Kjelsås, N-0411 Oslo, Norway,

[2]Division of Natural Science, Purchase College, State University of New York, Purchase, NY 10577, USA

[3]School of Medicine, University of Connecticut, 263 Farmington Avenue Farmington, CT 06030, USA

[4]Department of Ecology and Evolutionary Biology, University of Connecticut at Stamford, Stamford CT 06901, USA

ABSTRACT

Three species of *Porphyra*, one winter species *P. leucosticta* (Type C, *sensu* Neefus et al. 2000), one eurythermic species *P. suborbiculata* (Kjellm.), together with a summer species, *P. purpurea* (Roth) C Agardh., were collected from Long Island Sound (LIS), U.S.A. when the algae were in exponential growth phase. Multiple uptake experiments in two nutrient media, 3µmol [$NaNO_3$] with 0.5µmol $NaHPO_4$ (LCM = low concentration medium) and 30 µmol [$NaNO_3$] with 3µmol $NaHPO_4$ (HCM = high concentration medium), were run for 24 hours. Tissue samples from the field and from the uptake experiments, were analyzed for free amino acids (FAA), total carbon and nitrogen, and phycobiliprotein concentration. Differences in these tissue components reflected the environmental condition under which the different algae occurred in the field. *Porphyra purpurea* was the only species that showed significant positive response to the nutrient load (LCM & HCM), with increases in total N, FAA and pigment content. During nutrient uptake experiments, an increase in phycobiliprotein content was detected, while total N and FAA remained unchanged in the two other species *P. leucosticta* Type C and *P. suborbiculata*.

Keywords: FAA, *Porphyra*, CN, pigments, phycobiliproteins, nitrate, phosphate, uptake experiment, ecophysiology.

INTRODUCTION

Seaweeds have been utilized all over the world as a valuable source of food, especially in Asia, where different species are an important part of the daily diet (Jensen 1993; Noda 1993). *Porphyra* species (mainly *P. yezoensis* Ueda, *P. tenera* Kjellm. and *P. haitanensis* Chang and Zheng Baofu) dominate the world seaweed market and are one of the most valuable aquaculture industries in Asia (Shin and Miura, 1997; Tseng, 1993), estimated at US$1.46 billion per year (McHugh, 2002). The specific free amino acids (FAA) composition is believed to give "hosi-nori" (dried nori) its good taste, with high contents of alanine (Ala), glutamic acid (Glu) and aspartic acid (Asp) giving a pleasant flavor (Noda and Horiguchi, 1975; Yoshie et al., 1993). The concentration of different FAA as well as total FAA content in seaweeds varies within and among species (Fleurence 1999; Norziah and Ching, 2000; Wong and Cheung 2000, 2001); with seasons (Chapman and Craigie 1977; Saitoh et al., 1975) and environmental conditions (Jones et al., 1996; Lapointe et al., 1984; Vona *et al.* 1992; Wong and Cheung 2000, 2001). Nutrient source as well as concentration also affects the FAA composition in algae (Bird et al. 1982; Horrocks et al., 1995; Vona et al., 1992). Compared to ammonium (NH_3) as a nutrient source, nitrate (NO_3^-) increased the Glu and Ala contents in *Gracilaria edulis* (Gmelin) Silva, (Jones et al., 1996). Even though the FAA pools in red seaweed have frequently been studied, very few studies have focused the short-term change in FAA.

FAA appears to function as a nutrient store. Under nitrogen (N) limited growth, FAA are catabolized to supply cells with N (Bird et al., 1982). During periods when available in surplus, N can be stored, either in vacuoles or intracellular, as nitrate and ammonium, but most likely as FAA and proteins (Bird et al., 1982; Martínez and Rico 2002; Oohusa et al., 1977). Thomas and Harrison (1985) found that intracellular NO_3 and NH_3 in *P. perforata* composed only 10% of the pool of FAA. Storage of these components also can function as osmolites (Cole and Sheath, 1995). In the tribouxiophyceaen *Chlorella vulgaris* M. Beijerinck, the FAA acid content increased approximately 3-fold within 24 h when exposed to either NO_3^- or NH_3 (Reisner et al., (1960). During times of excess N supply, red algae can also store N in form of phycobiliproteins (PBP) (Fredriksen and Rueness, 1989; Martínez and Rico, 2002). In rhodophyta phycobilisomes (PBS) serves as antennae that supply the electron transport chain of the chlorophyll. The light energy is absorbed in the phycoerythrin (PBE) on the outside of the PBS in the green spectra and funnelled in to chl.a (PSII) via a cascade of excitation energy steps. When the light energy is reached PSII it goes through the electron flow system and produces ATP and NADPH by a two-step process called non-cyclic electron flow (Falkowski and Raven, 1997). Phycobilisomes are globular units constructed of phycobiliproteins like PBE, phycocyanin (PBC) and allophycocyanin (APC) and are loosely attached to the thylakoid (Algarra and Rüdiger 1993; Algarra et al., 1990; Falkowski and Raven 1997; Gantt, 1995). In the inner core of a PBS, a few APC is closely attached to Photosystem II (PSII) of the photosynthetic unit (PSU) within the thylakoid membrane. Surrounding APC's are several PBC's with subsequently a dominating portion of PBE-units speading out from the PBC's - like a needle on a pin-cushion (Gantt, 1995). PBE is the dominat phycobiliprotein in a PBS, up to 85% and can easily be disassociated from the PBS under different light and nutrient conditions (Algarra and Rüdiger, 1993). With less PAR during winter, algae adapt to the conditions by: a) either an increase in the number of Photosynthetic Units (PSU); or b) increases the harvesting antenna (PBS); or c) both mechanisms. PBE content usually increases under low light condition i.e. the size of the light harvesting phycobilosomes increases (Cole and Sheath 1995; Waaland and Cleland, 1974). Among phycobiliproteins, most focus has been on PBE, PBC and APC, for which ratios in planktonic producers are known to change during starvation and photo-acclimation (Algarra and Rüdiger, 1993). No such variations in the ratios between the pigments were found for the benthic red alga *Chondrus crispus* Stackh. (Chopin et al., 1995), however, the total phycobiliprotein content was significantly correlated with nutrient concentration in the media.

As *Porphyra* sp. usually inhabits mid-littoral zone from the arctic to the tropics (Yoshida et al., 1997 Silva 1999), the species within the genus are also tolerant to variability to many environmental factors (i.e. salinity, temperature, light etc.), all which are known to affect the physiological parameter being analyzed (Yarish and Edwards, 1982). Hence, in the present study, the FAA composition, PBE, PBC and APC as well as total carbon (C) and nitrogen (N), were analyzed in 3 species of *Porphyra* (*P. purpurea* (Roth) C Agardh., *P. leucosticta* Type C, *sensu* Neefus (2000) and *P. suborbiculata* Kjellm.) from Long Island Sound (LIS), U.S.A. New molecular sequencing of *Porphyra leucosticta* Type C, *sensu* Neefus (2000) suggest that it is to be *Porphyra rosengurtii* (Dr. C. Neefus pers.comm). The aim was to examine whether the content of these compounds varied according to the different environmental conditions under which these algae occurred.

MATERIAL AND METHODS

Three species of *Porphyra* were collected from Long Island Sound (LIS), U.S.A. *Porphyra purpurea*, was collected from East of Port Jefferson, New York (N:40°57'53, E:-73°02'5) in Aug 2001. *Porphyra suborbiculata* and *P. leucosticta* Type C were collected in Dec. 2001 at Cove Island Park, Stamford, CT (N:41°02'35, E:-73°30'1). The algae were kept in aerated autoclaved seawater (ASW) at 5° or 10°C (15°C only for *P. purpurea*) from LIS with circulated water for a week before acclimatization for two days at an appropriate test temperature in ASW. Circular disks of either 20 mm or 14 mm in diameter depending on the size of the blades, were then punched. After an additional day in aerated ASW the disks were transferred into artificial seawater (Tropic Marine® Dr. Biener Aquarientechnik, Germany-TM), with no inorganic nitrogen or phosphorus compounds (detection level ≤ 0.7 μM and ≤ 0.3μM, respectively). The algae were starved for 5 days at 100 μmol m^{-2} s^{-1}, before the uptake experiments. To prevent or delay any spore formation in the sporophyte, the light attenuation was reduced from 100 to 50 μmol m^{-2} s^{-1} during the uptake experiments measured with a Li-COR radiometer and a LI-190 Quantum Sensor (Li-COR, Biosciences, Nebraska, U.S.A.). This is at least 3 times higher than the species compensation point (E_c) which varies between 1 and 17μmol photon m^{-2} s^{-1} with the highest E_c at high temperatures. E_k values, an index of light adaptation that describes the irradiance at the onset of photosynthesis saturation were found in the range of 17 to 134 μmol photon m^{-2} s^{-1}, hence 50 μmol m^{-2} s^{-1} during the uptake seemed adequate light condition. The uptake experiments were performed in two series of aerated ASW, one low concentration medium (LCM) augmented with 0.5 μmol $NaHPO_4$ and 3μmol $NaNO_3$ and one high concentration medium (HCM) with 3μmol $NaHPO_4$ and 30 μmol $NaNO_3$. Experimental temperatures were 5° C and 15° C for the winter species *P. suborbiculata* and *P. leucosticta* with short day light regime

(8:16; L:D) and for the summer species *P. purpurea* 5° C and 25° C with short day and long day light (16:8; L:D), respectively. Samples of incubated medium were analyzed for inorganic N and P by the Environmental Research Institute, University of Connecticut, using a Four Channel Auto Analyzer equipped with High-Sensitivity Seawater Cartridges (Lachat - QuikChem AE Ion Analyzer, Hach Company, Loveland, Colorado). Samples for FAA and PBP were taken after 24hrs of uptake. Two (d = 20 mm) or 4 (d = 14 mm) disks were used for each sample for PBP analysis and double amount of disks were used for FAA analysis (~ 60 mg FW depending on species). PBP samples were frozen for later analysis and FAA samples were dried for 48 hrs at 55° C. Four replicate tests flasks were used for each nutrient medium, however, they represented pseudo-replicates for temperature since all flasks were contained within the same environmental chamber. Replicate tissue samples for FAA were pooled, from which four mg dry weight (DW) were ground by hand in a mortar with liquid nitrogen and extracted according to a modified procedure by Harada et al., 1990 and Noda et al., 1975. One ml of boiling 75% ethanol (HPLC grade, Sigma, St. Louis, MO, U.S.A.) was added to the samples in an Eppendorf tube and placed on a shaker for 15 min, followed by centrifugation for 6 min at 1500 rpm. The supernatant was washed into a new tube by adding 0.5 ml petroleum ether, vortexed for 10 sec and the supernatant was transferred into a new tube for a second wash. The samples were filtered through a 0.2µl PVDF membrane filter. Samples and standards were derivatized using 6-aminoquinolyl-N-hydroxysuccinimidyl carbamat (AQC) as a derivatizing agent, according to Waters AccQ Tag Chemistry Package Manual (Millipore Corporation, Milford, MA. U.S.A.). Samples were determined in triplicate runs with a Perkin-Elmer HPLC with a Series 200 Pump and Autosampler, a Series 200 Vacuum Degasser, custom-built column heater, a 785 UV Detector, LC 240 Fluorescence Detector, and a PeakSimple Data Chromatography System (SRI Instruments, Torrance, CA, USA). Phycobiliproteins were extracted in 0.1 M phosphate buffer according to Beer and Eshel, 1985. Quantities of PBE and PBC were calculated according to Beer and Eshel, 1985 and APC were calculated according to Rosenberg and Ramus, 1981. Total C and N content in the samples were determined in triplicates for each replicate sample with a Perkin-Elmer Series II 2400 CHNS/O Analyzer. Dried samples (48 hrs at 55° C) were ground by hand in a mortar. Statistical differences between the values were tested, when appropriate, by non-parametric t-test (MINITAB, Minitab Inc.). Pearson correlation analyses were performed to test correlation among different components and a one way ANOVA (Holm-Sidak Multipel Comparison Procedures) were run to test if there were significant differences among test-groups.

RESULTS

Field Collection

Porphyra purpurea was collected once during summer and *P. leucosticta* and *P. suborbiculata* twice, both during winter and early spring. The environmental conditions were quite different during the two seasons with 10-13 µM ambient nitrate concentrations during winter (October-February) and with nitrogen being a limiting factor for growth during mid summer in July/August at concentrations below 0.1 µM NO_3^-. The surface water temperature measured bimonthly at Cove Island, Stamford, CT varied from –0.5° to 5° C in the wintertime (January-March) and from 20° to 32°C in the summer time. In addition a surplus of PAR (Photosynthetic Active Radiation) exists during summer as opposed to low PAR during the winter season. Hence, one would expect such difference to influence the FAA content in the different seaweeds.

Total free amino acid content showed significant differences among all the species (field samples) (ANOVA, $p \leq 0.001$) with much higher total FAA in the winter species than in the summer species *P. purpurea* (Table 1; Fig. 1). There were, however, distinct differences among the *Porphyra* species in the most abundant FAA (Fig. 1). Even though all the species contained alanine (Ala), glutamic acid (Glu), taurine (Tau, an amino acid derivative) and aspartic acid (Asp), only *P. leucosticta* Type C had a significantly higher content than the other species (ANOVA, $p \leq 0.001$) and Ala, Glu, Tau and Asp in *P. leucosticta* Type C, exceeded all 1000 mg/100gr DW. *Porphyra leucosticta* Type C contained more than double the amount of Asp as the other species. *Porphyra suborbiculata* had high concentrations of Ala (2281 mg/100 gr DW), Tau (2119 mg/100 gr DW) and Glu (1300 mg/100gr DW), while *P. purpurea* contained a high concentration of Glu only (1322 mg/100gr DW) and significantly less Tau (605 mg/100 gr/DW) than the other species (ANOVA, $p \leq 0.001$).

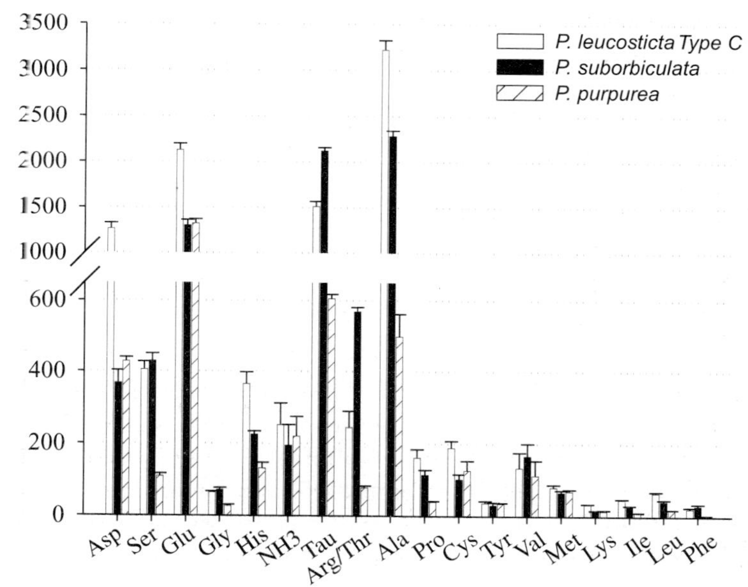

Fig. 1 Free amino acid content (FAA, incl. NH_3 and taurine -Tau) in *Porphyra leucosticta* (Type C), *P. suborbiculata* and *P. purpurea* from the field in Long Island Sound (New England-USA). Error bars are standard error.

Table 1 Total sum of free amino acids (FAA)(mg/100gr DW), percent total carbon (C), percent nitrogen (Tot. N) and C:N ratio in *Porphyra leucosticta* (Type C), *P. suborbiculata* and *P. purpurea* from the field in Long Island Sound (U.S.A)

	Sum FAA (mg/100g DW)	%C	%N	CN-ratio
P. leucosticta Type C	10 205	35.98 ± 0.50	5.81 ± 0.06	6.19 ± 0.03
P. suborbiculata	8 145	38.57 ± 0.20	4.46 ± 0.02	8.65 ± 0.01
P. purpurea	3 844	38.04 ± 0.91	3.07 ± 0.40	12.54 ± 1.42

The three species of *Porphyra* from LIS all contained less than 500 mg/100 gr DW for the rest of the FAA, with the exception of arginine (Arg)/threonine (Thr) in *P. suborbiculata*, which had 569 mg/100 gr/DW (i.e., two to seven times the content in *P. leucosticta* and *P. purpurea*, respectively). *Porphyra purpurea* also had much less serine (Ser), glycine (Gly), histidine (His), proline (Pro), Isoleucine (Ile), leucine (Leu) and phenylalanine (Phe) than the other two species, but the amount of these and the other FAA were small in all species. The total FAA content in the species correlated significantly with the percentage of total nitrogen in the species ($r = 0.98$, $p = 0.017$) (Table 1, Fig. 2).

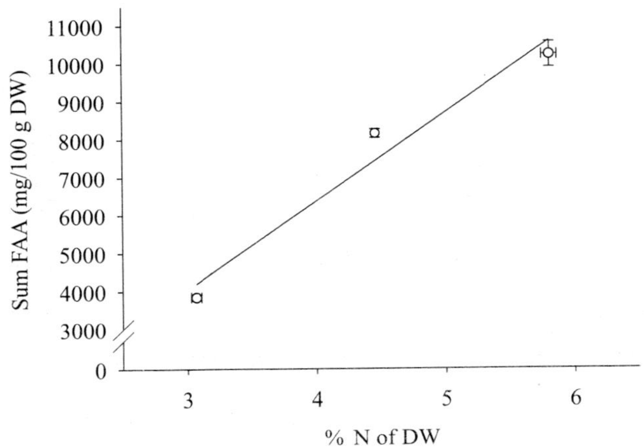

Fig. 2 Total sum of free amino acids (FAA)(mg/100gr DW and percent nitrogen (Tot. N) in *Porphyra leucosticta* (Type C), *P. suborbiculata* and *P. purpurea* from the field in Long Island Sound (New England-USA). The bars are standard error bars (ref. Tab. 1).

The phycobiliproteins in the three species from LIS showed the similar pattern as for FAA, i.e. higher pigment content (PBS) in the winter species than in the summer species *P. purpurea* (Fig. 3). Among the winter species the phycobiliprotein content (PBE + PBC + APC) in *P. leucosticta* was more than double the content of *P. suborbiculata* (23 and 9 mg/g DW, respectively; Fig. 3). The difference between summer and winter species were more pronounced for APC than PBE, which also reflects the small PBE/APC ratio of 2.4 in *P. purpurea*, as opposed to 3.2 in *P. leucosticta* and 4.7 in *P. suborbiculata*.

Experimental Nutrient Uptake

The FAA content of both *P. suborbiculata* and *P. leucosticta*, showed a reduction from the field values, to the start of the uptake experiments (0 hr), as opposed to the FAA in *P. purpurea*, which showed a slight increase in total FAA content in 5°C during the precondition treatments (Fig. 4). The precondition treatment in 5°C resulted in a drop of total FAA in both winter species, but more so for *P. suborbiculata* than in *P. leucosticta* Type C (24% and 35% reduction, respectively). The decline in FAA was less in 5°C than in 15°C. *Porphyra leucosticta* showed a drop in all the major FAA from the field concentration to the initial concentration (0 hrs) and the reduction was larger in 15°C (47%) than in 5°C (24%) (Fig. 4). Most dramatic were the declines in Ala contents which

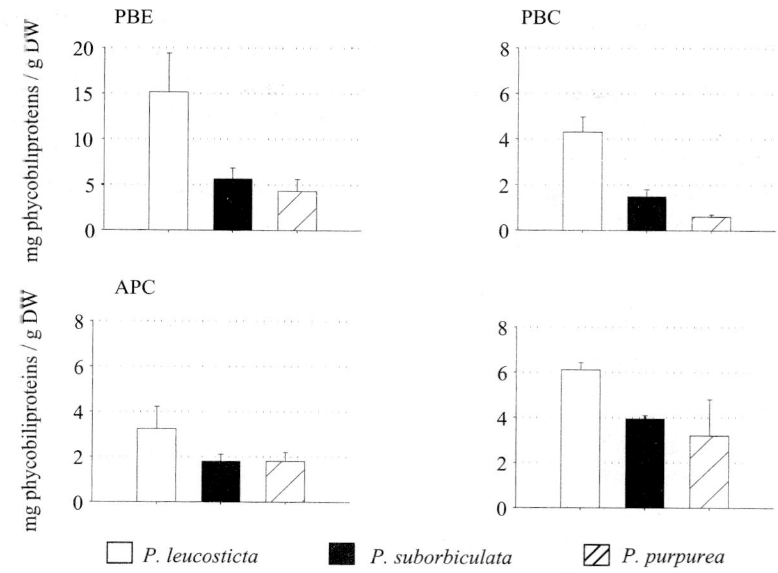

Fig. 3 Phycoerythrine (PBE), phycocyanine (PBC) and allophycocyanobilin (APC) in *Porphyra leucosticta* (Type C), *P. suborbiculata* and *P. purpurea* from the field in Long Island Sound (New England-USA). Ratio PBE+PBC/APC. Note the different scale on Y-axis for PBE and PBC. Error bars are standard error.

dropped from 3218 mg/100g DW to 1150 and 1394 mg/100g DW in 5°C and 15°, respectively (Fig. 2; Fig. 5A). In 15°C, all FAA dropped (except for Ile), but in 5°C treatment, only Ala, Asp, Ser & His showed a decrease from field samples.

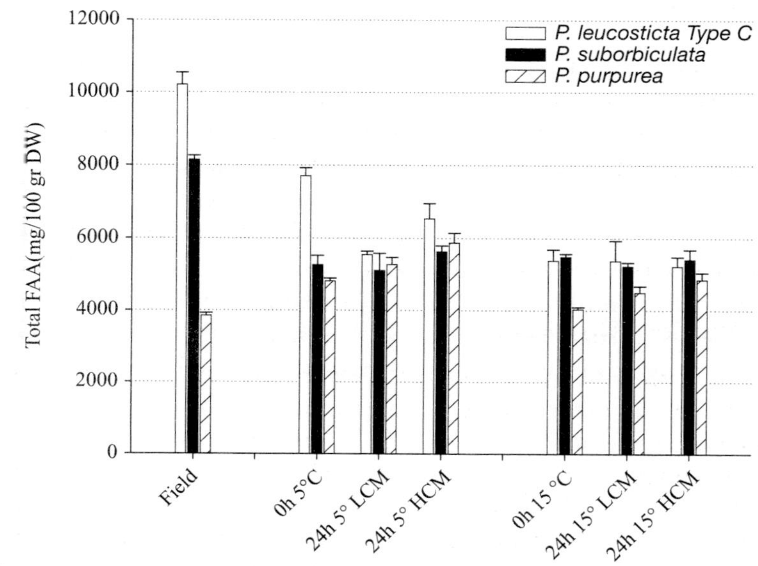

Fig. 4 Total content of free amino acids (FAA) in three *Porphyra* species from LIS. Field = total FAA content from field samples, Start indicated content after pretreatment in 5°C and 15°C (25°C for *P. purpurea*), LCM = low concentration medium, HCM = high concentration medium.

After the uptake period there was no detectable change in total FAA content at 15°C in *P. leucosticta* Type C (Fig. 4), even though there were minor changes in the specific FAA composition (Fig. 5A). In 5°C there was still a decline in total FAA content. However, the decline was less in HCM than in LCM (Fig. 4). In 5° only Ala increased during the experiment (Fig. 5A) and the increase was similar in LCM and HCM. However, in 15° Glu was the only FAA that showed an increase, even though it was not detectable in the total FAA content of *P. leucosticta* Type C (Fig. 4; Fig. 5A). All other FAA declined, but less so in HCM than in LCM. For the other winter species (*P. suborbiculata*), the FAA content was similar to that of *P. leucosticta* Type C, but the pretreatment procedure was not as dependent on temperature as in *P. leucosticta* Type C (Fig. 4; Fig. 5B). The significance of this presumed difference cannot be tested due to pseudoreplication of temperature. The drop in total FAA content in *P. suborbiculata* between 15°C and 5°C was about the same, 33% and 35% respectively. After the short-term uptake experiment, the total FAA content did not change (Fig. 4), only minor variation were observed among the FAA composition in *P. suborbiculata* (Fig. 5B). The total FAA content in *P. purpurea* remained at the same level in 15°C during preconditioning, while at 5°C there total FAA increased. *Porphyra purpurea* was the only species that showed a distinct increase in the total content of FAA at both temperatures after the uptake experiments. The increase in total FAA was higher in HCM than in LCM, and the levels were higher at 5°C than at 25°C (Fig. 1). There were an increase in most of the FAA and a decline in NH_3 at both temperatures (Fig. 5C).

Total N in *P. leucosticta* decreased with both temperatures at 5° and coincided with the decrease in FAA content in the alga at 5°C. Total N content of *P. suborbiculata* decreased in both temperatures and nutrient levels (Fig. 6B), unlike *P. leucosticta* Type C where no decrease occurred at high nutrient and high temperature (Fig . 6B, A). In *P. purpurea* total tissue N coincided with the total FAA content within the species, however, the total FAA at 5°C was higher than at 15°C, opposite of the total N content.

The phycobiliproteins in *P. leucosticta* Type C decreased during preconditioning, and more so in 15°C than in 5°C (Fig. 7A). However, at 15°C there was an increase at both nutrient mediums for all phycobiliproteins. In 5°C the phycobiliproteins showed no significant change (Fig. 7A). The phycobiliproteins in *P. suborbiculata* dropped in high nutrient and high temperature, but increased in low temperature and high nutrient. *Porphyra purpurea*, a summer species, showed patterns similar to those of *P. leucosticta* Type C (Fig. 7A, C), with an increase in all phycobiliproteins at 25° in both nutrient mediums, but not so at 5°. *Porphyra leucosticta* Type C was the only species that showed a decline in all phycobiliproteins preconditioning (Fig. 7A), opposite of the other two species (Fig. 7B, C). The increase in phycobiliprotein content in *P. leucosticta* Type C and *P. purpurea* only occurred at high temperature and not at low temperature, opposite to *P. suborbiculata* where a decline was recorded at similar conditions (Fig. 7).

DISCUSSION

Field Collection

The amino acid content in the three species of *Porphyra* was within the range found in Asian species (Harada et al., 1990; Nisizawa and Oofusa, 1990; Noda 1993; Noda and Horiguchi, 1975;

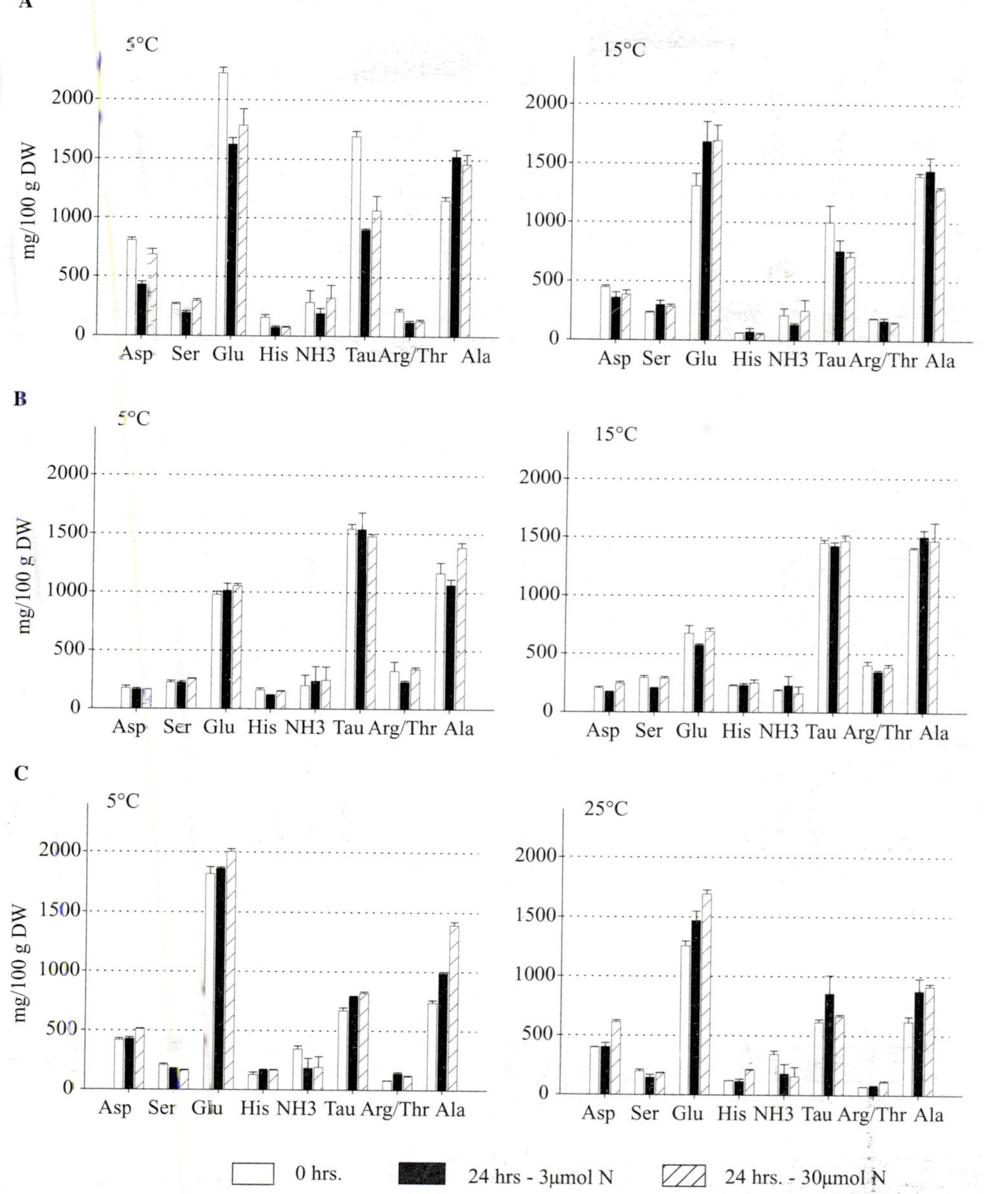

Fig. 5 FAA in *Porphyra* sp. from L.I.S. during uptake of nitrate and phosphate during a 24hrs period. *A. Porphyra leucosticta* (Type C) in 5 and 15°C. *B. P. suborbiculata.* in 5 and 15°C C. *P. purpurea* in 5 and 25°C. Error bars are standard error.

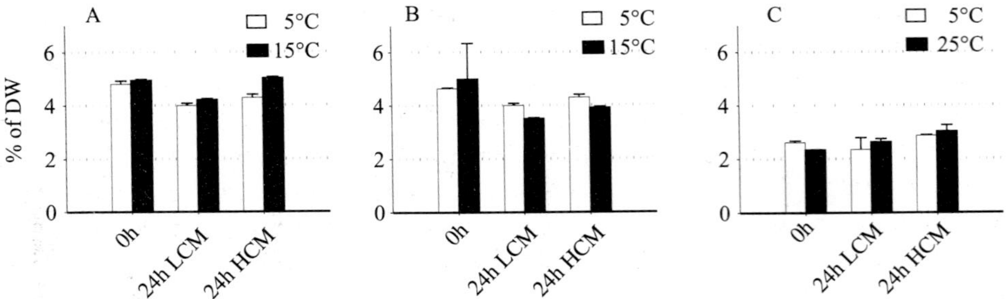

Fig. 6 Total N in three species of *Porphyra* from L.I.S during 24 hrs uptake in two concentration of nitrate and phosphate; A) *Porphyra leucosticta* (Type C), B) *P. suborbiculata*, C) *P. purpurea*. Error bars are standard deviation.

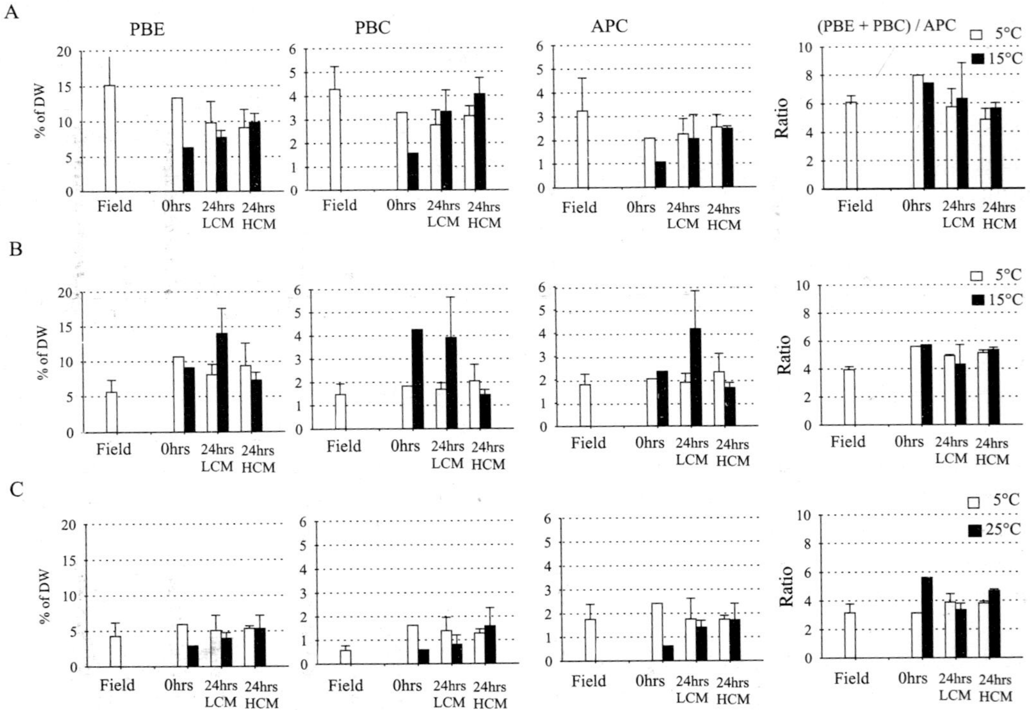

Fig. 7 Phycobiliproteins: Phycoerthythrin (PBE); Phycocyanin (PBC); and Allophycocyanobilin (APC) in three species of *Porphyra* from LIS during 24 hrs uptake in two concentrations of nitrate and phosphate (3N = 3μmol nitrate, 30N = 30μmol nitrate). A. *Porphyra leucosticta* (Type C) in 5 and 15°C. B. *P. suborbiculata*. in 5 and 15°C C. *P. purpurea* in 5 and 25°C. Error bars are standard deviation. Note different axis among PBE and PBC, APC.

Shin and Miura 1995, 1997), and Atlantic species (Fleurence, 1999; Munda and Gubensek, 1986). However, the content in the three species showed considerable variation in FAA content. The differences appear to reflect the different environmental conditions under which they occur.

The FAA content also varies between and among Asian species of *Porphyra* and Harada et al. (1990) showed that especially Ala varied in dried laver during the winter months. However, the FAA composition in *Porphyra* from LIS differed somewhat from the commercial Asian species. *P. leucosticta* Type C resembled most the dried Asian *Porphyra yezoensis* and some of its mutants, in its FAA content and composition (Harada et al., 1990; Noda and Horiguchi, 1975; Shin and Miura, 1997). The other two species had considerably less of one or two of the major FAA and the total FAA content was lower than in the dried Asian sheets of *Porphyra* (Shin and Miura, 1997). Total N content in the three species from the field correlated positively with the FAA content in these algae.

The nutrient-deprived *P. purpurea* (summer ambient nitrate concentration 0.1μM) had much lower FAA content than the other species, and only Glu was found in high amounts (>1000 mg/100 gr DW). Among the species Ala, Glu and Tau dominated and *P. leucosticta* had in general higher FAA content that *P. suborbiculata*. A possible cause for the differences is their location within the littoral zone and their access to available nutrients and light. *Porphyra suborbiculata* grows from higher up in the midlittoral, up to the supralittoral fringe, as opposed to *P. leucosticta* which grows the middle to low mid-littoral zone. *Porphyra leucosticta* Type C is immersed for longer periods that *P. suborbiculata* and thereby exposed to less light, but more nutrients than *P. suborbiculata*, resulting in the observed differences in FAA (longer nutrient exposure) and phycobiliproteins (due to low light adaptation and to some extent also high nutrient concentrations).

As shown in Fig. 3, the pigment content in *P. leucosticta* Type C was twice as high as in *P. suborbiculata*. Both alga had, with respect to *P. purpurea*, larger "antenna" i.e. high PBE + PBC content and relatively less of the closely PSII-attached APC (Algarra and Rüdiger, 1993). However, the antenna (PBE + PBC) was much larger in *P. leucosticta* Type C than in *P. suborbiculata*, probably due to less light exposure. *Porhyra suborbiculata* does not require "large antennas" for light harvesting as *P. leucosticta* Type C does, due to *P. leucosticta* Type C's more frequent and longer inundation in the lower littoral zone. The pigment concentration in *P. purpurea* was lowest of all the species as it grows during late summer with high irradiances, hence, the need for antenna's in the summer species is less. The pigment contents within the algae were positively correlated with total N in the species, however, not significant due to few replicates ($r = 0.94$, $p = 0.057$).

Experimental Nutrient Uptake

The uptake experiment showed that during the first 24 hrs, the FAA content in *P. leucosticta* Type C did not increase, but instead, decreased, especially in 5°C. The decreases in FAA are also supported by a decrease in the total N in the alga (Fig. 6A). This indicates either that the FAA syntheses, under the specific precondition procedure followed, were not initiated during the first 24 hrs of nutrient uptake or that a loss of FAA from the cells exceeded the formation rate of FAA. Total FAA content in *P. leucosticta* Type C at the start point of the experiment in 5°C (Fig. 4) was very high but not so in 15°C. This does not coincide with the almost equal to N content in the two tissues (Fig. 6A) during startup of the experiments if there were a linear correlation between total FAA content and total N content as shown in Fig. 3. Nitrogen could have been stored as different components within the cell at the two temperatures and the breakdown of FAA in high temperature still occurred but not so at low temperature (5°C). Tau, Asp, and Glu were the FAA that contributed to the high total FAA content in *P. leucosticta* Type C, all of which have been reported to be the predominant FAA in *Porphyra* species (Inoue et al., 1973). Asp is the only FAA with a distinct umami taste (the fifth taste quality defined by Toko (2000)). Prior to preconditioning the algae were stored in ASW. During N starvation at 5°C the loss of FAA from the algae both from internal breakdown and external loss, happened much slower in 5°C than in 15°C. The respiration rate in *P. leucosticta* Type C at 5°C was only one-tenth of that at 15°C (0.3 μmol $O_2 \cdot g^{-1} DW \cdot min^{-1}$ vs. 3.0 μmol $O_2 \cdot g^{-1} DW \cdot min^{-1}$, respectively; 2004 in prep). Our experiments cannot explain the drop in the three amino acids Ala, Glu and Tau in 5°C, unless they all were used in rebuilding proteins necessary in facilitating an active uptake process in *P. leucosticta* Type C effective at 5°C but not at 15°C. Active nitrate uptake in *P. leucosticta* Type A was more pronounced in 5°C than at 15°C at the nitrate concentrations between 3 μM -30 μM (Pedersen et al., 2004), which indicates that such an assimilation could have taken place. The increase in total FAA for *P. purpurea* after preconditioning in 5°C (i.e., during N starvation)(Fig. 1), was probably an effect of a potential uptake during storage in ASW prior to preconditioning and starvation. As for *P. leucosticta* Type C the loss of FAA from *P. suborbiculata* was less in 5°C than in 15°C due to higher metabolic turnover at 15°C than at 5°C. Hence, starvation did not reduce the total FAA content in 5°C as much as in 15°C. In *P. purpurea* the content of almost all FAA (Fig. 5C) increased at both temperatures and both nutrient mediums, indicating that amino acid synthesis in this summer species was initiated earlier than in the two winter species. Different isoenzymes could have been involved resulting in different responses among the different species. One should, however, not exclude the possibility that the early FAA synthesis in *P. purpurea* could be an effect of an uptake of nutrients from the storing media (ASW) prior to the starvation period and the uptake experiments.

The pigment content in *P. leucosticta* Type C and *P. purpurea* showed similar patterns as the content increased during the 24 h experiment in 15°C, but not so in 5°C (Fig. 7A,C). However this was not the case for *P. suborbiculata* (Fig. 7C). As described by Falkowski and Raven, 1997, deactivation and reactivation of phycobiliproteins can be regulated by phosphorylation. The observed increase in phycobiliproteins in *P. leucosticta* Type C and *P. purpurea* at high temperatures, especially in 30μmol N (Fig. 7A, B, C), can be a result of phosphorylation of inactive free pigment phycobiliproteins, induced by a change in PAR when the algae were moved from 100 μmol/m^2/s during preconditioning and starving, to 50 μmol/m^2/s at uptake condition. In colder water (5°C), the process was slower and there seemed to be a slightly loss of phycobiliproteins in both *P. leucosticta* Type C and *P. purpurea.* In this study the algae were exposed to low light (50 μmol/m^2/s) only for 8 hrs and to dark for 16 hrs. One reason for the reduction in phycobiliproteins at 5°C, can be that the algae in high nutrient regime (30 μmol N) produced or reactivated more PBE than in the low nutrient regime (3 μmol N), due to higher nutrient availability. During night the deactivation of PBE occurred and after 16 hrs in dark, the process in low nutrient medium, could have exhausted the pool of PBE obtained during the 8 hrs light period, resulting in a decrease in active PBE in the overall 24 hr experimental period. In high nutrient medium, the pool of newly produced or light-reactivated PBE might have exceeded the amount of PBE deactivated during the dark period. Hence, after 24 hrs the PBE increased in high nutrient and decreased in low nutrient. However, the reduction in total N in *P. leucosticta* Type C does not coincide with change in phycobiliproteins (over 24 hrs), but more so with FAA content. The change in FAA was 1-4% of the change in total N, whereas the change in PBE which is most prominent of the phycobiliproteins, represented only on average 0.0008% of the change in total N content. These results shows that the FAA content is a more important component of the total N content in the *P. leucosticta* Type C than phycobiliproteins are as described by Thomas and Harrison (1985). The patterns for *P. suborbiculata* did not coincide with the stenothermic *P. leucosticta* Type C. There was an increase in PBE and APC at low temperature and low nutrient levels, unlike in *P. leucosticta* Type C. This might be due to competitive processes between the nutrient assimilation and pigment or FAA formation and degradation, which also is dependent upon irradiance (Algarra and Rüdiger 1993; Falkowski and LaRoche, 1991) (Fig. 7B). The general increase in phycobiliproteins in *P. suborbiculata* in LCM might be a response to the reduction in light from preconditioning at 100 $\mu mol/m^2/s$, to the light condition during the experiments (50 $\mu mol/m^2/s$). The change in light might have rebuilt or reactivated the phycobiliproteins in the alga as the increase in PBE happened in low nutrient conditions and not in high nutrient conditions at low irradiance, similar to the conditions described by Algarra and Rüdiger, 1993.

Porphyra purpurea is a summer species and usually occurs under nitrogen limited conditions, hence, its affinity to and ability to assimilate NO_3, would expect to surpass the same processes in the other species which usually are found in higher nutrient concentration during winter. In *P. purpurea* we found that the total FAA content was 1.9% of the total N content. The increase in total N (10%) during the uptake experiments was reflected in the increase in FAA (15%), which support these assumptions. *Porphyra purpurea's* responses in FAA content, pigment proteins and total N to nutrient and temperature, resembled more so *P. leucosticta* Type C than *P. suborbiculata*. This might indicate that the different species acclimatization to vertical occurrence in the littoral zone, with its differences in exposure, plays an important role in the species ability to utilize nutrients.

CONCLUSION

The results from the field samples coincide with the ambient environmental conditions under which these algae occur. The nutrient condition in the sea during winter is reflected in the high nitrogen content in the winter algae — *P. leucosticta* Type C and the eurythermic *P. suborbiculata*. The difference in pigmentation and FAA content in *P. leucosticta* Type C and *P. suborbiculata* can be explained by the difference in their habitats — one high up in the littoral zone and one low in the littoral zone. During the uptake experiments the three species responded differently. *P. purpurea* which occurs in low nutrient condition during summer, was the only species that showed: a) an uptake of nitrogen (nutrient dependent) and b) an initiation of FAA synthesis. Both *P. purpurea* and *P. leucosticta* Type C both showed a reactivation or reassembling of phycobiliproteins at low temperatures and high nutrient conditions. This suggests that *P. purpurea* has high affinity to N and that its enzymes (like nitrate reductase or nitrite reductase) are more efficient or more easily activated than in *P. leucosticta* Type C and *P. suborbiculata.* The FAA content in species did not change in response to the increase in nutrients (*P. leucosticta* Type C in HCM might be an exception). However, the phycobiliproteins increased within *P. leucosticta* Type C and *P. suborbiculata*, although differently among them. As the two species were transferred from 100 μmol/m^2/s irradiance during preconditioning to 50 μmol/m^2/s during uptake experiment, the observed differences might be due to different competitive processes within the two species when it come to the photoindependent and the photodependent process of building up PBE and PBS under low and high nutrient levels.

ACKNOWLEDGEMENTS

We would like to thank R. Hamilton (Greenwich High School, Greenwich CT) for invaluable assistance with the HPLC analyses. Special thanks to S. Baver and J.E. Heitert at Environmental Research Institute, University of Connecticut for analyzing the nutrient samples. We also would like to thank Dr. J. Zertuche-Gonzalez and Dr. D. Sahoo for their valuable discussions on FAA and pigments. We thank Perkin-Elmer Analytical division for donating analytical instruments to the Marine Biotechnological Laboratory, UCONN, Stamford. We also acknowledge support of the State of Connecticut Critical Technology Program, State of Connecticut, and the Connecticut Sea Grant College Program.

REFERENCES

Algarra, P. and Rüdiger, W. 1993. Acclimation processes in the light harvesting complex of the red alga *Porphyridium purpureum* (Bory) Drew et Ross, according to irradiance and nutrient availability. *Plant Cell. Environ.* 16: 149-159.

Algarra, P., Thomas, J. C. and Mousseau, A. 1990. Phycobilisome heterogeneity in the red alga *Porphyra umbilicalis. Plant Physiol.* 92: 570-576.

Beer, S. and Eshel, A. 1985. Determining phycoerythrin and phycocyanin concentrations in aqueous crude extracts of red algae. *Aust. J. Mar. Freshwat. Res.* 36: 785-792.

Bird, K. T., Habig, C. and DeBusk, T. 1982. Nitrogen allocation and storage patterns in *Gracilaria tikvahiae* (Rhodophyta). *J. Phycol.* 18: 344-348.

Chapman, A. R. O. and Craigie, J. S. 1977. Seasonal growth in *Laminaria longicruris*: Relations with dissolved inorganic nutrients and internal reserves of nitrogen. *Mar. Biol.* 40: 197-205.

Chopin, T., Gallant, T. and Davison, I. 1995. Phosphorus and Nitrogen Nutrition in Chondrus-crispus (Rhodophyta) — Effects on total phosphorus and nitrogen content, carrageenan production, and photosynthetic pigments and metabolism. *J. Phycol.* 31: 283-293.

Cole, K. and Sheath, R.G. 1995. Biology of the red algae. Cambridge Universtiy Press, Cambridge.

Falkowski, P.G. and LaRoche, J. 1991. Acclimation to spectral irradiance in algae. *J. Phycol.* 27: 8-14.

Falkowski, P.G. and Raven, J.A. 1997. Aquatic photosynthesis. Blackwell Science, Inc., Malden, Massachusetts.

Fleurence, J. 1999. Seaweed proteins: Biochemical, nutritional aspects and potential uses. *Trends food Sci. Technol.* 10: 25-28.

Fredriksen, S. and Rueness, J. 1989. Culture studies of *Gelidium latifolium* (Grev.) Born et Thur. (Rhodophyta) from Norway. Growth and nitrogen storage in response to varying photon flux density, temperature and nitrogen availability. *Bot. Mar.* 32: 539-546.

Gantt, E. 1995. Pigmentation and photoacclimation. In: Biology of the red algae (Eds. K. M. Cole and R. G. Sheath), pp. 203-219. Cambridge University Press, Inc.

Harada, K., Osumi, Y., Fukuda, N., Amano, H. and Noda, H. 1990. Changes of amino acid compositions of nori, *Porphyra* spp. during storage. Nippon Suisan Gakkaishi/*Bull. Jap. Soc. Sci. Fish.* 56: 607-612.

Horrocks, J. L., Stewart, G. R. and Dennison, W. C. 1995. Tissue nutrient content of *Gracilaria* spp. (Rhodophyta) and water quality along an estuarine gradient. *Mar. Freshwat. Res.* 46: 975-983.

Inoue, A., Asakawa, S., Saito, Y. and Yoshikawa, K. 1973. On the changes of extractive amino acid contents in the infected purple laver, red alga Porphyra. *J. Fac. Fish. Anim. Husb. Hir. Univ.* 12: 61-72.

Jensen, A. 1993. Present and future needs for algae and algal products. In: 14. Int. Seaweed Symp. Eds A. R. O. Chapman M. T. Brown, and M. Lahaye), Brest (France), *Hydrobiol.*, 15-23.

Jones, A. B., Dennison, W. C. and Stewart, G. R. 1996. Macroalgal responses to nitrogen source and availability: Amino acid metabolic profiling as a bioindicator using *Gracilaria edulis* (Rhodophyta). *J. Phycol.* 32: 757-766.

Lapointe, B. E., Tenore, K. R. and Dawes, C. J. 1984. Interactions between light and temperature on the physiological ecology of *Gracilaria tikvahiae* (Gigartinales: Rhodophyta). 1. Growth, photosynthesis and respiration. *Mar. Biol.* 80: 161-170.

Martínez, B. and Rico, J. M. 2002. Seasonal variation of P content and major N pools in *Palmaria palmata* (Rhodophyta). *J. Phycol.* 38: 1082-1089.

McHugh, D. J. 2002. Prospects for seaweed production in developing countries., pp. 28. FAO Fisheries Circular., Rome, FAO.

Munda, I. M. and Gubensek. F. 1986. The amino acid content of some benthic marine algae from the northern Adriatic. *Bot. Mar.* 29: 367-372.

Neefus, C. D., Mathieson, A. C., Yarish, C., Klein, A., West, A., Teasdale, B. and Hehre, E. J. 2000. Five cryptic species of Porphyra from the Northwest Atlantic. *J. Phycol.* 36: suppl.

Nisizawa, K. and Oofusa, T. 1990. Aroma and Taste of Dried Purple Laver (Hosi-Nori). In: Introduction to Applied Phycology (Ed. I. Akatsuka), SPB Academic Publishing, The Hague, The Netherlands. pp. 191-198.

Noda, H. 1993. Health benefits and nutritional properties of nori. *J. Appl. Phycol.* 5: 255-258.

Noda, H. and Horiguchi, Y. 1975. Studies on the flavor substances of 'Nori', the dried laver *Porphyra tenera*. 1. Dimethyl sulfide and dimethyl-Betha-propiothetin. *Bull. Jap. Soc. Sci. Fish.* 41: 481-486.

Norziah, M. H. and Ching, C. Y. 2000. Nutritional composition of edible seaweed *Gracilaria changgi*. *Food Chem.* 68: 69-76.

Oohusa, T., Araki, S., Sakurai, T. and Saitoh, M. 1977. Physiological studies on diurnal biological rhythms of Porphyra. 2. The growth and the contents of free and total nitrogen and carbohydrate in the thallus cultured in the laboratory. *Bull. Jap. Soc. Sci. Fish.* 43: 251-254.

Pedersen, A., Kraemer, G. and Yarish, C. 2004. The effect of temperature and nutrient concentrations on nitrate and phosphate uptake in different species of Porphyra from Long Island Sound (USA). *J. Exp. Mar. Biol. Ecol.* 312: 235-252.

Reisner, G. S. Gering, R. K., and Thompson, J. F. 1960. The metabolism of nitrate and ammonia by Chlorella. *Plant physiol.* 35: 48-52.

Rosenberg, G., and Ramus, J. 1981. Ecological Growth Strategies in the Seaweeds *Gracilaria foliifera* (Rhodophyceae) and *Ulva* sp. (Chlorophyceae):. *Bot. Mar.* Yale University. pp. 583-589.

Saitoh, M., Araki, S., Sakurai, T. and Oohusa, T. 1975. Variation in content of photosynthetic pigments, total nitrogen, total free amino acids and total free sugars in dried lavers obtained at different culture grounds and harvesting times. *Bull. Jap. Soc. Sci. Fish.* 41: 365-370.

Shin, J. A. and Miura, A. 1995. Genetic Improvement of Eating Quatlity of Dried Sheet of Porphyra by Using the Recombinant Wild-Type in *P. yezoensis* (Bangiales, Rhodophyta). *Korean J. Phycol.* 10: 109-115.

Shin, J. A. and Miura, A. 1997. Inheritance Mode of Some Characters of *Porphyra yezoensis* (Bangiales, Rhodophyta) Crispness and Free Amino Acid Content in Two Recombinant Wild-types Bred by Reciprocal Crosses. *Algae* 12: 313-318.

Silva P. C. 1999. Porphyra from the Index Nominum Algarum. http://ucjeps.berkeley.edu/rlmoe/porphyra.html. Univ. Herbarium, UC Berkley. *Rev:* 12/20/99.

Thomas, T. E. and Harrison, P. J. 1985. Effect of nitrogen supply on nitrogen uptake, accumulation and assimilation in *Porphyra perforata* (Rhodophyta). *Mar. Biol.* 85: 269-278.

Toko, K. 2000. Taste sensor. Sensors and Actuators B 64: 205-215.

Tseng, C. K. 1993. Notes on mariculture in China. Elsevier, Amsterdam (Netherlands).

Vona, V., Martino Rigano, V., Espisito, S., Martino, C. and Rigano, C. 1992. Growth, photosynthesis, respiration, and intracellular free amino acid profiles in the unicellular alga *Cyanidium caldarium* . Effect of nutrient limitation and resupply. *Physiol. Plant.* 85: 652-658.

Waaland, S. D. and Cleland, R. E. 1974. Cell repair through cell fusion in the red alga *Griffithsia pacifica. Protoplasma.* 79: 185-196.

Wong, K. H. and Cheung, P. C. K. 2000. Nutritional evaluation of some subtropical red and green seaweeds. Proximate composition, amino acid profiles and some physico-chemical properties. *Food Chem.* 71: 475-482.

Wong, K. H. and Cheung, P. C. K. 2001. Nutritional evaluation of some subtropical red and green seaweeds. *In vitro* protein digestibility and amino acid profiles of protein concentrates. *Food Chem.* 72: 11-17.

Yarish, C., and Edwards, P. 1982. A field and cultural investigation of the horizontal and seasonal distribution of estuarine red algae of New Jersey. *Phycologia* 21: 112-124.

Yoshida, T., Notoya, M., Kikuchi, H. and Miyata, M. 1997. Catalogue of species of Porphyra in the world, with special reference to the type locality and bibliography. *Nat. Hist. Res.*, Special Issue 3: 5-18.

Yoshie, Y., Suzuki, T., Shirai, T. and Hirano, T. 1993. Free amino acids and fatty acid composition in dried nori of various culture locations and prices. *Nippon Suisan Gakkaishi/Bull. Jap. Soc. Sci. Fish.* 59: 1769-1775.

17

Phycodiversity of River Pandu, Uttar Pradesh in Relation to Anthropogenic Stress

Divya Tiwari
Department of Botany, A.N.D.N.N. Mahila Mahavidyalaya, Kanpur-208002
Uttar Pradesh, India

ABSTRACT

Pandu flowing on the southern outskirts of Kanpur is severely subjected to anthropogenic stress all along its stretch in Kanpur. Municipal and industrial wastes of the southern Kanpur, especially of the Kanpur Industrial Area, are discharged into it without any treatment. Physicochemical and phycological attributes of the river were studied to assess the pollution stress on the river and its consequences on the phyco profile of the river. The study revealed that discharge of wastes into the river adversely affected its water quality and algal profile. Only 33 genera and 46 species were recorded. Algal profile of the river at different stations lacked consistency of results due to fluctuating quality and quantity of wastes discharge. Low diversity and similarity index values reported are indicative of an unhealthy and stressed ecosystem.

Keywords: Anthropogenic stress, waste effluents, river pandu, algae.

INTRODUCTION

The river Pandu flows near the southern part of Kanpur city, Uttar Pradesh, India. The river runs through 120 km and meets the main Ganga in Fatehpur. Pandu river which was known for its crystal clear water two decades back has turned black and stinks now due to heavy pollution load. Before Pandu river enters Kanpur, its water colour is green but due to the high industrial and domestic sewage in the Kanpur city, the water colour changes to black. The biggest source of pollution comes from the Panki Thermal Power Plant. The plant uses around 3000 tonnes of coals and churns out 40 tonnes of fly ash every day into the river. This ash is dumped in the pond due to the lack of proper maintenance and finds its way to the Pandu river. Moreover many municipality and domestic sewages were also drained into the river Pandu. The present work is on the physico-chemical parameters and the phycodiversity of the river Pandu.

MATERIALS AND METHODS

Material sampling was done at five major outfalls and sampling stations of the river. The major outfalls are Thermal Power Plant drain, Panki Municipal drain, Ganda Nala, Halwa Khanda drain and Central Ordinance Depot (COD) drain. Sampling stations on the river were Tikra village, Panki Industrial Area, Mardanpura, Meharbanpurwa, Hamirpur Road Bridge and Sher Shah Suri bridge. Effluent, water and algal samples were collected and analyzed for qualitative and quantitative estimation of the pollutants and algae as per standard procedure (APHA, 1989). Algal cell number was estimated using Sedgewicck-Rafter Counting cell (SR Counting Cell) and results have been expressed as number of individuals per liter after applying the appropriate correction factor. Algal identification was done with the help of pertinent literature on algae (Desikachary, 1959, Smith, 1950, Tiffany and Britton, 1952). Algal diversity was computed using standard indices (Shannon and Weaver 1963, Margalef, 1958). Taxonomic similarity was computed following Sorenson (1948).

RESULT AND DISCUSSION

Results of the physicochemical investigation of the waste effluents and river water are depicted in Tables 1and 2. River is a shallow river with a depth ranging from 1-2.5 meters but also has a few interspersed deeper zones of about 4 meters or so. Except a few small portions where the riverbed

Table 1 Physicochemical characteristics of the effluents (*n* = 4)

Parameter	Panki TPP drain	Panki Municipal drain	Ganda Nala	Halwakhanda drain	C.O.D. drain	IS:2490, 1981
Present status	untapped	untapped	untapped	untapped	untapped	-
Colour	Murky	Murky black/Reddish brown	Murky	Black	Murky black	*
Odour	Foul	Organic/pungent	Faecal foul	Faecal foul	faecal foul	*
pH	7.7	8.6	7.9	7.9	8.3	
Temp. oC	34	36.2	37.4	33	32.4	
Transp.	Nil	Nil	Nil	Nil	Nil	
Total solids	1699	1308	1254	1112	760	
Susp.solids	1540	183	251	200	70	100
D. O.	Nil	Nil	Nil	Nil	Nil	
B.O.D.5	141	202	229	180	130	30
C.O.D.	304	440	475	302	271	250
Nitrate	ND	ND	ND	ND	ND	
Ammonia	143.4	205	107	151.7	136	50
Urea	Nil	I.6	Nil	Nil	Nil	
Chloride	125	175	285	178	140	1000.0
Phosphate	1.6	2.4	4.4	1.8	2.2	5
Oil and Grease	+	+	+	+	+	-

All values in mg/l except colour, odour, pH, temperature(oC) and transparency (cm)
All efforts should be made to remove colour and odour as far as possible

Table 2 Physicochemical characteristics of the river water at different stations (n = 12)

Parameters	Sampling Stations					
	S1	S2	S3	S4	S5	S6
Colour	Murky	Murky Black	Murky Black	Murky	Muddy	-
Odour	Nil	Foul	Faecal foul	Faecal foul	Nil	-
Temperature	25.3	27.8	25.7	26.8	25.6	27.1
Transparency	17.2	2.3	6.2	7.3	15.4	13.7
Total solids	121	934	587	830	417	469
pH	7.8	8.2	8.5	8.3	8.1	8.2
D.O.	5.5	2.4	1.8	2.3	3.7	4.1
B.O.D.$_5$	10.3	39.5	45.2	67.1	44.4	30.0
C.O.D.	20.3	220	165	144	69	23.7
Chloride	25.5	80.7	62.6	75.4	54.9	45.0
Total hardness	135	366	457	599	387	517
Ammonia	0.46	3.64	18.6	9.95	8.92	2.0
Nitrate	0.15	0.05	0.02	0.09	0.13	1.28
Phosphate	0.24	1.72	1.76	1.86	1.44	1.22
Sulphate	0.78	5.3	4.4	12.5	3.66	4.68

All values in mg/l except temperature (°C), transparency (cm) and pH

is rocky and composed of small pebbles, it has a muddy bottom. River discharge fluctuated in between 3240-5450m^3 per hour. River flows with a slow velocity that ranged from <0.01m^{-s} to 0.07 m^{-s}. Peak velocity was noticed in September and minimum in June. Algal profile of the river is depicted in Table 3 whereas quantitative data pertaining to its abundance, species number, diversity values are given in Table 4. Table 5 depicts the coefficient of similarity values between the algal profile at different stations.

Colour of the waste effluents varied widely ranging from light brown, red brown, blue, blackish brown to light black or murky. This may be attributed to the use of different coloured compounds in diverse industrial processes. The qualitative as well as the quantitative variations in the effluent characteristics may be attributed to varied chemical and technical processes used in different industries, effluent of different industries in different drains and also to the relative proportion of the municipal and industrial wastes carried by them. A critical perusal of the data pertaining to effluent characteristics with prescribed ISI standards (IS:1981) for wastes to be discharged into surface waters revealed that most of the parameters are either beyond or approaching the permissive limits. Discharge of these wastes into the river adversely affected the river water quality as is obvious from the water quality gradients depicted in Table 2. Coloured water, low transparency, low dissolved oxygen coupled with significantly high values of BOD,COD, hardness, chloride, ammonia and phosphate all indicated the oxygen stressed polluted condition of the river. Except station 1 all the down stream stations are heavily polluted.

The impact of this anthropogenic stress is also evident on the algal community of the river (Tables 3-5). Temperature an important factor affecting chemical reactions occurring in an aquatic ecosystem as well as on the soil water interphase is well within the biokinetic range. Low water transparency as a result of huge waste discharge adversely affected the euphotic zone of the river and consequently the algal population and diversity. Total solids (suspended and dissolved)

Table 3 Algal profile of the river at different stations

Algae	S1	S2	S3	S4	S5	S6
Cyanophyceae						
Anacystis sp.	+	–	–	–	–	+
Merismopedia minima	+	–	–	–	–	+
Microcystis sp.	–	–	–	+	–	+
Nostoc sp.	+	–	–	–	–	–
Oscillatoria tenuis	–	+	+	+	+	–
Oscillatoria limosa	–	+	–	+	–	+
Oscillatoria princeps	+	–	–	+	+	–
Oscillatoria sancta	+	–	–	+	+	+
Oscillatoria curviceps	+	–	–	–	–	–
Phormidium tenue	+	+	+	+	+	–
Phormidium uncinatum	+	+	+	+	+	+
Chlorophyceae						
Actinastrum sp.	+	–	+	+	–	+
Ankistrodesmus sp.	–	+	–	+	+	–
Chlamydomonas sp.	–	–	+	+	+	+
Chlorella vulgaris	+	+	–	+	+	–
Cladophora glomerata	–	+	+	+	+	–
Closterium sp.	+	–	+	+	+	–
Coelastrum sp.	+	–	–	–	–	–
Hydrodictyon reticulatum	+	–	+	–	–	+
Pandorina morum	–	+	+	+	+	–
Pediastrum boryanum	+	–	–	–	–	–
Pediastrum duplex	+	–	–	–	–	–
Pediastrum simplex	+	–	–	–	–	+
Pediastrum tetras	+	–	–	–	–	+
Scenedesmus quardicauda	+	+	+	+	–	+
Scenedesmus abundans	+	–	–	–	–	+
Spirogyra sp.	+	–	+	+	+	+
Stigeoclonium tenue	+	+	+	+	–	+
Zygnema sp.	+	–	–	–	+	+
Bacillariophyceae						
Achnanthes sp.	+	–	–	–	–	–
Asterionella sp.	+	–	–	–	–	+
Cyclotella sp.	+	–	–	–	+	+
Cymbella sp.	+	–	–	–	–	+
Diatoma elongatum	+	–	–	–	–	+
Fragilaria pinnata	+	–	–	–	–	–
Fragilaria capucina	+	–	–	–	–	–
Melosira granulatum	+	–	+	+	+	–
Navicula cryptocephala	–	+	+	+	+	+
Navicula cuspida	+	+	–	–	–	+
Nitzschia palea	–	+	+	+	+	+
Nitzschia acicularis	+	–	–	+	+	+
Pinnularia appendiculata	+	–	–	+	+	–
Synedra sp.	–	+	–	+	+	–
Tabellaria floculosa	+	–	–	+	–	–
Euglenophyceae						
Euglena sp.	–	+	–	–	+	–
Phacus sp.	+	+	+	+	+	+

Table 4 Algal individuals-l (n=24), species number and diversity values at different stations

	S1	S2	S3	S4	S5	S6
No of individuals	174	63	73	199	339	336
No of species	35	16	16	25	22	25
Shannon-Weaver Index	4.41	1.73	1.68	2.49	2.09	2.96
Margalef index	3.95	1.06	1.18	2.08	1.44	1.93

Table 5 Coefficient of taxonomic similarity among different stations

STATIONS	S1	S2	S3	S4	S5	S6
S1	—	0.27	0.39	0.5	0.44	0.66
S2	—	—	0.63	0.63	0.58	0.41
S3	—	—	—	0.73	0.63	0.51
S4	—	—	—	—	0.68	0.54
S5	—	—	—	—	—	0.44

contributed by carbonates, bicarbonates, chlorides, sulphates, phosphates and nitrates of Ca, Mg, Fe, Mn and other compounds exert a pronounced effect on the osmoregulatory phenomenon of aquatic phototrophs. H+ ion concentration serves as an important index of nature and intensity of pollution. The pH of the river water at different stations ranged from 7.8-8.5 and was suitable for propping the algal growth.

Dissolved oxygen of the river water at any point is a tell-tail sign of the nature and intensity of pollution. Low D O (<4ppm) and high B O D (>2ppm) at downstream stations 2-6 is an indication of anthropogenic pollutants on the river. At times nil DO was reported at stations 2 and 4 in the months of May and June respectively indicating that at these stretches anoxygenic conditions prevailed for a short while. Invariably high ammonia and low nitrate values in the river water further insinuate the oxygen stressed condition of the river. River is rich in nutrients as is obvious from the reported values of ammonia and phosphate (Table 2). Despite the nutrient rich water, low algal diversity and population of the river may be attributed to the presence of toxic materials in water. Maximum species number was recorded at station 1(35) which drastically declined at station 2(16) and 3(16) and could not reach at par to station 1 till station 6. Chlorophyceae out numbered the other algal groups in species representation but so far as the numerical strength is concerned cyanophyceae and bacillariophyceae dominated the other groups.

Algal diversity values computed for the present study also indicate the adverse impact of effluent discharged on river algae (Table 3). Maximum values were recorded at station 1. Values declined sharply at stations 2 and 3 and remained quite low at rest of the downstream stations. It is evident from the table that not only the species number but algal population too declined sharply at stations 2 and 3 as compared to station 1. At rest of the downstream stations it was slightly better than station 1. This may be attributed to certain obvious reasons such as low transparency and reduced illumination as a consequence of residual fly ash discharged by Panki Thermal Power Plant drain upstream of station 2 and discharge of coloured industrial effluent from Panki municipal drain upstream of station 3 and presence of heavy metals in the effluents

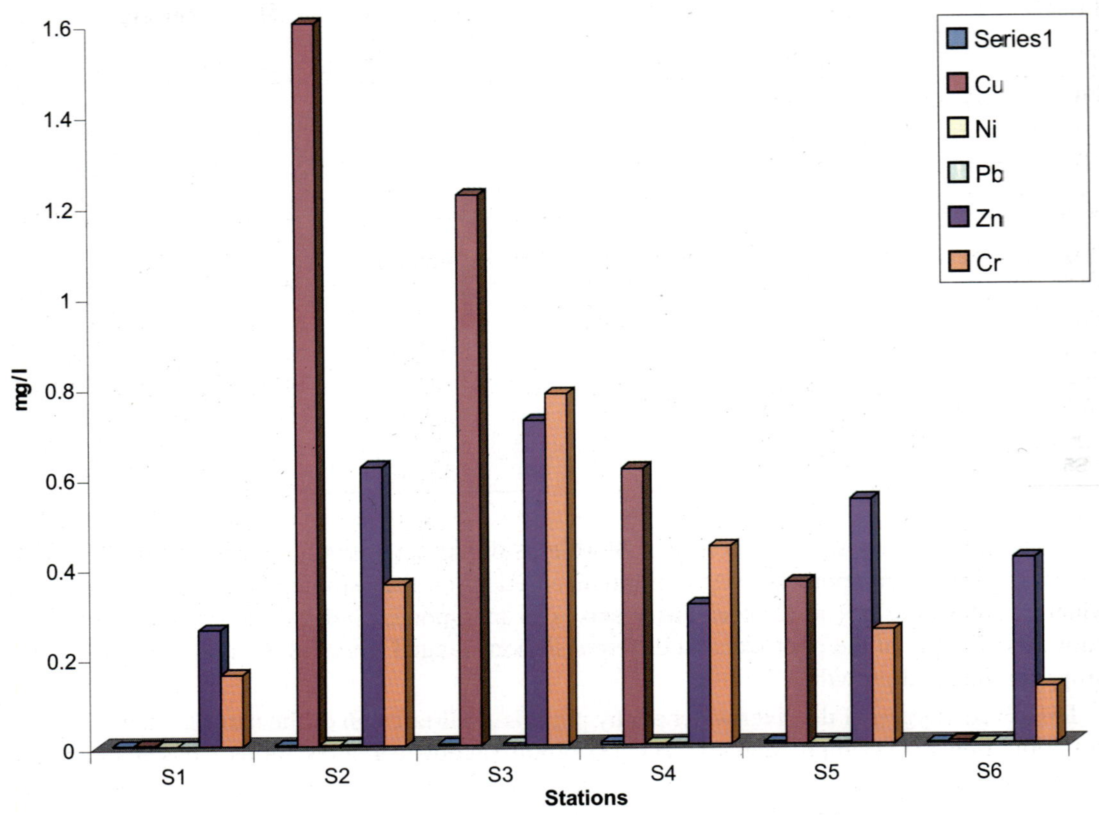

Fig. 1 Concentration of heavy metals in river water.

(Fig.1). These factors probably acted synergistically declining the algal growth and diversity. Albeit domestic sewage is also discharged via these drains in to the river but probably it growth-stimulating effect on algal population was nullified by the presence of heavy metals coupled with low transparency of water at these stations. As river flowed downstream, toxicants got diluted in due course and natural removal of pollutants enhanced the euphotic zone. Further addition of sewage through other drains enriched the river water with nutrients. Thus reduced toxicity, improved transparency and nutrient enriched water triggered the algal growth at stations 4,5 and 6 reaching a maximum of 335 individuals^{-1} at station 5 (Table 4). But despite this the algal diversity as well as algal population in the river was quite low a compared to other perennial rivers. Out of total 46 species, 8 were reported exclusively at station 1 and were never reported from any other station. This shows that discharge of anthropogenic wastes in the river severely disturbed the ecosystem resulting in disappearance of more than 50% species at the very next downstream station 2 and 3. At station 4 a few of them reappeared but the species number could not reach at par to station 1 till station 6. Highest taxonomic similarity (0.73) was recorded between stations 3and 4 indicating that these stations had similar physicochemical ambience. Lowest taxonomic similarity was recorded at stations 1 and 2 indicating that anthropogenic stress exerted a pronounced effect on algal spectrum of the river.

ACKNOWLEDGEMENTS

The authors are thankful to thePrincipal A.N.D.N.N. Mahila Mahavidyalaya, Kanpur for the facilities provided.

REFERENCES

A. P. H. A. 1989. Standard methods for the examination of water and waste water. 17th Ed. American Public Health Association, Washington D.C.

Desikachary, T. V. 1959. Cyanophyta. Indian Council of Agricultural Research, New Delhi.

IS: 2490 1981. Tolerance limits for industrial effluents. Indian Standard Institution, New Delhi.

Margaled, R. 1958. Information theory in ecology. *Gen. Systems.* 3:36-71.

Shannon, C. E. and Weaver, W. 1963. The mathematical theory of communication. University of Illinois Press, Urbana.

Smith, G. M. 1950. The fresh water algae of the United States. McGraw-Hill Book Company, Inc. New York.

Sorensen, R. 1948. A method of establishing groups of equal amplitudes in plan sociology based on similarity of species content and its application to analysis of vegetation on Danish Commons. K. Dan. Vid. Selsk. *Bio. Skr.* 5:1-35.

Tiffany, L. H. and Britton, M. E. 1952. The algae of Illinois. The University of Chicago Press.

18

Use of *Lyngbya limnetica* (Blue-green Alga) in the Treatment of Cattle Shed Wastewater

Sreelaxmi M. Murthy
Department of Biology and Environmental Systems,
International Baccalaureate Diploma Programme, Jamnabai Narsee School,
Juhu, Mumbai-400049,
Maharashtra, India

ABSTRACT

Decentralized wastewater treatment facilitates recycling and reuse of water at the source. Cattle sheds in Mumbai represent one such example where on site treatment can be practised. Removal of nutrients and reduction of BOD by intensive algal culture can provide a low cost treatment method. Algae from the local habitat being naturally acclimatized to the climatic conditions do not require special culturing requirements and hence may provide a system with least maintenance that works naturally. The algal sample from the local habitat showed mixture of green and blue-green algae; out of which *Lyngbya limnetica* showed the ability to remove nutrients successfully from the wastewaters during the pilot study. The nutrient removal efficiency of *Lyngbya limnetica* from cattle shed effluent diluted in the ratio of 1:1 was investigated by batch-culture method. *Lyngbya limnetica* showed positive growth response indicated by increase in its fresh biomass. Autochthonous alga *Chlorella* developed in both control as well as test sets. The pH of the sample increased to 11.43 by the end of the experimental period. Phosphates and ammonium nitrogen with other forms of nitrogen were removed effectively from 6th day onwards. BOD values remained below 1 mg/l. The maximum reduction of the total nitrogen concurred with that of ammonium nitrogen, indicating ammonium nitrogen as the most prevalent form of nitrogen occurring in the cattle shed wastewater. Test set showed a maximum of 83% reduction in ammonia levels; a maximum of 62% reduction in phosphate phosphorus by 10th day. Control set showed a maximum of 74% reduction in ammonia levels by 12th day and a maximum of 43% reduction in phosphate phosphorus by 6th day. Nutrients were effectively removed from 6th day onwards, by 10th day maximum nutrient removal and minimum BOD levels could be achieved. The test set with *Lyngbya* and *Chlorella* showed higher efficiency in removal of nutrients than the control set containing only *Chlorella*. These findings may help to explore the use of *Lyngbya limnetica* in onsite-wastewater treatment.

Keywords: *Lyngbya limnetica*, phosphates, nitrates, cattle shed, wastewater.

INTRODUCTION

Microalgae play an important role in the treatment of wastewater especially in systems such as waste stabilization ponds, oxidation ditches and intensive lagoon systems. High rate algal pond can be envisaged as an adaptation of waste stabilization pond in which treatment efficiency is increased and pond area is reduced by optimizing conditions for algal growth. (Fallowfield et al., 1996). Literature reports indicate use of algae in the treatment of different types of wastes such as domestic sewage (Patil 1991, Craggs et al., 1997); animal wastes, (Canizares, et al., 1994,Yang and Nagano, 1985); food industry such as tomato concentrate industry (Rodrigues and Oliveira, 1987); cheese factory anaerobic effluent (Blier et al., 1995). Algae were successfully employed to remove nitrogen from inorganic wastewaters especially that of nitrogenous industry (Hendricks et al., 1980, Przytocka et al., 1984). Fischerella has been reported to show ability to remove the nutrients from heated effluent tolerating the temperature of 55°C. (Radway et.al., 1994). Filamentous algae such as *Stigeoclonium* (Pretorius and Hensman, 1985) *Phormidium bohneri* (Lessard et al., 1994; Dumas et al., 1998); *Oscillatoria* (Hashimoto et al., 1989) are shown to be effective in nutrient removal as well as convenient to harvest during the post-treatment stage owing to their ability to aggregate. Immobilisation of algae is another technique exploited to bypass the harvesting of algae in algal wastewater treatment systems. *Chlorella vulgaris* immobilized in carrageenan (Lau et al., 1998), in calcium alginate beads (Tam-n and Yong Lou 1994); *Oscillatoria* immobilized in calcium alginate (Singh et al., 1992) have been to shown to remove the nutrients from wastewater efficiently. Thus, algal technology engineered to suit the specified use has a tremendous potential to renovate, recycle and recover the water and valuable resources dissolved in it from all categories of wastewaters.

Privatization, decentralization and reduction-recycle-reuse at source are some novel strategies that are gaining ground worldwide for effective management of the wastewaters and their use. In expanding urban settlements centralized wastewater treatment systems may not cope up with increasing volume of waste generated. Decentralized wastewater treatment system facilitates reduction in huge costs incurred for the construction and maintenance of the sewer network including the required pumping stations; demand for fresh water to flush the wastes; and facilitates reuse of water at the source itself. (Lier and Lettinga, 1999). 'Johkasous' are onsite wastewater treatment plants used for domestic wastewater treatment in unsewered areas in Japan (Ohmori et al., 2000). In such onsite wastewater treatment units immobilized filamentous algae can be integrated to exploit the potential of harnessing solar energy for removal of nutrients, and reduction of BOD. India being a tropical country has the facility to avail solar energy adequately. Algae growing in a locality are well acclimatized to that area. Such algae if employed for treatment; culturing conditions and maintenance can be made simple. Based on this hypothesis, ability of locally growing algae to remove the nutrients from cattle shed effluent was investigated.

MATERIALS AND METHODS

Culturing of algae: Algal sample from the local site was inoculated in two types of nutrient media (1) Knop's medium containing- $Ca(NO_3)_2.4H_2O$, KNO_3, KH2PO_4, $MgSO_4$. $7H_2O$, $FeCl_3$. (2) Chu-10 medium, containing $Ca(NO3)_2$, KH_2PO_4, $MgSO_4$.7H2O, Na_2CO_3, Na_2SiO_3, $FeCl_3$ taken in 100 ml sterilized conical flasks. These two nutrient media were selected to give a different range of nutrient composition for growth of different forms of algae. The flasks were kept in a chamber of artificial illumination of 12 h light and 12 h dark. After a period of three months flasks containing knop's medium showed growth of green alga *Chlorella*, while the flasks containing Chu-10 medium showed growth of the above-mentioned blue-green algal forms.

Blue-green algae were isolated through microscopic observation and subculture in Chu-10 medium. The cultured algae were identified with the help of a monograph on Cyanophyta by Desikachary (1959). Continuous subculturing for more than six months led to the cultures 2 predominant with *Chroococcus* sp., *Nostochopsis* sp., *Lyngbya limnetica*, *Microcystis* sp.. Culture from each type of alga was inoculated onto agar plates containing Chu-10 medium. *Lyngbya limnetica* spread as filaments and *Nostochopsis* grew as colonies on the agar plates. The filaments from these colonies were inoculated into liquid Chu-10 medium. These cultures yield pure cultures of *Lyngbya limnetica* and *Nostochopsis* sp. All the cultures were kept under artificial illumination 12 h L : 12 h D light regime set in a chamber fitted with a timer and with an average light intensity of 600-800 lux. The intensity of light was measured using LX-101 Lutron Luxmeter.

Waste water sample collection: The water samples were collected from the outlets of municipal drainage which consult of the water used for washing the cattles as well as cattle shed and its excreta. The samples were collected for over a period of one-year

Sample preparation: The cattle shed wastewater was dark yellow in color containing a little quantity of sediments and straw pieces and was less dense. The sample was filtered through filter paper and diluted in the ratio of 1:1 using tap water. The sample was diluted to reduce the intensity of colour and allow light penetration. The diluted sample was immediately analyzed for temperature, pH, dissolved oxygen, orthophosphate, ammonia nitrogen, nitrite nitrogen and nitrate nitrogen by standard methods (APHA 1980)

The experiments were set up at room temperature conditions in duplicate sets of each control set (without *Lyngbya*) and test set (with *Lyngbya*). Each experiment was set for a total period of 12 days. 100 ml of diluted sample was taken in each 250 ml sterilized conical flask and 0.5 gm wet weight of *Lyngbya limnetica* was inoculated into the sample. Every alternate day one of the flasks was taken, pH recorded, Orthophophates, ammonia nitrogen, nitrite nitrogen, nitrate nitrogen and BOD were analysed. The wet weight (fresh biomass) of *Lyngbya* was determined. Microscopic examination was also carried out to observe the type of microorganisms growing in the sample.

Methods for Analysis of Water Sample

Each sample had been analyzed for the selected parameters by the following methods

pH: pH of the sample was measured by using digital pH meter. The pH meter was standardized using standard buffers of pH 4.00, 7.00, and 9.2 as per the manufacturer's manual. Then the glass electrode was immersed in the sample. The pH of the sample was directly read from the display.

BOD: For each analysis of the sample two sets of BOD bottles were filled with 10 ml of sample diluted with dilution water. One set of the bottles was kept for incubation for 5 days. The dissolved oxygen content of the second set of the bottles was immediately analyzed by azide modification method.(D_1). After completion of 5 days incubation, the D_0 of the sample was determined (D_2). From the values of D_1 and D_2 BOD of the sample was computed.

Ammonia-Nitrogen: The ammonia nitrogen present in the sample was determined by Indophenol method. 10 ml of the sample diluted to 50 ml with deionised water was taken in a glass stoppered test-tube, to which 2 ml of phenol solution, 2 ml of nitroprusside and 5 ml of oxidizing solution. After thorough mixing, the flask is allowed to stand at room temperature for 1-2 h. The extinction is read at 640 nm against deionised water blank.

Nitrite nitrogen: 10 ml of the sample diluted to 50 ml was taken in a conical flask to which 1 ml of sulfanilamide solution was added. After 2 min 1.0 ml of NED-dihydrochloride solution was added and mixed immediately. After 10 min the absorbance was measured at 543 nm.

Nitrate nitrogen: 10 ml of the sample diluted to 50 ml with distilled water was evaporated to dryness in an oven. Using a glass rod, the residue is rubbed thoroughly with 2.0 ml phenoldisulphonic acid reagent to insure the dissolution of all the solids. This was diluted with 20 ml distilled water and 6-7 ml of liquor ammonia was added. Finally the contents were made to 50 ml, and the absorbance of the yellow colour developed was read at 420 nm.

Phosphate phosphorus: (Vanadate molybdate method) 10 ml of the sample diluted to 35 ml was taken in a conical flask to which 10 ml vanadate molybdate reagent was added. The final volume is made up to 50 ml. After 10 min the absorbance was measured at 420 nm.

RESULT & DISCUSSION: The results of all the nine experiments carried out showed a similar trend with respect to algal growth and nutrient uptake except in case of the experiments conducted during December, January and February i.e. winter season. The mean values of all the experiments indicated the following:

Table 1 Composition of cattleshed samples collected during the experimental periods

Expt. No.	Testing Order	Days	pH Value	Phosphate mg/l	Ammonia-Nitrogen mg/l	Nitrite - Nitrogen- mg/l	Nitrat Nitrogen mg/l	Total Nitrogen mg/l	BOD mg/l
1	Control-May 2001	Day of Collection	8.14	1.43	5.11	0.02	0.13	5.26	1.94
2	Control-20/4/01	Day of Collection	7.90	4.30	7.86	0.03	0.05	7.94	1.69
3	Control-19/04/01	Day of Collection	8.80	8.61	4.18	0.06	0.06	4.30	1.66
4	Control-16/02/01	Day of Collection	8.30	8.97	3.08	0.08	0.15	3.31	2.83
5	Control-22/01/01	Day of Collection	7.75	1.08	2.88	0.04	0.07	2.99	3.02
6	Control-19/01/01	Day of Collection	8.00	5.74	8.52	0.03	0.10	8.64	1.58
7	Control-27/12/00	Day of Collection	7.00	3.59	4.85	0.06	0.05	4.96	4.20
8	Control-21/11/00	Day of Collection	8.50	7.17	3.93	0.05	0.03	4.01	1.59
9	Control-18/10/00	Day of Collection	7.90	4.30	7.86	0.03	0.05	7.94	1.77

Table 2 Mean values of the test parameters of all the nine experiments

Days	Phosphate (mg/l) control	Phosphate (mg/l) Test	Toatal nitrogen (mg/l) control	Toatal nitrogen (mg/l) Test	pH control	pH Test	Chlorella (mg) control	Chlorella (mg) Test	Lyngbya (gm)
0	5.02	5.02	5.48	5.48	8.03	8.03	not detectable	not detectable	0.50
2	3.71	2.43	4.41	3.73	8.27	8.43	not detectable	not detectable	0.55
4	3.39	2.18	3.91	2.36	8.42	9.00	0.50	0.66	0.62
6	1.95	1.29	3.28	1.22	8.67	9.35	2.72	3.42	0.73
8	2.15	1.24	2.18	1.09	9.42	9.63	3.37	4.04	0.89
10	2.11	1.83	1.86	0.84	9.55	9.83	3.86	1.29	1.18
12	2.07	1.87	1.37	0.36	10.05	10.04	2.10	0.30	1.50

pH: On 4th day pH value had scaled up to 9.00 in case of test set against 8.42 ascontrol set. By 12th day pH values had reached to 10 in both control as well as test set. (Table 2). Increase in the pH was recorded in all the experimental sets. (Rodrigues and Oliviera 1987; Wood, et al., 1989; Patil 1991; Canizares et al., 1994; and Blier et al., 1995). Such increase in the pH is the result of the algal photosynthesis, which involves removal of CO_2 from the medium. (Buhr and Miller 1983; Rodrigues and Oliviera 1987; Wood et al., 1989; Patil 1991; and Canizares, et al., 1994). Rise in pH also indicates that the medium is poorly buffered. (Blier et al., 1995). Statistical regression analysis based on the mean values of all the experiments revealed positive correlation between pH and growth of algae. In the control set the P-value was 0.47306 that indicates significant correlation between the growth of *Chlorella* and rise in pH. The coefficient of determination R^2-0.577889 implies that growth of *Chlorella* explains approximately 58% rise in pH. In case of test set, where *Chlorella* was growing along with *Lyngbya*, it explains only 19.75 of rise in pH. (P value = 0.317911 & R^2 = 0.19741). At the same time a strong relation between pH and *Lyngbya* had been exhibited; the P value being 0.009155 & R^2 = 0.772643. *Lyngbya* explains approximately 77% rise in pH. Thus statistical analysis substantiates that algal photosynthesis leads to increase in the pH of the medium.

BOD: During December, January & February the BOD values in the sample were between 2.00-4.00 mg/l, while during other months the values were below 2mg/l (Table:1). By 10th day BOD values have fallen down to less than 1 mg/l in both control as well as test sets (Table 2). However the percentage reduction in control was 72 in case of control set and 67 in case of test set by 10th day (Fig: 2)

Nitrite, Nitrate and Ammonia Nitrogen: 65% of Nitrite nitrogen and 68% of nitrate nitrogen reduced in control set on 10th day. 74% of nitrite nitrogen and 64% of nitrate nitrogen reduced in test set on 12th day. Test set showed a maximum of 83% reduction in ammonia levels by 10th day and showed 94% reduction of ammonium nitrogen on 12th day as against 74% in case of Control set by 12th day. (Fig. 1, Fig. 2 and Fig. 3).

The reduction in nitrite and nitrate nitrogen showed inverse relationship during the experiments. During winter months when algal growth was slow, reduction pattern of these two forms of nitrogen pointed to the occurrence of bacterial metabolism within the culture. However these two forms constituted less than 1mg/l and Ammonium nitrogen was the predominant form

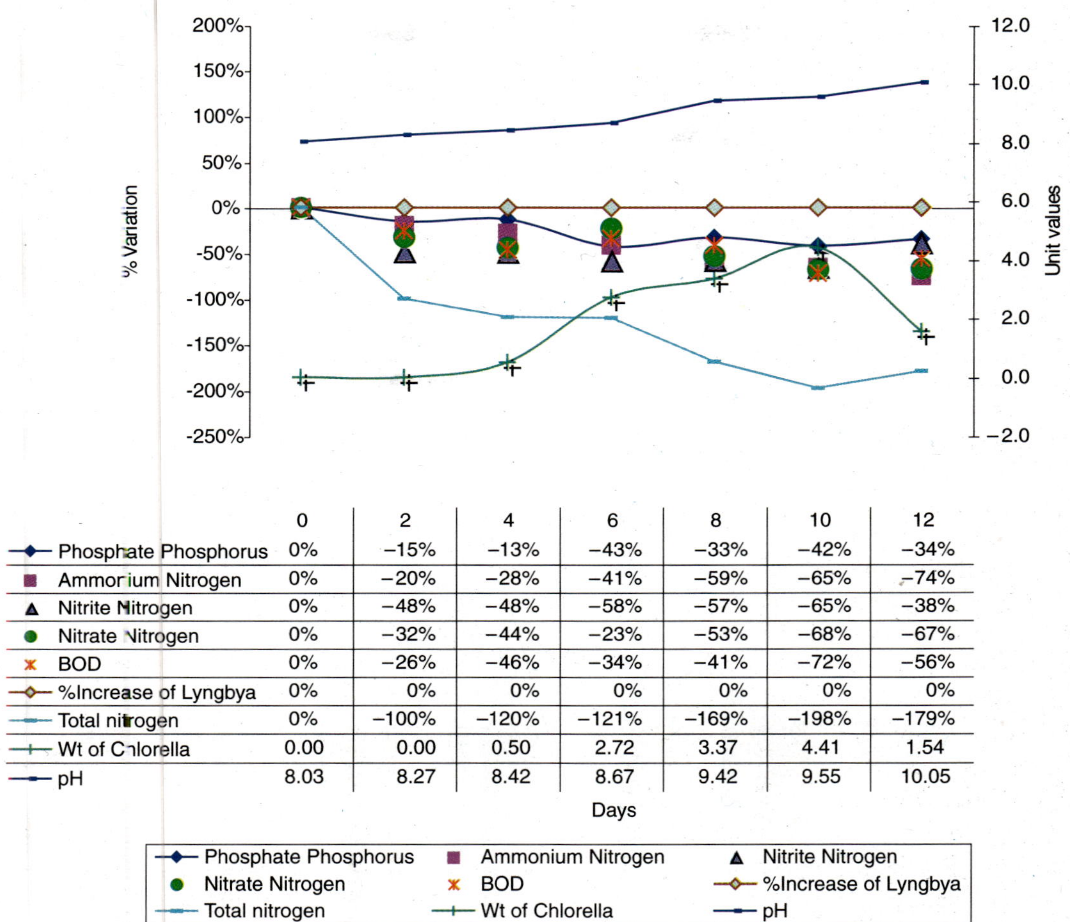

	0	2	4	6	8	10	12
Phosphate Phosphorus	0%	−15%	−13%	−43%	−33%	−42%	−34%
Ammonium Nitrogen	0%	−20%	−28%	−41%	−59%	−65%	−74%
Nitrite Nitrogen	0%	−48%	−48%	−58%	−57%	−65%	−38%
Nitrate Nitrogen	0%	−32%	−44%	−23%	−53%	−68%	−67%
BOD	0%	−26%	−46%	−34%	−41%	−72%	−56%
%Increase of Lyngbya	0%	0%	0%	0%	0%	0%	0%
Total nitrogen	0%	−100%	−120%	−121%	−169%	−198%	−179%
Wt of Chlorella	0.00	0.00	0.50	2.72	3.37	4.41	1.54
pH	8.03	8.27	8.42	8.67	9.42	9.55	10.05

Fig. 1 Relationship between various parameters and algal growth—Control set (based on mean values of all expts).

of nitrogen in the sample (Table 1). Reduction of ammonium nitrogen in the control as well as test set coincided with the active growth of *Chlorella* as well as *Lyngbya* and a pH range of 8.85-10.00 (Fig. 1, Fig. 2 & Fig. 3). Therefore the reduction of ammonium nitrogen can be attributed to the algal uptake and volatilization due to high pH. Neither in the control set nor in the test set removal of ammonium nitrogen was accompanied by increase in nitrite and nitrate nitrogen content. So the removal of ammonium nitrogen through nitrification appears to be unlikely. (Craggs et al., 1997).

Total nitrogen: From 6th day onwards total nitrogen reduced by more than 100%. On 10th day control showed 198% reduction as against 202% in case of test set. Maximum percentage of reduction in total nitrogen concurred with that of ammonium nitrogen (Fig. 4). It is in accordance with the fact that the sample contained predominantly ammonium nitrogen. The efficiency of the test set over the control set may be due to the active uptake by the green alga besides that of *Chlorella*. The high pH resulted due to algal photosynthesis also contributes to the removal of

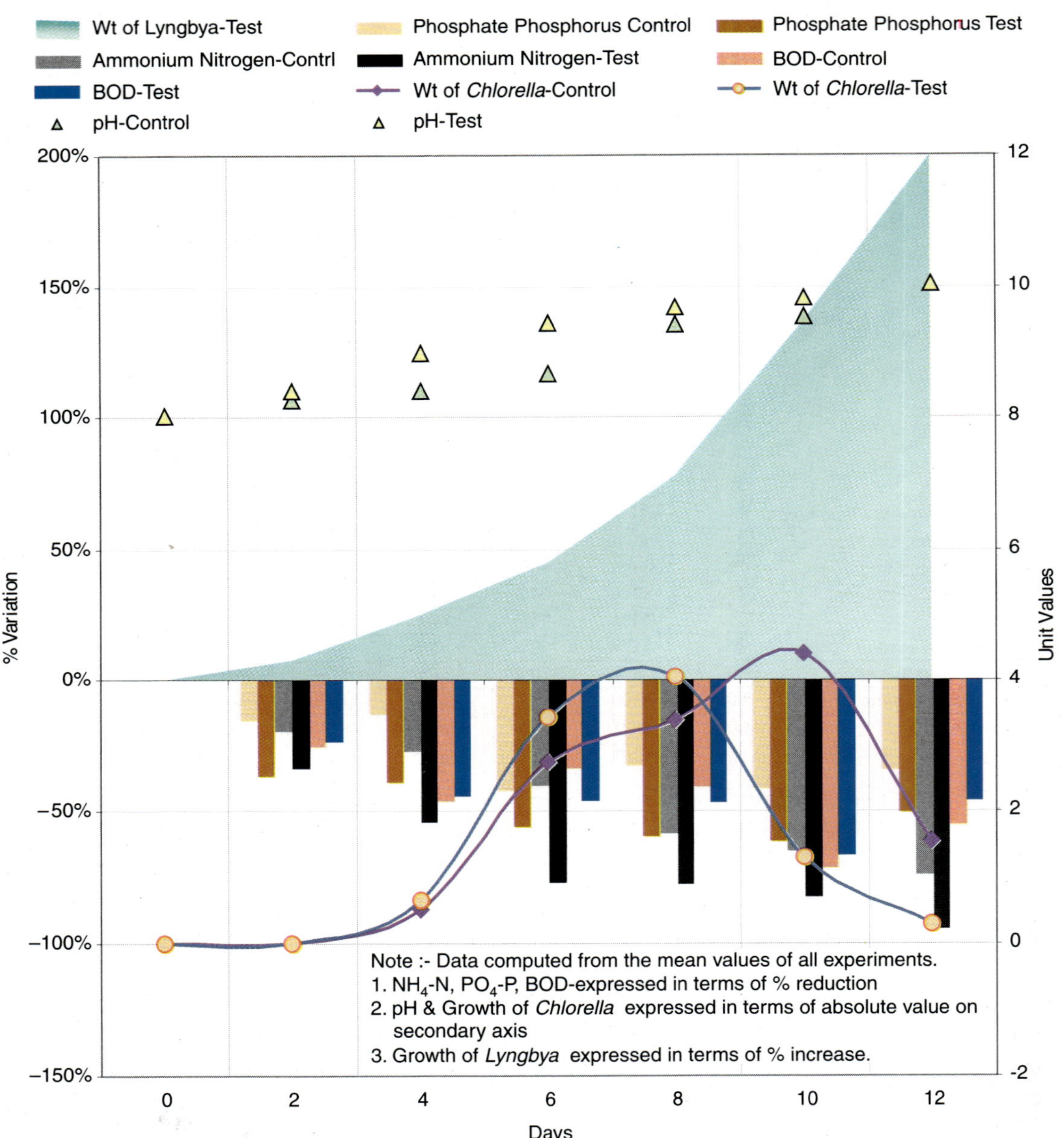

Fig. 2 Comparison of nutrient removel efficiency of algae in control set vs test set.

ammonium nitrogen.

Regression analysis of pH and total nitrogen showed a negative correlation. The P value for control set being 0.000202 and for the test set being 5.32 E-0; and the coefficient of determination $R^2 = 0.9701$ in case of test set and 0.9491 in case of control set. This confirms the removal of

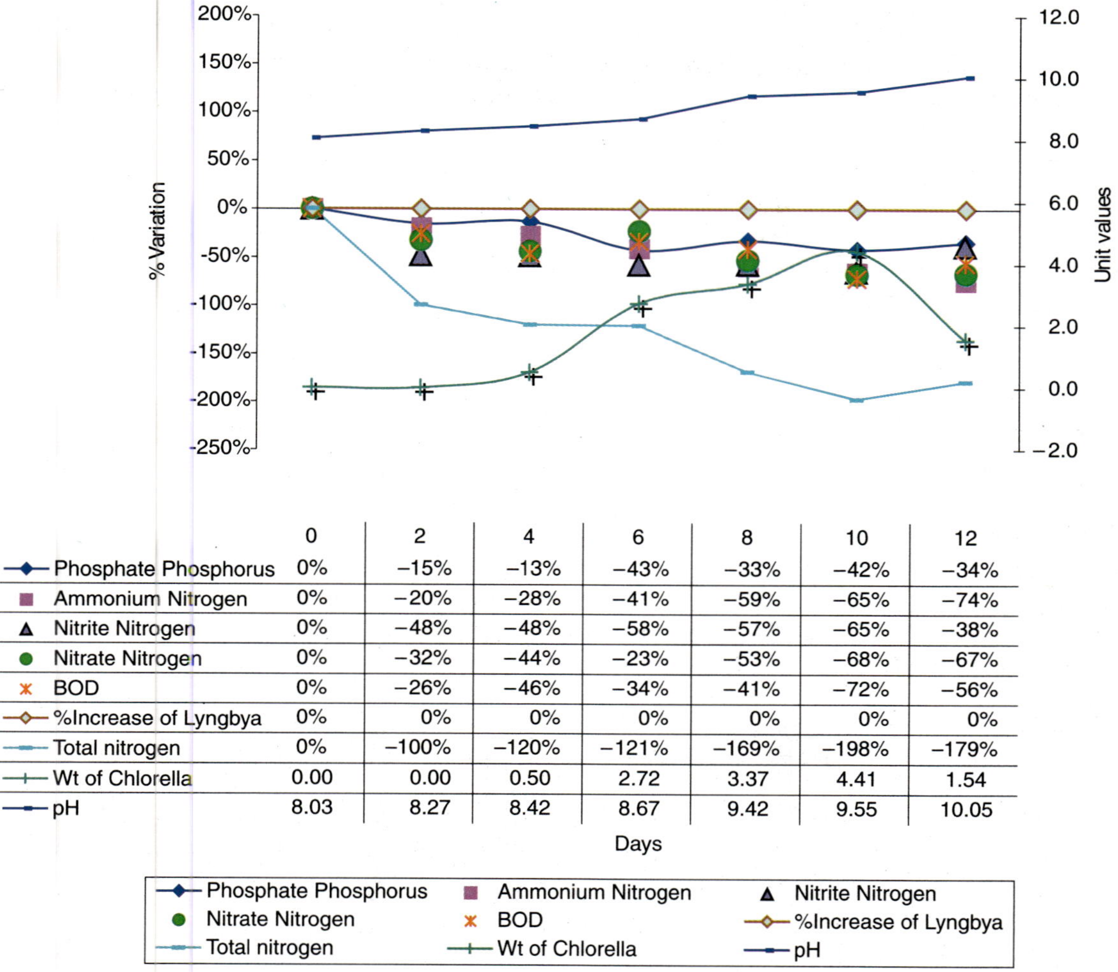

	0	2	4	6	8	10	12
Phosphate Phosphorus	0%	−15%	−13%	−43%	−33%	−42%	−34%
Ammonium Nitrogen	0%	−20%	−28%	−41%	−59%	−65%	−74%
Nitrite Nitrogen	0%	−48%	−48%	−58%	−57%	−65%	−38%
Nitrate Nitrogen	0%	−32%	−44%	−23%	−53%	−68%	−67%
BOD	0%	−26%	−46%	−34%	−41%	−72%	−56%
%Increase of Lyngbya	0%	0%	0%	0%	0%	0%	0%
Total nitrogen	0%	−100%	−120%	−121%	−169%	−198%	−179%
Wt of Chlorella	0.00	0.00	0.50	2.72	3.37	4.41	1.54
pH	8.03	8.27	8.42	8.67	9.42	9.55	10.05

Fig. 3 Relatonship between varior parameters and algal growth—test set (Mean value of all experiments.)

Ammonium nitrogen by volatilization due to high pH. Growth of *Chlorella* showed negative correlation with total nitrogen, both in case of test set as well as control set the P values being 0.392783 and 0.021464 respectively. Contribution of *Chlorella* towards the removal of nitrogen can be explained up to 68% ($R^2 = 0.68186$) in case of control set while in the test set it showed 14% contribution ($R^2 = 0.1488$) In test set presence of *Lyngbya* may be the influential factor for the low uptake of nitrogen by *Chlorella.* The P value 0.035112 for *Lyngbya* and total nitrogen indicates significant relationship between the growth of *Lyngbya* and removal of total nitrogen. This in conjunction with the negative correlation ($R^2 = 0.618602$) implies nitrogen uptake by *Lyngbya* in substantial amount i.e. 61.86%.

Thus, the removal of both total nitrogen from the cattle shed effluent can be attributed to

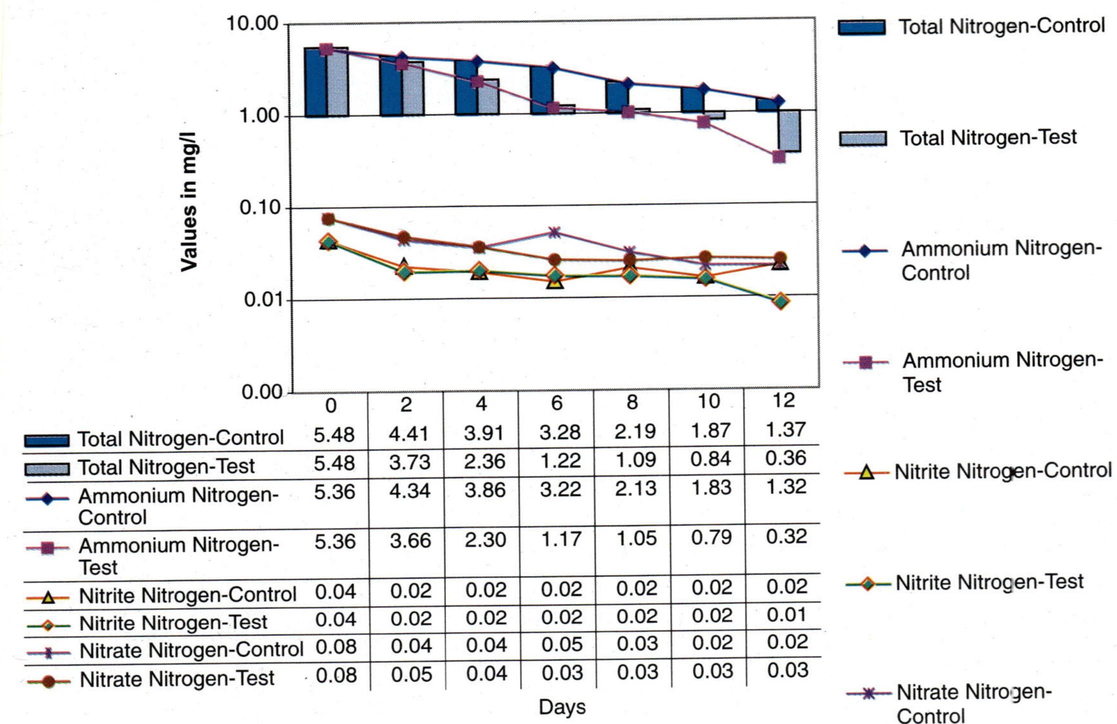

	0	2	4	6	8	10	12
Total Nitrogen-Control	5.48	4.41	3.91	3.28	2.19	1.87	1.37
Total Nitrogen-Test	5.48	3.73	2.36	1.22	1.09	0.84	0.36
Ammonium Nitrogen-Control	5.36	4.34	3.86	3.22	2.13	1.83	1.32
Ammonium Nitrogen-Test	5.36	3.66	2.30	1.17	1.05	0.79	0.32
Nitrite Nitrogen-Control	0.04	0.02	0.02	0.02	0.02	0.02	0.02
Nitrite Nitrogen-Test	0.04	0.02	0.02	0.02	0.02	0.02	0.01
Nitrate Nitrogen-Control	0.08	0.04	0.04	0.05	0.03	0.02	0.02
Nitrate Nitrogen-Test	0.08	0.05	0.04	0.03	0.03	0.03	0.03

Fig. 4 Reduction in nitrogen forms contol vs test.

Lyngbya as well as autochthonous alga *Chlorella* besides volatilization due to high pH. It is reported that in waste stabilization ponds apart from nitrfication-denitrification, algal uptake of ammonium, nitrate and nitrite forms of nitrogen also contribute to the removal of nutrients (Loi and Lam 1997).

Phosphate phosphorus: Phosphate phosphorus in the sample was in the range of 1.43 - 8.97 mg/l during the experimental period. Maximum reduction was recorded in both control as well as test set by 10th day. The test set showed maximum reduction 62% and control set showed maximum reduction of 42%. Removal of phosphates phosphorus from the medium can be attributed to precipitation at high pH as well as the algal uptake. The pH of the sample in the experimental runs was in the range of 9.00-11.43. Orthophosphates precipitation is known to occur at elevated pH of more than 8.5. (Hartley et al., 1997). Removal of phosphates from the waste water due to precipitation at high pH is well documented. (Craggs et al., 1997, 1993; Mesple et al., 1995; Doran and Boyle 1979). Regression analysis based on the mean values of all the experiments indicated significant correlation between the pH and phosphate content. (P value for control set is 0.0031792 and 0.035951 for test set.) The coefficient of determination R^2 in case of control set is 0.6355 and in case of test set is 0.618476. This implies that high pH value explains 61-63% reduction of phosphates. Negative correlation indicated that the progress in the growth of *Chlorella* was accompanied by reduction in the amount of orthophosphates. This confirms the uptake of phosphates by *Chlorella*. However the uptake of phosphates by *Chlorella*

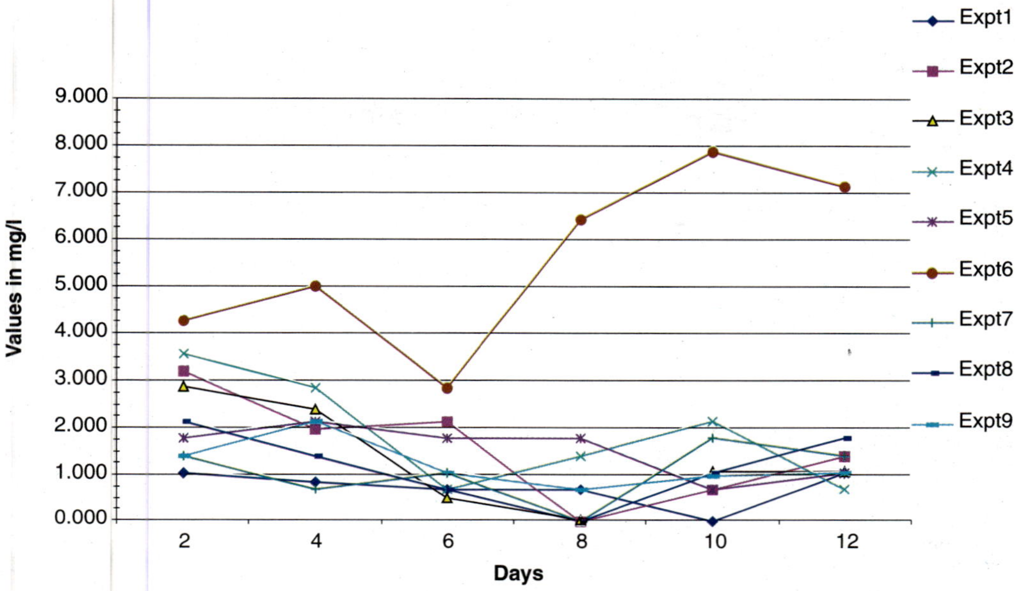

Fig. 5 Trend of phosphate reduction for all test experiments.

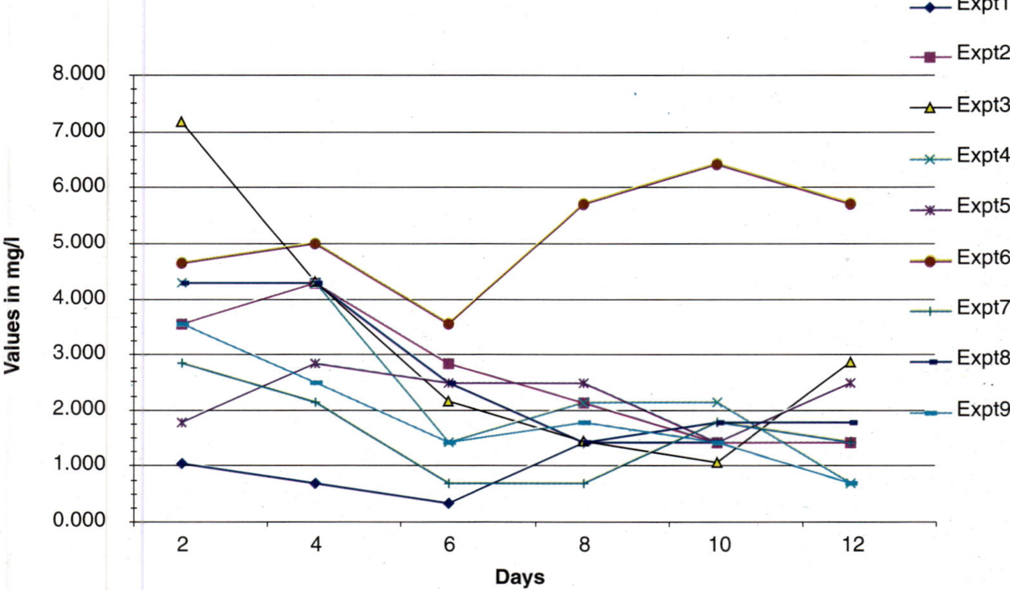

Fig. 6 Trend of phosphate reduction for all control experiments.

in the control set was more significant than that of the test set. This may be due to the competitive influence exerted by *Lyngbya*. The growth of *Lyngbya* showed a negative correlation with the reduction of phosphates, but with relatively less significance, P = 0.296949. The coefficient of determination $R^2 = 0.213199$, implies 21.3% of phosphate phosphorus removal by *Lyngbya.*

Algal Growth:

Chlorella: The growth of *Chlorella* developed indigenously in the culture in both control as well as test set. It showed a distinct pattern of growth curve with lag phase, log phase followed by senescent phase. *Chlorella* is reported to be a predominant alga capable of surviving in wastewaters. (Fruend et al., 1993; Cohen and Anton 1993). It is also known for its nutrient uptake

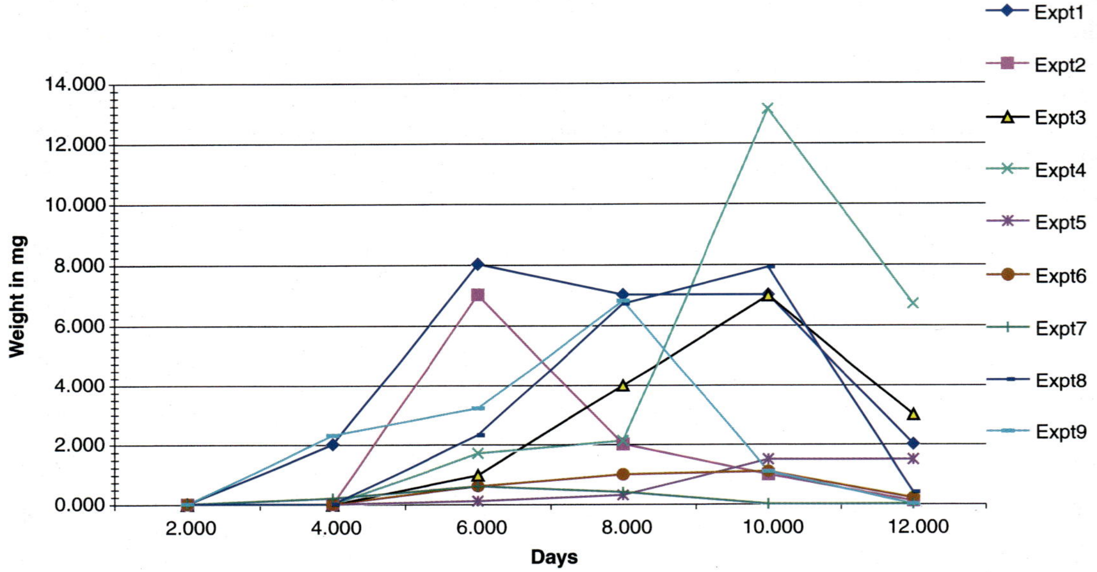

Fig. 7 Trend of weight of *Chlorella* for all control experiments.

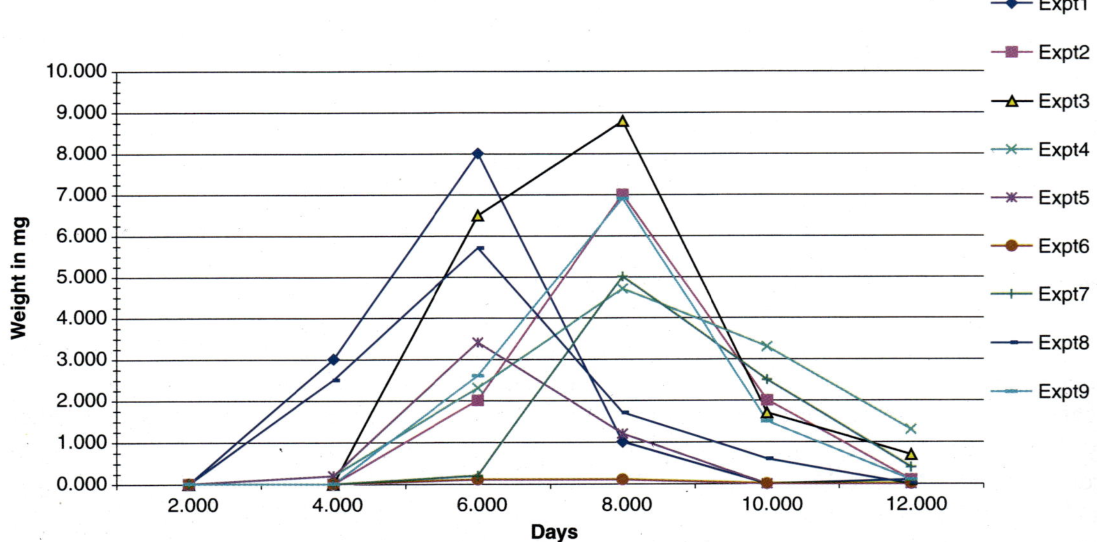

Fig. 8 Trend of weight of *Chlorella* for all test experiments.

ability. (Lau et al., 1996; Travieso et al., 1996).

Lyngbya limnetica: *Lyngbya limnetica* grew well in all the experimental sets. It could out compete the autochthonous alga *Chlorella*. Towards the later stage of the experimental period, *L. limnetica* grew favorably and dominated the culture. This observation points towards the ability of *L. limnetica* to withstand competition. Gantar et al., (1991) reported the succession of indigenous alga over inoculated ones while investigating the efficiency of pig water treatment by the green alga *Scenedesmus quadricauda* and blue-green alga *Spirulina*. This may be attributed to the fact that *Lyngbya* being blue-green alga had tolerance to alkaline pH and had the ability to utilize even traces of CO_2 present in the sample. The maximum tolerable pH without effecting the activity of *Chlorella* is around 9.00). At pH 11.5 *Chlorella* vulgaris was not detected (Mayo and Noike 1994). As the selected alga *Lyngbya limnetica* was collected from the same locality as that of the effluent; natural acclimatization might have enabled the blue green alga to out compete *Chlorella*.

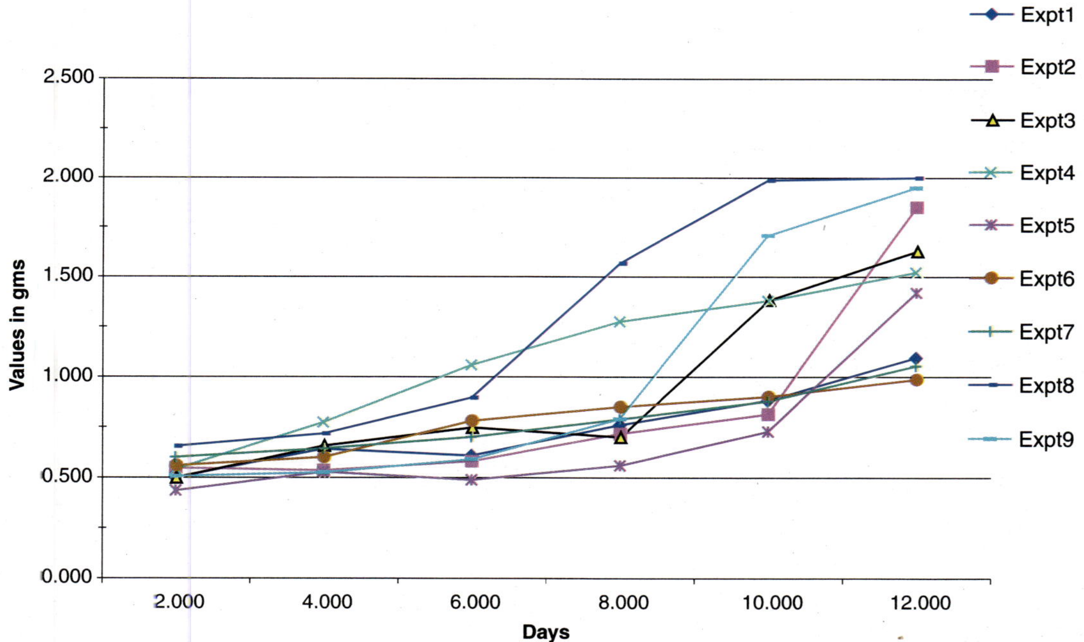

Fig. 9 Trend of *Lyngbya* growth for all test experiments.

Lyngbya limnetica showed sustained growth in the cattle shed effluent. 64% reduction in phosphate phosphorus by 10th day and 94% reduction in ammonium nitrogen. by 12th day were achieved. This observation is augmented by the reports of nutrient removal by other cyanophyceae members from secondary effluent of a dairy (Lincoln et al., 1996), by *Phormidium* from fish farm effluent (Dumas et al., 1998), by sp. of *Oscillatoria* from primary sewage (Craggs et al., 1997). Efficient phosphate removal from the wastewaters by immobilized *Oscillatoria* was reported by Singh et al., (1992). Cromar et al., (1992) reported Cyanophycean members as an integral component of high rate algal ponds and *Lyngbya* sp. was one of them.

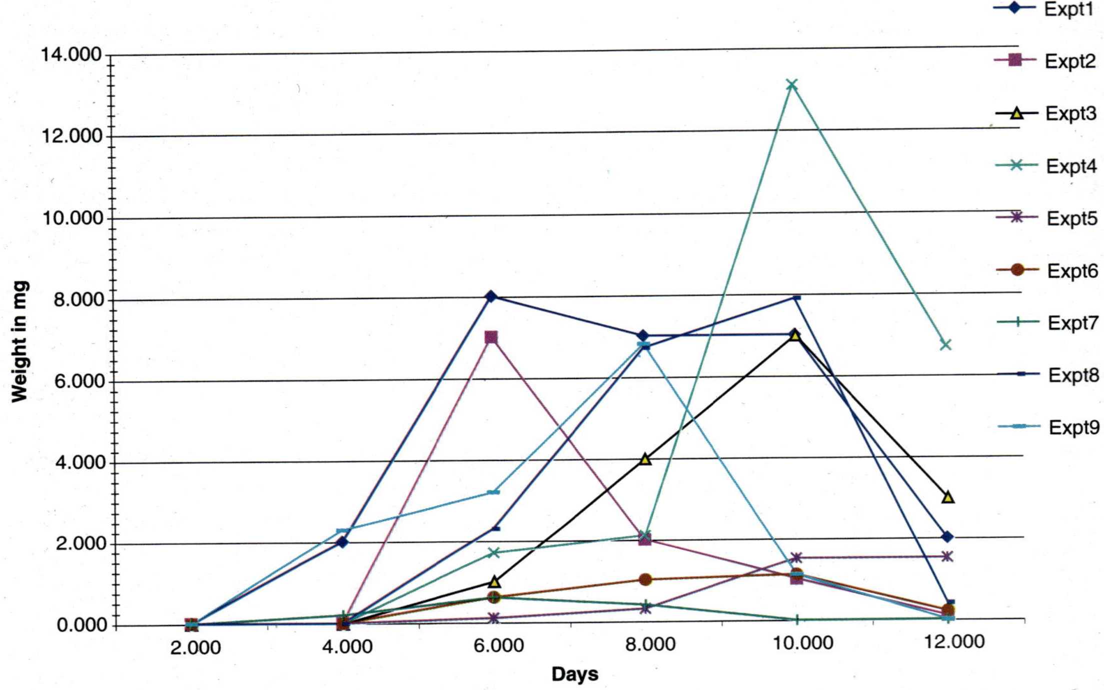

Fig. 10 Trend of weight of *Chlorella* for all control experiments.

CONCLUSION

Lyngbya limnetica showed positive growth response in cattle shed effluent and could effect the nutrient removal as well as reduction in BOD. Hence *Lyngbya limnetica* can be successfully used to explore designing of a compact on-site algal wastewater treatment plant for small holdings like cattle shed.

ACKNOWLEDGEMENTS

I would like to thank my teacher Late Dr. Mrs S.M. Joshi for her guidance through the work. I wish to extend my thanks to Dr. Rajan for his guidance in statistical analysis and Mr. Shashank for his assistance in spreadsheet software.

REFERENCES

APHA, AWWA & WPCF 1980 Standard Methods for the Examination of Water and Wastewater. 15th edition U.S.A.

Blier, R., Laliberte, G. and Noue, J, De La. 1995. Tertiary treatment of cheese factory anaerobic effluent with *Phormidium bohneri*, Micro *actinium pusillum. Biores. Tech.* 52: 151-155.

Buhr, B. O. and Miller, S. B. 1983. A dynamic model of the high-rate algal bacterial waste water treatment pond. *Water Research* 17: 29-37.

Canizares, R. O., Villimueva, Ramos A., Corona, A. I., Monroy, O., Torre M. De La, Gomez Lojero, C. and Travieso, L. 1994. Phormidium treatment of anaerobically treated swine waste water. *Water Research* 28 (90): 1891-1895.

Cohen Iris and Anton, F. Post 1993. The heterotrophic connection in photoautotrophic *Chlorella vulgaris* dominant in waste water oxidation ponds. *Wat. Sci. Tech.* 27(7-8): 151-155.

Craggs R., J., Mcauley, J. Paul and Smith, J. Valerie 1997. Wastewater nutrient removal by marine microalgae grown on a corrugated raceway. *Wat. Res.* 31 (7): 1701-1707.

Cromar, N. J., Martin, N. J., Christofi, N., Read, P. A. and Fallowfield, H. J. 1992. Determination of nitrogen and phosphorus partitioning within components of biomass in a high rate algal pond—Significance for the coastal environment of the treated effluent discharge. *Wat. Sci. Tech.* 25(12): 207-214.

Desikachary, T. V. 1959. Cyanophyta. Indian Council of Agricultural Research. New Delhi, India.

Doran, D. Michael and Boyle, William C. 1979. Phosphorus removal by activated algae.*Wat. Res.*13: 805-812.

Duams, A., Laliberte, G., Lessard, P., and Noue, J. De La 1998. Biotreatment of fish farm effluent using the cyanobacterium *Phormodium bohneri. Aquacul. Engin.* 17: 57-68.

Fallowfield, H. J., Cromar, N. J. and Evison, L. M. 1996. Coliform die-off rate constants in a high rate algal pond and effect of operational and environmental variables. *Wat. Sci. Tech.* 34 (11): 141-147.

Fruend, C., Romen, E. and Post, F. A. 1993. Ecological physiology of an assembly of photosynthetic microalgae in wastewater oxidation ponds *Wat. Sci. Tech.* 27 (7): 143-149.

Gantar, M., Obreht, Z. and Dalmacija, B. 1991. Nutrient removal and algal succession during the growth of *Spirulina platensis* and *Scenedesmus quadricauda* on swine waste water. *Biores. Techno.* 36: 167-171.

Hartley, A. M., House, W. A., Callow, M. E. and Leadbeater, B. S. C. 1997. Co-precipitation of phosphate of phosphate with calcite in the presence of photosynthesizing green algae *Wat. Res.* 31(9): 2261-2268.

Hashimoto, S. and Furukawa, K. 1989. Nutrient removal from secondary effluent by filamentous algae. *J. Ferment. Bioengin.* 67 (1): 62-69.

Hendricks, F. Bosman, J. 1980. The removal of nitrogen from an inorganic indusrial effluent by means of intensive algal culture. *Prog. Wat. Tech.* 12: 651-665.

Lau, P. S., Tam, N. F. Y. and Wong, Y. S. 1996. Wastewater nutrient removal by *Chlorella vulgaris* optimization through acclimation. *Environmental Technology.* 17 (2): 183-189.

Lau, P. S. Tam, N. F. Y. and Wong, Y. S. 1998. Operational optimization of batchwise nutrient removal from wastewater by carragenan immobilized Chlorella vulgaris. *Wat. Sci. Tech.* 38(1):185-192.

Lessard, Paul, Proux, Daniel. and Joel, De La. Noue 1994. Nutrient removal using cyanobacterium (phormidium bohneri) experimental results with a batch reactor. *Wat. Sci. Tech.* 24 (5): 75-83.

Lier. Jules, B., and Lettinga, G. 1999. Appropriate technologies for effective mangement of industrial and domestic wastewater. The decentralized approach. *Wat. Sci. Techno.* 40 (7): 171-183.

Lincoln, E. P., Wilkie, A. C. and Frenhc, B. T. 1996. Cyanobacterial process for renovating dairy waste water. *Biomass Bioenergy.* 10 (1): 63-68.

Loi, P. C. C. and Lam, P. K. S. 1997. Major pathways for nitrogen removal in waste water stabilization ponds *Water, Air and Soil Pollution.* 99 (1-2): 125-136.

Mayo, A. W., and Noike, T. 1994. Response of mixed cultures of *Chlorella vulgaris* and heterotrophic bacteria to variation to pH. *Wat. Sci. Tech.* 30 (8): 285-294.

Mesple, F., Troussellier, C. and Bontoux, J. 1995. Difficulties in modeling phosphate evolution in a high rate algal pond. *Wat. Sci. Tech.* 31 (12): 45-54.

Ohmori, M.,Yahashi, T., Furukawa, Y., Kawamuna, K., Yamamoto, Y. 2000. Treatment performance of newly developed johkasous with membrane separation. *Wat. Sci. Tech.* 41 (10-11): 197-204.

Patil, H. S. 1991. The role of *Ankistrodesmus falcatus* and *Scenedesmus quadricuada* in sewage purification. *Biores. Techno.* 37: 121-126.

Pretorius, W. A. and Hensman, L. C. 1985. The selective cultivation of easily harvestable algae using Cross-Flow Micro screening. *Wat. Sci. Tech.* 17: 791-892.

Przytocka-Jusiak Magdalena, Duszota marek, Matusiak Kazimierz and Mycielski Roman 1984. Intensive culture of *Chlorella vulgaris*/AA as the second stage of biological purification of nitrogen–indutrial waste water. *Wat. Res.*18 (1): 1-7.

Radway, J. C.,Weissman, J. C. 1994. Nutrient removal by thermophilic *Fischerella* (Mastigocladus laminosus) in a simulated algae culture process. *Biores. Techno.* 50: 227-233.

Rodrigues, A. M. and Oliviera, S. J. F. 1987. treatment of wastewater from the tomato concentrate industry in high rate algal pond. *Wat. Sci. Tech.* 19 (12): 43-49.

Singh, A. K., Dwivedi Alka and Kumar, H. D. 1992. Phosphorus uptake by immobilized cells of alkalitolerant cyanobacterium *Oscillatoria amphibia* Ag. Ex. Gomont. *Environmental Technology* 13: 89-94.

Talbot, P., J. de. La. Noue 1993. Tertairy treatment of wastewater with *Phormidiium bohneri* (schmidle) under various light and temperature conditions. *Water. Res.* 26 (1): 153-159.

Tam-n, F. Y., Lau P. S. Yong, 1994. wastewater inorganic N and P removal by immobilized *Chlorella vulgaris*. *Wat. Sci. Tech.* 30 (6): 369-374.

Travieso, L., Bentez, F., Weiland, P., Sanchez, E., Dugeyron and Domiguez, A. R. 1996 Experiments on immobilization of microalgae for nutrient removal in wastewater treatments. *Biores. Techno.* 55: 181-186.

Wood, A., Scheepers, J. and Hills, M. 1989. Combined artificial wetlands and high rate algal ponds for wastewater treatment and protein production. *Wat. Sci. Tech.* 21: 659-668.

Yang, P. Y. and Nagano, S. Y. 1985. A potential treatment alternative for swine wastewater in the tropics. *Wat. Sci. Tech.* 17: 819-831.

19

Studies on Biodegradation Potential of Linear Alkyl Benzene Sulphonate by *Nostoc muscorum* Ag. ex. Born et Flah

Dilip Kumar Rathore[1*], Preeti Sharma[2], Praveen Kumar Chandel[3], Sandhya Tyagi[1] and Krishna Chandra Soni[3]

[1]Laboratory of Anatomy and Ecophysiology of Medicinal Plants and Cyanobacterial Research

[2]Microbial Research Laboratory

[3]Laboratory of Biomolecular Technology

Department of Botany, University College of Science, Mohanlal Sukhadia University, Udaipur-313001 (Rajasthan)

ABSTRACT

Cyanobacteria (blue-green algae) are prokaryotic organisms with oxygen evolving photosynthetic process as well as atmospheric nitrogen fixation property. They occur in a wide range of ecologically extreme habitats which prove their very high order capacity of adaptability. Algae is known for its ability to degrade and use organic substances natural or synthetic as source of nutrients or very rarely for energy, out of which Linear Alkyl Benzene Sulphonate (LABS) is a very important detergent compound used worldwide. LABS is an anionic surfactant and is biodegradable. It enters lakes, rivers and oceans as component of household sewage (around 1 to 20 ppm) and makes conditions stressful for algae. Many cyanobacterial organisms respond to this and degrade LABS during their normal growth cycle. In the present study, we have studied growth rate percentage, biodegradation potential and heterocyst frequency of pure cultures of a cyanobacterium *Nostoc muscorum* Ag. ex. Born et Flah using 3, 10, and 15 µg/ml LABS concentration at regular time interval of five days starting from 0 to 30 days. Heterocyst frequency of exponentially growing cultures were determined by calculating an average of five fields at five days interval. *Nostoc muscorum* was found to degrade LABS efficiently up to 88.33%, 96.22% and 96.89% at 3, 10 and 15 µg/ml concentration respectively after 25 days which would help toward pollution monitoring and maintenance of water reservoirs used for municipal supply.

Keywords: Biodegradation potential, cyanobacteria, detergent, growth rate percentage, surfactant, water self-purification.

INTRODUCTION

Cyanobacteria are the simplest members of group - algae, single celled, forming most diverse group of photosynthetic prokaryotes. They are referred to in literature by various names, chief among which are Cyanophyta, Cyanobacteria, Blue-green algae, Myxophyta and Cyanochloronta. Cyanobacteria are common in all types of habitats ranging from aquatic to terrestrial ones, often in environments where no other vegetation can exist. They can live in extremes of temperatures ranging –60°C to 85°C. The majority of blue-greens are aerobic photoautotrophs. At the onset of nitrogen limitation, certain cells evolve into heterocysts, which convert nitrogen gas into ammonium, which is then distributed to the neighbouring cells of a filament (Echlin,1966; Fay, 1983). The filamentous heterocystous cyanobacteria like our test organism, *Nostoc muscorum*, are endowed with property of being photosynthetic as well as aerobic nitrogen-fixers. *Nostoc* species form olive-green coatings and are more frequent about the water line. Cyanobacteria can cope up stressful conditions such as waters polluted with linear alkyl benzene sulphonate (LABS).

LABS are the most widely used surfactant in detergents and cleaning products, in both liquid and powder preparations and for household and industrial use. LABS are anionic surfactants with molecules characterized by a hydrophobic and a hydrophilic (polar) group. LABS are non-volatile compounds produced by sulphonation of linear alkyl benzene (LAB) with sulphur tri-oxide (SO_3) or sulphuric acid (H_2SO_4). The corresponding sulphonic acid is subsequently neutralized with an alkali such as caustic soda. The product is generally used in detergents in the form of the sodium salt. After use such detergent compounds are discharged into the environment in waste water and household sewage. For anionic surfactants in general, the most important compartments are sewage water treatment plants, surface waters, sediment- and sludge-amended soils, and estuarine and marine environments (Brunner et al., 1988; De Henau et al., 1986).

This compound LABS is biodegradable (McAvoy et al., 1993). It was introduced in 1960's as more biodegradable replacements for highly branched alkyl benzene sulphonates. Studies on aerobic biodegradation of LABS can be divided into primary and ultimate. Primary degradation of LABS occurs during the initial reactions in the metabolic pathway, and products are often shorter-chain homologues (Okpokwasili and Nwabuzor, 1988). The ultimate degradation of LABS is that of the entire molecule to its biodegradation end-products CO_2, H_2O and NH_4. Both biodegradation (primary and ultimate) and adsorption occur, resulting in decreased environmental concentrations and bio-availability. Reduction in chain length and loss of the parent structure both result in compounds that are less toxic than the parent compound (Nielsen and Huddleston, 1981; Berna et al., 1989). The rate of biodegradation depends on temperature. Biodegradation is rapid between 10° and 25°C; at lower temperature, biodegradation kinetics are reduced (Hollis et al., 1976).

LABS can be analyzed by non-specific methods. The primary method is assay as methylene blue-active substances (MBAS). The procedure developed by Manivasakam, 1996 is considered to be the best for determination of MBAS in aqueous samples. The sensitivity of this procedure is such that levels of 0.01 to 0.02 mg/l MBAS can be determined. The methylene blue reaction responds to any compound containing anionic center and a hydrophobic center, because such compounds tend to form an extractable ion pair when they combine with cationic dyes such as methylene blue.

Our findings have sought to identify the species capable of degrading detergent compounds. Surfactants produce negative and sometimes stimulatory effects also on cyanobacteria, diatoms, plant seedlings and green algae. Scanty work and literature is available on the effects and biodegradation of LABS on algal (oxygenic) organisms (Kerterz et al., 1994; Mampel et al., 1998), yet lot of literature is available in terms of LABS degradation, analytical studies, chemical nature, etc (Matthijs and De Henau, 1987; Schleheck et al., 2004).

In this paper we have examined the growth rate percentage, biodegradation potential and heterocyst frequency of *Nostoc muscorum* Ag. ex. Born et Flah using LABS at 3, 10 and 15 μg/ml concentration.

MATERIALS AND METHODS

In the present study we have taken sodium dioctyl sulpho succinate (Manoxol OT) as the reference material of LABS to observe its effect on growth of test organism, *Nostoc muscorum* under culture conditions. Initially growth experiments were set using various concentrations of sodium dioctyl sulpho succinate and it was found that higher concentrations than 20 μg/ml were inhibitory to the growth of test organism but concentration lower than this supported measurable growth which was also used to calculate biodegradation potential of test organism in terms of its growth rate percentage and total chlorophyll-*a*. The cultures were grown in Allen and Arnon's nitrate free medium (Kaushik, 1987). Prior to experiments, the cultures were made bacteria-free by following standard manuals and axenic cultures of the test organism were maintained and were used as inoculums for further experiments. The average temperature maintained was 28 ± 2°C and white fluorescent light at a distance of 30 cm and intensity of 1800 lux was provided to growing cultures. Culture media and glassware were autoclaved every time before use at 1.1 kg/cm^2 for 15 minutes atleast. The growth period of the cyanobacteria was taken as 18 days. Exponentially growing cultures were homogenized by sonication and used for inoculation in culture tubes (125 × 25 mm). The cyanobacterial sample was centrifuged keeping the supernatant for determination of LABS concentration using the process as indicated and pellet was kept for pigment extraction at 4°C for 24 hours and then chlorophyll absorbance was recorded at 630 nm, 645 nm, and 665 nm by UV-VIS spectrophotometer systronics-117 using following formula (Kaushik, 1987).

Total growth (Chl.a) = $C_a \times V_a/V_c$ mg/l

Where,

C_a = amount of Chl-*a*

V_a = Volume of 80 % acetone

V_c = Volume of culture concentrate

$(C_a = 11.6\ OD_{665} - 1.31\ OD_{645} - 0.14\ OD_{630})$

And the growth rate percentage was calculated using this formula-

Growth Rate Percentage = $dx/dt \times 1/x \times 100$

Where,

dx = change in growth (final - initial growth)

dt = time (days)

x = initial growth

For determination of concentration of sodium dioctyl sulpho succinate, Longwell and Maniece Methylene Blue Method (Manivasakam, 1996) was used which is a quite sensitive test for quantitative measurement and gives blue colour by reacting with anionic surfactants. Absorbance of the samples were recorded at 652 nm using control as blank. The amounts were determined using standard curves and the results were expressed in mg surfactant µg/ml as sodium dioctyl sulpho succinate into the culture medium, obtained as supernatant liquid after centrifugation. Heterocyst frequency of exponentially growing cultures were determined by calculating an average of 5 fields under microscope at five days interval. Percentage heterocyst frequency as represented indicates number of heterocysts per 100 vegetative cells.

RESULTS

Initial growth experiments were set to see the effect of different concentrations of LABS in the range of 0.1 µg/ml to 20 µg/ml using test organism *Nostoc muscorum* for growth period of 18 days. It was observed that there was negative correlation between growth and concentrations of LABS used. At 3 µg/ml, growth rate decreased at a rapid pace and it exhibited the same trend upto 10 µg/ml of LABS concentration. Beyond this level, maximum growth rate was exhibited at 15 µg/ml of LABS and decline thereafter regularly and at highest level of LABS tested i.e. 20 µg/ml no growth was recorded due to toxicity of LABS (Fig. 1). At second stage taking 15 µg/ml as sub-lethal concentration of LABS, growth rate percentage and biodegradation potential of test organism was recorded at regular interval of 5 days. It was found that at the concentration of 3, 10 and 15 µg/ml, the growth rate percentage increased significantly upto 10 days and later decreased gradually upto the last day of observation, i.e. upto 30th day (Fig. 3, 4 and 5). In terms of LABS degradation, the amount of remaining LABS was least after 30 days in 3 µg/ml and in 10, 15 µg/ml concentration it was so after 25 days that was indicative of degradation potential of *Nostoc muscorum* with reference to time (Fig. 3, 4 and 5).

An increase in heterocyst frequency was observed in control as well as all the concentrations tested, being highest after 15 days in control, after 10 days in 3, 15 µg/ml; and 5 days in 10 µg/ml. A gradual decrease was then observed up to 20-25th day in all the concentrations including control, after which a rapid increase in heterocyst frequency was seen up to the last day of observation i.e., 30th day (Fig 2).

DISCUSSION

Wide differences in the ability to degrade surfactants may occur within a particular genus as evidenced by our results. Also, high concentration of surfactant may increase degradation to a greater degree than growth. Also, at toxic levels of surfactants no degradation may occur and inoculum is killed.

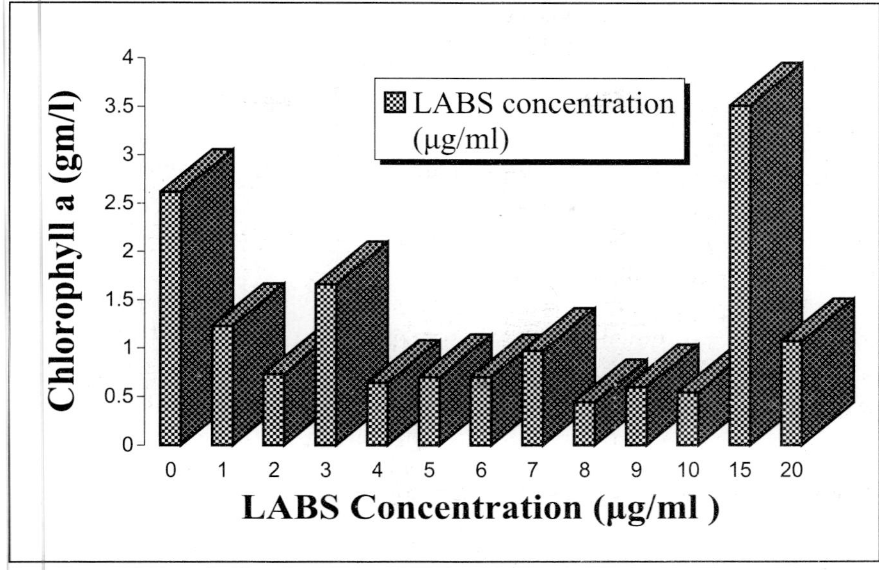

Fig. 1 LABS concentration and algal growth.

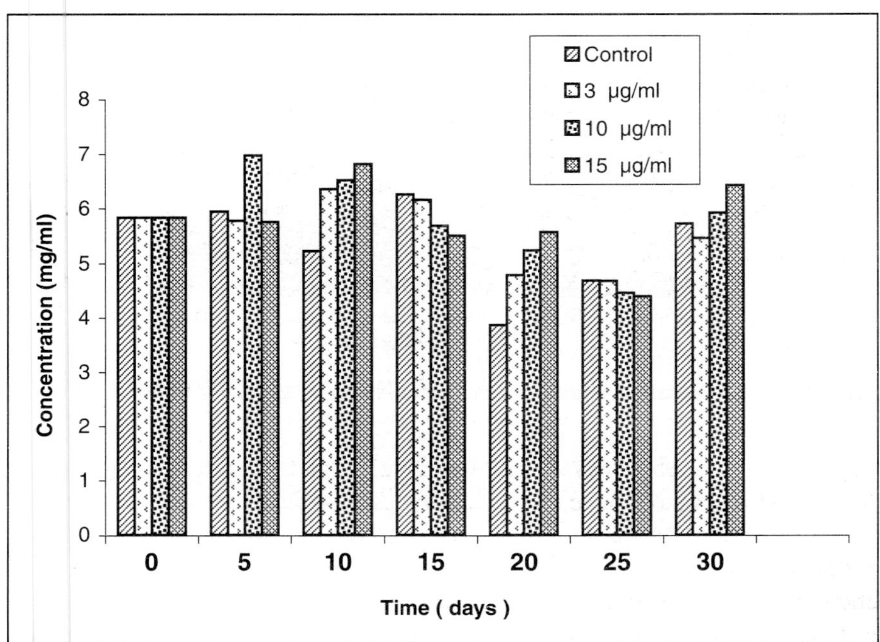

Fig. 2 Heterocyst frequency of *Nostoc muscorum*.

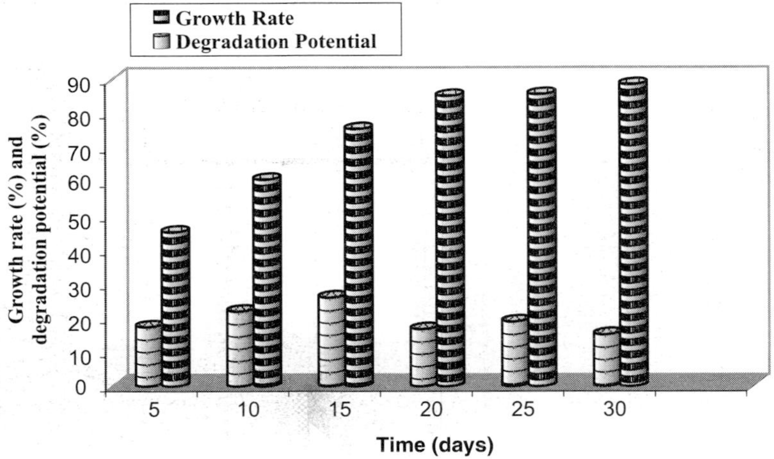

Fig. 3 Growth rate percentage and degradation potential at 3 μg/ml LABS concentration.

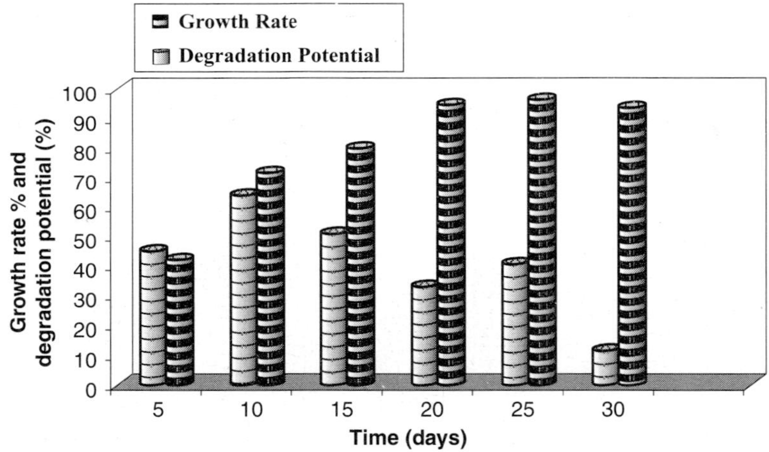

Fig. 4 Growth rate percentage and degradation potential at 10 μg/ml LABS concentration.

Heterocyst frequency show more or less similar trend in control as well as in the LABS concentration tested throughout the experiment. It indicates that heterocyst frequency is not much affected by the presence of LABS. The cyanobacteria have an ability to degrade a detergent as well as nitrogen-fixing. The nitrogen fixation using spent detergent as an energy source in conjunction with biochemically complementary species is an attractive idea from the ecological point of view. *Nostoc muscorum* was found to degrade LABS efficiently up to 88.33%, 96.22% and 96.89% at 3, 10 and 15 μg/ml concentrations respectively after 25 days which would help toward pollution monitoring and maintenance of water reservoirs.

So, it can be used as an agent for water self-purification processes in municipal water bodies contaminated with detergents, chemicals and solvents. The results obtained are under laboratory conditions but their relative importance in soil, water and sewage need to be assessed.

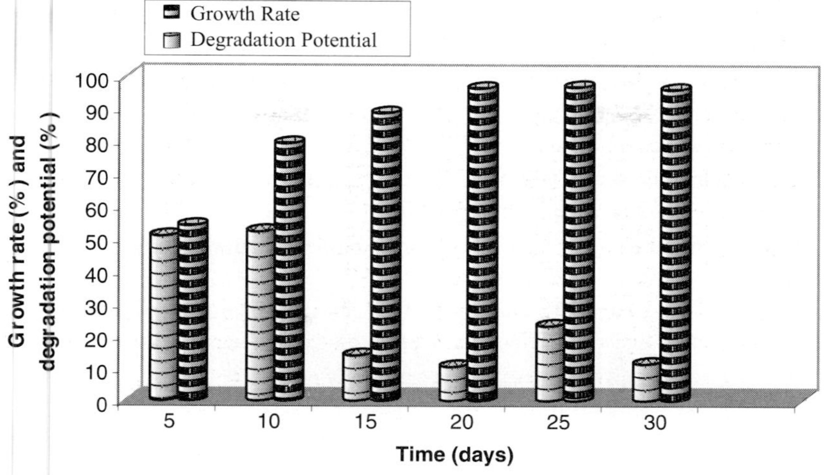

Fig. 5 Growth rate percentage and degradation potential at 15 μg/ml LABS concentration.

ACKNOWLEDGEMENTS

We are grateful to CSIR for their financial help. The authors are also thankful to the Department of Botany, University College of Science, Udaipur for their support.

REFERENCES

Berna, J. L., Ferrer, J., Moreno, A., Prats, D., and Ruiz, B. F. 1989. The fate of LAS in the environment. *Tenside Surfactants Deterg*. 26: 101-107.

Brunner, P. H., Capri, S., Marcomini, A., and Giger, W. 1988. Occurrence and behaviour of linear alkyl benzene sulphonates, nonylphenol diethoxylates in sewage and sewage sludge treatment. *Water Res,* 22: 1465-1472.

De Henau, H., Matthijs, E., and Hopping, W. D. 1986. Linear alkylbenzene sulphonate (LAS) in sewage sludges, soils and sediments : Analytical determination and environmental safety considerations. Int *J. Environ. Anal. Chem.* 26: 279-293.

Echlin, P. 1966. The blue-green algae. *Scientific American.* 214 (6): 74-81.

Fay, P. 1983. The Blue-greens (Cyanophyta-Cyanobacteria). Edward Arnold (Pubs.), Baltimore, MD.

Hollis, C. G., Philipps, W. K. and Johnson, J. E. 1976. Thermal effects on biodegradation of linear alkyl benzene sulfonate. Dev Ind Microbiol. 17: 345-350.

Kaushik, B. D. 1987. Laboratory methods for blue-green algae. Associated Publishing Company, New Delhi, India, pp. 31-62.

Kertesz, M. A., Kalbener, P., Stockinger, H., Beil, S. and Cook, A. M. 1994. Desulfonation of linear alkyl benzene sulfonate surfactants and related compounds by bacteria. *Appl. Environ. Microbio.* 60: 2296-2303.

Mampel, J., Hitzler, T., Ritter, A. and Cook, A. M. 1998. Desulfonation of biotransformation products from commercial linear alkyl benzene sulfonates (LAS). *Environ. Toxicol. Chem.* 17: 1960-1963.

Manivaskam, N. 1996. Physico-chemical Examination of Water, Sewage and Industrial Effluents. Pragati Prakashan, Meerut, India. 201-203 pp.

Matthijs, E. and De Henau, H. 1987. Determination of LAS. *Tenside Surfactants Deterg*. 24: 193-199.

McAvoy, D. C., Eckhoff, W. S. and Rapaport, R. A. 1993. Fate of Linear alkyl benzene sulfonate in the environment. *Environ Toxicol Chem.* 12: 977-987.

Nielsen, A. M. and Huddleston, R. L. 1981. Ultimate biodegradation of linear alkyl benzene sulfonate alkyl and ring carbon. *Dev Ind Microbiol.* 22: 415-424.

Okpokwasili, G. C. and Nwabuzor, C. N. 1988. Primary biodegradation of anionic surfactants in laundry detergents. *Chemosphere.* 17 : 2175-2182.

Schleheck, D., Knepper, T. P., Fischer, K. and Cook, A. M. 2004. Mineralization of individual congeners of linear alkyl benzene sulfonate by defined pairs of heterotrophic bacteria. *Appl. Environ. Microbio.* 70 (7): 4053-4063.

20

Astaxanthin Production from Green Alga *Haematococcus* under Autotrophic and Heterotrophic Conditions

R. Sarada, B. Sandesh Kamath, R. Vidhyavathi, B.R. Brinda and G.A. Ravishankar
Plant Cell Biotechnology Department
Central Food Technological Research Institute, Mysore, India-570 020

ABSTRACT

Haematococcus is one of the potent sources for production of astaxanthin, a ketocarotenoid. Owing to its high antioxidant activity, astaxanthin gained nutraceutical and pharmaceutical importance besides its use as pigmentation source in farmed aquaculture. *Haematococcus* is capable of growth under autotrophic and heterotrophic conditions. In the present study, a comparison was made using different nitrogen sources under autotrophic and heterotrophic conditions. Various amino acids such as sulfur containing, hydroxy, aromatic, cyclic and hydrocarbon type were used to grow *H. pluvialis* under heterotrophic conditions. Among the amino acids, isoleucine, threonine showed high cell count compared to L-asparagine (control), while low cell count was observed in tyrosine, tryptophan and histidine did not support the growth. The biomass yields, total carotenoid and astaxanthin content and production differed among various amino acids. Autotrophic medium containing ammonium salts of chloride, carbonate, sulphate and acetate, urea and commercial salts such as suphala and diammonium phosphate(DAP) as nitrogen sources were used to grow *Haematococcus* in ambient and CO_2 enriched atmosphere. In the presence of CO_2, *H. pluvialis* exhibited enhanced growth in all the ammonium salts with a biomass yield of 2.6-2.8 g/l and astaxanthin productivity of 46-48 mg/l. Urea as nitrogen source did not support growth of *Haematococcus* and astaxanthin production.

Keywords: Ammonium salts, astaxanthin, autotrophic, *Haematococcus pluvialis*, total carotenoid, amino acids, heterotrophic.

INTRODUCTION

Haematococcus pluvialis, the freshwater green alga produces a ketocarotenoid — astaxanthin. In recent years, *H. pluvialis* has gained commercial importance owing to its ability to accumulate high concentration of astaxanthin (2-3% on dry weight basis) and its approval by FDA-USA and some European countries for its use in humans as dietary food supplement shows its potential applications (Lorenz and Cysewski, 2000). Astaxanthin production has been achieved under salt stress (Sarada et al., 2002), oxidative and light stress (Kobayashi et al., 1992; 1993), flagellated cell (Brinda et al., 2004). In the present study, evaluation of *Haematococcus* growth and astaxanthin production in autotrophic and heterotrophic conditions under the influence of various organic and inorganic nitrogen sources was carried out.

MATERIALS AND METHODS

Cultivation of Algae

Haematococcus pluvialis (SAG 19-a) was obtained from Sammlung von Kulturen, Pflanzen Physiologisches Institut, Universitat Gottingen, Gottingen Germany and was maintained in heterotrophic, as well as autotrophic liquid and slant cultures (Usha et al., 1999). Cultures were incubated at 25 ± 1°C temperature under continuous light intensity of 2.2058×10^7 erg m^{-2} s^{-1}. A 4-day old culture was used as inoculum for all the experiments.

Bold's basal medium (BBM, Kanz and Bold, 1969) consisting (g/L) of 0.074, K_2HPO_4; 0.175, KH_2PO_4; 0.073, $MgSO_4.7H_2O$: 0.249, $NaNO_3$; 0.024, $CaCl_2.2H_2O$; 0.025, NaCl; 0.045, EDTA; 0.005, $FeSO_4.7H_2O$ and 1 ml of trace elements per liter of distilled water. The trace elements consisted of salts of cobalt, copper, molybdenum, nickel, iodine at 0.05 µm, vanadium, tungsten, chromium at 0.01 µm, zinc at 0.1 µm and manganese at 1.0 µm level. The pH was adjusted to 7.0.

Heterotrophic medium (Kobayashi et al., 1991) containing (g/L) of 1.986, sodium acetate; 0.405, L-asparagine; 2.0, yeast extract; 0.2, $MgCl_2.6H_2O$; 0.01, $FeSO_4.7H_2O$; 0.02, $CaCl_2.2H_2O$.

Comparison of *Haematococcus* Growth in Autotrophic and Heterotrophic Media

Haematococcus culture was inoculated in both autotrophic medium with ambient carbon dioxide and also at 2% carbon dioxide atmosphere in two-tier flasks as detailed below and heterotrophic medium independently and the growth was monitored by following cell count at regular intervals.

Influence of ammonium salts on the growth of *Haematococcus* and astaxanthin production in autotrophic medium

Bold's basal medium (BBM) containing ammonium chloride (0.25 – 1.0g/l, equivalent to 4.67 – 18.68 mM in terms of nitrogen), urea (0.15 – 1.0 g/l or 4.99-33.267 mM) ammonium carbonate (0.15 – 1.0 g/l or 2.86 – 19.067 mM), ammonium sulphate (0.15 – 1.0 g/l or 2.27 –15.33 mM) and ammonium acetate (0.25 – 1.0 g/l or 3.24 – 12.96 mM) as nitrogen sources were independently prepared and *Haematococcus* cells in vegetative phase was inoculated so as to have an initial cell count of 12×10^4 cells/ml. BBM with sodium nitrate was considered as control. BBM contains 0.249 g/l of sodium nitrate, which is equivalent to 2.93 mM of nitrogen. The experiments were carried out in 150 ml conical flasks containing 50 ml medium. Effect of ammonium salts, urea and commercial salts such as suphala (nitrogen, phosphorus, potassium mixture), diammonium phosphate (DAP) was studied on *H. pluvialis* growth and astaxanthin production in presence of 2% carbon dioxide using two-tier flasks. While adding commercial salts, the concentration of nitrogen and phosphate were estimated using APHA (1989) and AOAC procedures (1999) accordingly the amounts were adjusted. Suphala was obtained from Rashtriya Chemicals and Fertilizers Ltd. Mumbai and DAP from Gujarat State Fertilizers Ltd. Gujarat, India.

The two-tier vessel consisting of two 250 ml narrow neck Erlenmeyer flasks (Husemann and Barz, 1977) was used for enriching CO_2 in the culture environment. The lower compartment of the flask contained 100 ml of 3 M buffer mixture ($KHCO_3/K_2CO_3$) at specific ratio, which generated a partial pressure of CO_2 at 2% in the two-tier flask (Usha et al., 2001). The upper chamber contained 50 ml of Bold's basal medium with different ammonia salts. A 10-day old autotrophic culture was used as inoculum for all the experiments. The mouths of the upper and lower compartments were sealed tightly using rubber caps and parafilm. All the flasks were incubated at 26 ± 1°C in a culture room under light intensity 2.2058×10^7 erg m^{-2} s^{-1}. After two weeks of growth, one set of flasks were harvested for cell count and biomass estimation while second set of flasks were incubated further for two weeks under high light 5.1468×10^7 erg m^{-2} s^{-1} for carotenoid accumulation. Data was represented as an average of three replicates in two repetitive experiments.

Influence of Different Amino Acids on Growth and Astaxanthin Production in *Haematococcus*

Heterotrophic medium (Kobayashi et al., 1991) containing amino acids such as leucine, isoleucine, valine, methionine, threonine, serine, tyrosine, tryptophan, histidine, and proline as nitrogen sources were independently prepared and inoculated at 10% level. The concentration of the individual amino acids was calculated in terms of nitrogen equivalent to L-asparagine in the heterotrophic medium. After 8 days of growth at the above-mentioned conditions, culture flasks were subjected to stress by adding 0.25% NaCl along with 14 mM sodium acetate and were incubated at high light intensity of 5.1468×10^7 erg m^{-2} s^{-1} for a period of 12 days. At the end of 12 days the cultures were harvested and biomass, total carotenoid and astaxanthin contents were estimated. All the experiments were carried out in triplicates.

Growth Measurement

Growth was monitored by cell count using haemocytometer and also on a dry weight basis. Known volume of culture was centrifuged, the pellet was washed with distilled water and the pellet was dried in oven at 60°C till constant weight was attained.

Estimation of Total Carotenoid and Astaxanthin

Carotenoids were extracted from known quantity of biomass with acetone and analyzed spectrophotometrically by the method of Lichtenthaler (1987) by measuring absorbance at 450, 470 and 480 nm. Astaxanthin was determined at 480 nm using an absorption coefficient, A1% of 2500 by the method of Davies (1976).

RESULTS

Growth Profile in Autotrophic and Heterotrophic Media

The growth phase of *Haematococcus* extended to 2-3 weeks at relatively slow growth rate under autotrophic conditions as shown in Fig. 1. The growth rate is significantly high in autotrophic medium when 2% carbon dioxide was supplemented and a 4-fold increase in cell count was obtained (Fig. 1). Under heterotrophic conditions the algal growth rate was fast and for a short duration (4-6 days). Maximum cell count of $41 - 44 \times 10^4$ cells /ml was observed on 5th day after inoculation. In heterotrophic medium due to limitation of nutrients, the cells enter into stationary phase and undergo encystment accompanied by carotenoid accumulation.

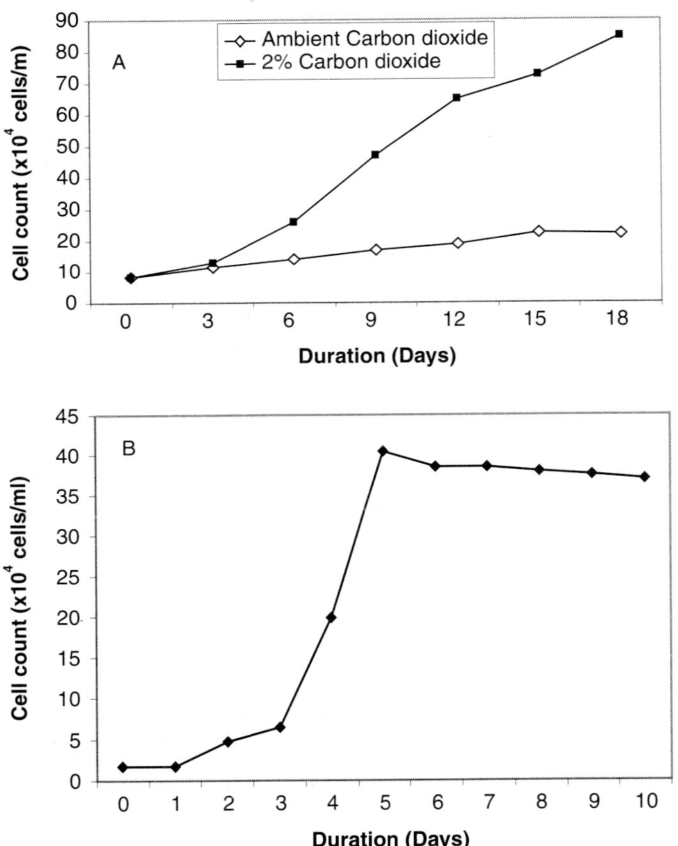

Fig. 1 Growth profile of *H. pluvialis* in autotrophic (A) and heterotrophic (B) medium.

Effect of Ammonia Salts on *H. Pluvialis* Growth

Growth of *H. pluvialis* under the influence of ammonium chloride, ammonium carbonate, ammonium sulphate and urea at ambient CO_2 are shown in Fig. 2.

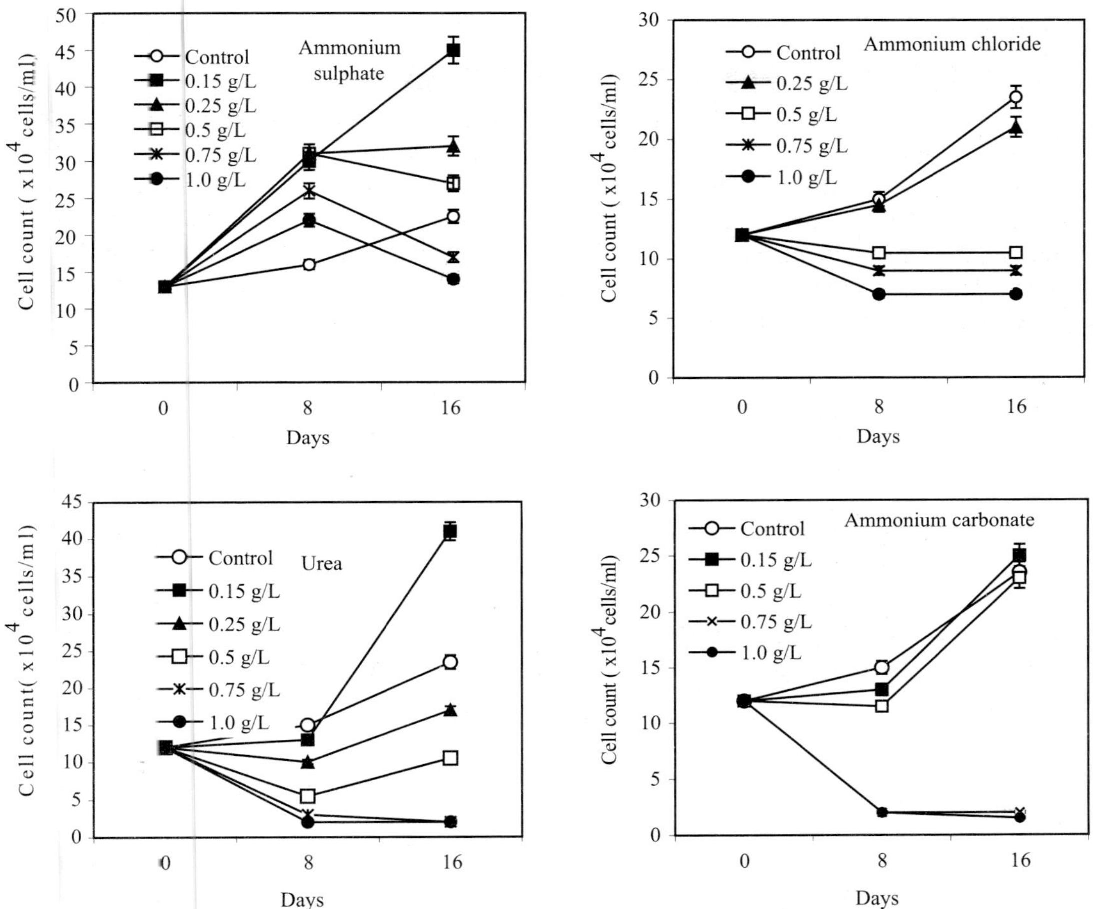

Fig. 2 *H. pluvialis* growth profile in different concentrations of ammonium salts at ambient CO_2 level.

The cells were found to be motile upto 48 hours at all concentrations of ammonium chloride tested, and later at higher concentrations, 50% of the cells lost motility. Only at 0.25 g/l concentration the growth in terms of cell count (21×10^4 cells/ml) was comparable to control and at higher concentrations the cell count per ml decreased (Fig. 2). Ammonium carbonate supported growth comparable to control at 0.5g/l concentration, while decrease in the cell count was observed at higher concentration.

Treatment of cells with ammonium acetate rendered them immotile and cell lysis was observed within 48 h. Therefore it was not found suitable for growth while in ammonium sulphate fed cultures the cell count was found to be 2- to 3-fold high compared to control at 0.15g/l concentration, however at higher concentrations, growth was inhibited. At 0.15 g/l level of urea, growth was comparable to control. At higher concentrations of urea (0.5 – 1.0 g/l), the cells lost motility and there was significant decrease in the growth.

Synergistic Effect of Ammonia and CO_2 on *H. Pluvialis* Growth and Astaxanthin Production

The results obtained for *H. pluvialis* growth with different ammonium salts in CO_2 enriched environment are shown in Table 1 and Fig. 3A. Cell count of 80 – 100 × 10^4 cells/ml was obtained with different ammonia salts (ammonium carbonate, ammonium acetate) which is 8 – 10 fold increase over initial cell count (Table 1) when compared to only 2-4 fold increase in cell count observed in the absence of CO_2 enriched medium (Fig. 1). The ammonia concentration in the *Haematococcus* culture was found to be decreased to 9% of the initial level, which shows 90% of NH_3 utilization. The cells grown in urea, turned brownish yellow colour after 6 days indicating early induction of carotenoid formation. The second set of flasks, which were exposed to high light, encysted and accumulated carotenoids.

Table 1 Total carotenoids an astaxanthin content in *H. pluvialis* cells grown in different ammonium salts in the presence of 2% CO_2.

	Cell count (× 10^4cells/ml)	Total carotenoids (% w/w)	Astaxanthin (% w/w)
BBM	80	2.27	1.97
Ammonium carbonate (0.15 g/l)	94	2.12	1.85
Ammonium acetate (0.25 g/l)	100	1.57	1.36
Ammonium sulphate (0.25 g/l)	55	0.43	0.36
Urea (0.15 g/l)	77	1.02	0.82
Suphala (1.0 g/l)	64	2.36	2.1
Diammonium phosphate (0.25 g/l)	82	2.12	1.87

Initial cell count – 12 × 10^4 cells ml^{-1}. Data recorded after 4 weeks

The encysted biomass yield obtained was found to be 2.6 – 2.8 g/l which was comparable to control (Fig. 3A) except in urea (0.15 g/l level) and ammonium sulphate (0.15 g/l level) where the biomass yield was significantly less. Reduction in biomass was to the order of 21% in ammonium carbonate. In the case of DAP there was 20% increase in biomass production compared to control. Total carotenoid content in all treatments was found to be in the range of 2.0 – 2.2 % (Table 1) except in ammonium acetate (1.57%) and urea (1.02%). However, the astaxanthin content was in the range of 85 – 88% of the total carotenoid in all the cultures.

Astaxanthin productivity which is sum of astaxanthin content and biomass yield was high in BBM and commercial salts (46 – 48g/l) followed by ammonium carbonate and ammonium acetate and least in urea (Fig.3B). Total carotenoid and astaxanthin productivities were found to 15 – 13% be more in DAP culture when compared to control and suphala cultures respectively.

Effect of Amino Acids on Growth and Astaxanthin Production in Heterotrophic Conditions

The initial cell count in the heterotrophic medium was maintained at 2.2 × 10^4 cells /ml. The cell count in different amino acids varied and the multiplication rate in 4 days duration ranged between 13-25 times in different amino acids studied. Growth in terms of cell count on 4th day

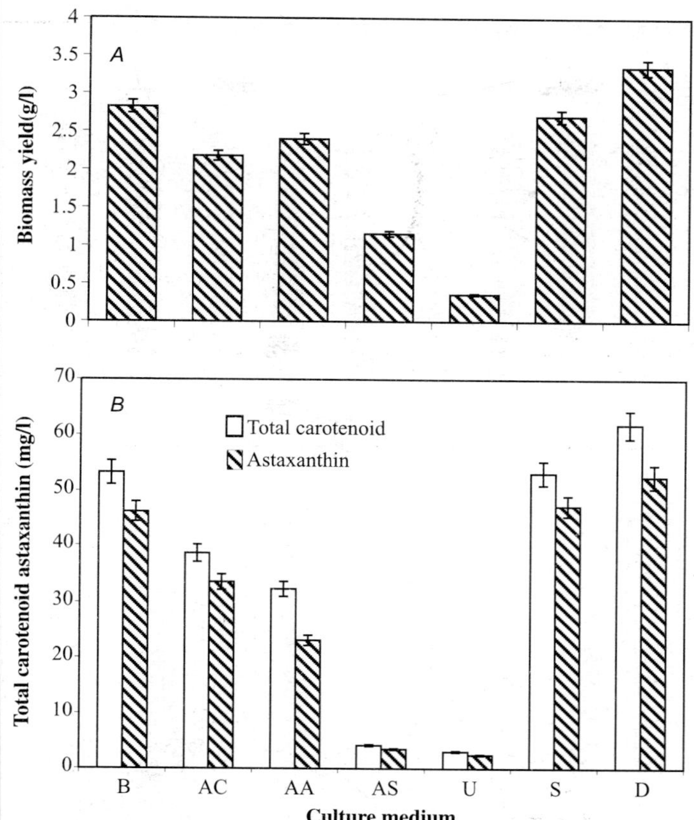

Fig. 3 Biomass yield of *H. pluvialis* (A) and total carotenoids and astaxanthin production (B) in presence of 2% CO_2 B-BBM, AC- Ammonium carbonate(0.15g/l), AA-Ammonium acetate(0.25g/l), AS-Ammonium sulphate(0.25g/l), U-Urea(0.15g/l), S- Suphala(1.0g/l) D- Diammonium phosphate(0.25g/l).

was found to be similar in leucine, methionine and 54.5 × 10^4 valine (45 to 49 × 10^4 cells/ml) while it was marginally high in isoleucine and threonine (53 to 54.5 10^4cells/ml) compared to L-asparagine (37.5 × 10^4 cells/ml; Table 2). Low cell count (29 to 30 × 10^4 cells /ml) was observed in tyrosine and tryptophan. The organism was not able to utilize histidine at the tested concentration and the cells were bleached. The biomass yield in all amino acids was in the range of 0.74 to 0.96 g/L except L-asparagine during the experimental period.

Total carotenoid and astaxanthin content and productions varied in amino acids. High content of total carotenoid and astaxanthin (1.41 %w/w; Table 2) was obtained in asparagine while lowest was observed in tryptophan followed by leucine and proline. Total carotenoid and astaxanthin production was found to be 7.8 to 7.7 mg/L in asparagine, isoleucine and threonine cultures (Fig. 4). Irrespective of the productivities of total carotenoids, all amino acid cultures showed high content of astaxanthin (90 – 95%, Table 2) under high light and stress conditions except for serine and tyrosine where it was 84%.

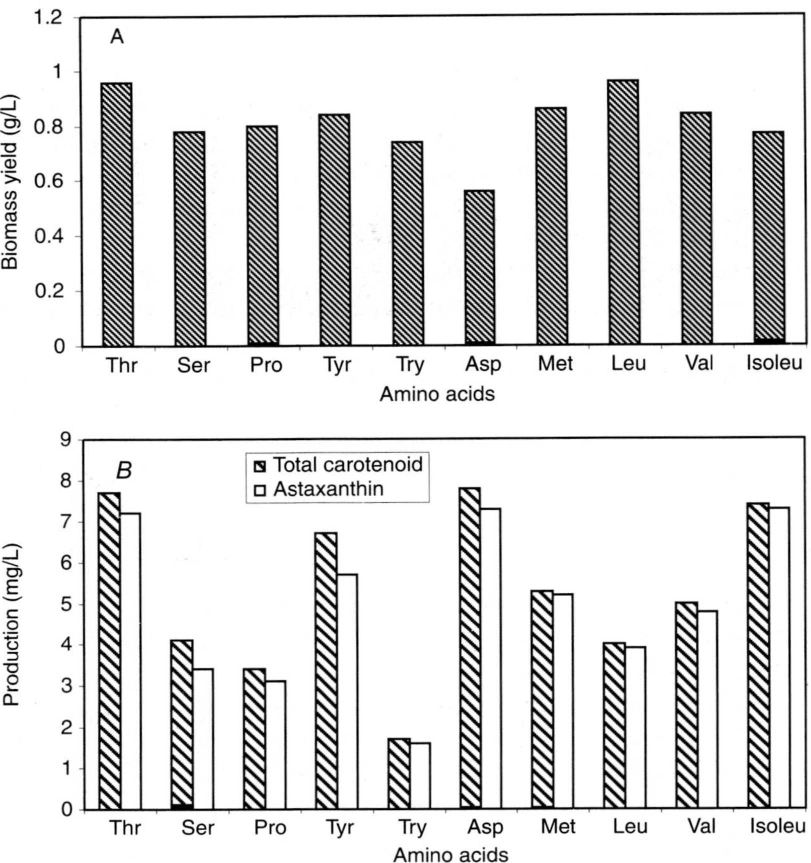

Fig. 4 Biomass yields (A) and Carotenoid production(B) in *H. pluvialis* grown in different amino acids under stress condition.

Table 2 Total carotenoids an astaxanthin content in *H. pluvialis* cells grown in different amino acids

	Cell count ($\times 10^4$ cells/ml)	Total carotenoids (% w/w)	Astaxanthin (% w/w)
Tyrosine	29	0.8	0.68
Tryptophan	30.5	0.23	0.22
Proline	43.5	0.42	0.39
Serine	51	0.52	0.44
Threonine	54.5	0.8	0.75
Asparagine	37.5	1.41	1.3
Methionine	45	0.62	0.6
Leucine	45	0.42	0.4
Valine	49	0.6	0.57
Isoleucine	53	0.96	0.95

DISCUSSION

Algae in general have the ability to use different organic and inorganic nitrogen compounds as nitrogen source. However some microalgae were shown to be sensitive to ammonia as sole nitrogen source and their growth may be inhibited (Abeliovich and Azov 1976). In the present study different ammonia salts were studied for their influenc in *Haematoccus pluvialis* growth since the information available was scanty and most of the reports (Fabregas et al., 2000, Olaizola, 2000) focused either potassium nitrate or sodium nitrate as nitrogen source under autotrophic conditions. Abeliovich and Azove (1976) reported unionized ammonia is more toxic to various photosynthetic organisms and ammonia among other weak electrolytes, causes swelling and osmotic lysis of the cells in the case of *Prymnesium parvum*. Swelling and cell lysis was also observed in the present study when ammonium acetate and urea were used in the medium at higher concentrations. Sarada et al. (2002) reported slow growth of *H. pluvialis* when ammonium nitrate was used as nitrogen source under heterotrophic conditions. There was a significant increase in cell count (8-10 fold) in cultures grown in presence of CO_2 compared to those in the absence (2-4 fold) with different ammonium salts (Fig. 1 and Table 1). The results indicated that *H. pluvialis* can utilize ammonia salts as nitrogen source in the presence of CO_2 for photoautotrophic growth and it could not utilize ammonium acetate without CO_2. It was of interest to know that the transfer of *Haematococcus* culture to medium containing DAP significantly increased the biomass yield (3.5 g/l) and carotenoid production (65 mg/l) with 2.2% astaxanthin content (Fig. 3B).

The increase in cell count and biomass yields shows that *Haematococcus* is able to utilize all the amino acids studied as nitrogen source. The differences in the biomass yields and carotenoid production in different amino acids in spite of high cell counts (especially in serine and proline) suggests the interrelationship of the various metabolic activities in the cell. Sarada et al., 2002 reported that stress is a cumulative factor of the physiological state of the culture and the culture conditions. Therefore the variations in the yields of biomass and carotenoids under stress conditions in different amino acids is as expected.

In general the amides, urea, glutamine and asparagine are good sources of nitrogen. Neilson and Larson (1980) reported that amino acids such as glycine, serine, alanine, glutamic acid and aspartic acid serve as nitrogen sources. Kirk and Kirk, (1978) showed the existence of specific amino acid carriers in various chlorophyceae and once taken up, all amino acids do not undergo same conversions inside the cells. The differences in the extent of astaxanthin formation observed in this study substantiates the above statement.

CONCLUSION

The results clearly indicated that the organism can tolerate ammonia salts in the range of 3mM - 4.7mM concentration in autotrophic conditions and almost all amino acids as nitrogen sources in heterotrophic conditions. The results also substantiated that cultivation of *Haematococcus* in autotrophic conditions with CO_2 supplementation resulted in higher yields of biomass, total carotenoid and astaxanthin. The significant increase in biomass and astaxanthin production achieved with DAP could be exploited for the batch production system while the stock culture can be maintained in basal medium. In this regard the present results substantiated the use of ammonia as nitrogen source suggesting the possible use of commercial nitrogen mixtures for cultivation of this algae to make the process cost effective.

ACKNOWLEDGEMENTS

The authors acknowledge the financial support from Department of Biotechnology, Government of India, New Delhi. The award of Senior Research Fellowship to BSK and RV by the Council of Scientific and Industrial Research (CSIR), New Delhi is gratefully acknowledged.

REFERENCES

Abeliovich, A., Azov, Y. 1976. Toxicity of ammonia to algae in sewage oxidation ponds. *Appl. Environ. Microbiol.* 31: 801-806.

AOAC 1999. Official methods of Analysis of AOAC international. 16th edn, Vol. 1, Chapter 2, ed. Patricia Cunniff. AOAC international SUITE 500, Maryland USA, pp. 5-12.

Usha, T., Sarada, R., Ravishankar, G. A. 2001. A culture method for microalgal forms using two-tier vessel providing carbon dioxide environment: Studies on growth and carotenoid production *World J. Microbiol. Biotechnol.* 17: 325-329.

Usha, T., Sarada, R., Ramachandra, Rao, S., Ravishankar, G. A. 1999. Production of astaxanthin in *Haematococcus pluvialis* cultured in various media. *Bioresource Technol.* 68: 197-199.

21

Saturated and Unsaturated Hydrocarbon Production from *Botryococcus braunii* sp. from Indian Freshwater Bodies and Culture Collection Centers

R. Sarada, C. Dayananda, A. Ranga Rao, T.R. Shamala[1], P. Srinivas[2] and G.A. Ravishankar

Plant Cell Biotechnology Department, [1]Food Microbiology Department,
[2]Plantation Products, Spices and Flavour Technology Department,
Central Food Technological Research Institute, Mysore, India-570 020

ABSTRACT

B. braunii is recognized as a renewable resource for production of hydrocarbons of fuel value. *B. braunii* strains collected from different places in South Indian freshwater bodies and culture collection centre were purified and established in autotrophic medium. Their growth profiles were recorded and the hydrocarbons produced were analysed by GC and GCMS and identified by their mass spectra, fragmentation pattern and NIST library. Accordingly the cultures were found to belong to A and B races and the hydrocarbon content varied with cultural and environmental conditions in the range of 20 to 50% (w/w) on dry weight basis. Their hydrocarbon production under controlled conditions was higher than that in out door conditions. The 'B' race produced C_{33} and C_{34} botryococcenes and higher than C_{34} and of which C_{34} constitutes 50-70% of total hydrocarbons. *B.braunii* strains of Indian origin were found to produce saturated hydrocarbons in the range of C_{20} to C_{32} and of which 66% of total hydrocarbons are in the range of C_{20} to C_{32}. The organisms produced fatty acids in the range of C_{12} to C_{24} of saturated and unsaturated ones. The B-race produced C18:1 as the major fatty acid followed by C16:0, while A race produced C16:0 as major followed by C18:1, C18:2, C18:3 and C18:0 fatty acids.

Keywords: *Botryococcus braunii*, biomass, hydrocarbons, botryococcenes, fatty acids.

INTRODUCTION

Botryococcus braunii is a colonial green microalga belongs to chlorophyceae, which produces large amounts of liquid hydrocarbons (Maxwell et al.,1968; Brown et al., 1969). *B. braunii* species are grouped into A, B and L races based on the type of hydrocarbons they produce (Metzger et al., 1985, 1986, 1990 and 2005). The B race produces two types of triterpenes called botryococcenes of $C_{30} - C_{37}$ as major hydrocarbons and small amounts of methyl branched squalene (Metzger et al., 1985; Wake and Hillen 1980; Achitouv et al., 2004). *B. braunii* is a promising renewable resource for the production of hydrocarbons and it has been reported that on hydrocracking, the distillate yields 67% gasoline, 15% aviation turbine fuel, 15% diesel fuel and 3% residual oil (Hillen et al., 1982). *Botryococcus braunii* is also known to produce large amount of fatty acids, the quantity and composition of which varies with species and also among the races (Metzger et al., 1990; Grice et al., 1998). The focus of present study is to evaluate growth, hydrocarbon and fatty acid profile of different strains of *B. braunii* to exploit the selected organism for its mass cultivation.

MATERIALS AND METHODS

Algal Strains

Botryococcus braunii strains were obtained from National Institute for Environmental Studies, Tsukuba, Japan. *B.braunii* (N-836); Sammlung von Algen Kulturen, Pflanzen Physiologisches Institut, Universitat Gottingen, Germany *B. braunii* (SAG 30.81) and University of Texas, U.S.A., *B.braunii* (LB 572). Indigenous strains were collected from Kodaikanal and Mamallapuram and the strains are designated as CFTRI-Bb-1 and CFTRI -Bb-2. The stock cultures were maintained both in agar slants and liquid cultures of modified Chu 13 medium (Largeau et al., 1980). The cultures were grown in Erlenmeyer flasks of 150 ml capacity, containing 40 ml modified Chu 13 medium for a period of 4 weeks. The culture flasks were inoculated with 20% (v/v) inoculum and incubated at 25 ±1°C under 1.7647 erg m^{-2} s^{-1} light intensity with 16h: 8h light and dark cycle. The effect of bicarbonate on *B.braunii* strains were studied at 0.25-1.0 gL^{-1}.

Biomass estimation

The algal cultures were harvested and the cells were washed twice with distilled water after centrifugation at 5000 rpm for 10 min. The dry weight of algal biomass was determined gravimetrically and expressed as g L^{-1} (Frenz et al., 1989).

Hydrocarbon extraction

Hydrocarbon was extracted by homogenizing the dry biomass with n-hexane for 30 min and supernatant recovered after centrifugation was evaporated to complete dryness under the stream of nitrogen gas. Hydrocarbon content was measured gravimetrically and expressed as dry weight percentage (Maxwell et al., 1968; Sawayama et al., 1992).

Hydrocarbon Analysis by GC/GC-MS

Hydrocarbon extract was purified by column chromatography on silica gel. The hydrocarbon sample was analyzed on GC/GC-MS (PerkinElmer Instruments, USA) using Elite-5 (cross bond 5% diphenyl-95% dimethyl polysiloxane) capillary column (30 m × 0.25 mm ID 0.25 μm film thickness, PerkinElmer Instruments, USA) with temperature gradient from 130°C (5min)/8°C/to 200°C (2 min)/5°C/to 280°C (15 min) at a flow rate of 1 ml/min carrier gas. The injector and detector port temperatures were maintained at 240°C and 250°C respectively (Dayananda et al., 2005). The mass spectra were recorded under electron impact ionization at 70 eV electron energy with a mass range from 40-600 at a rate of one scan /s. Hydrocarbons were identified by their $M^{+\bullet}$ ions and comparison of the mass spectra with those of the standards as reported in literature (Wake and Hillen 1980). Comparison of the botryococcenes major daughter ions and their relative abundances afforded a high degree of (>96%) match. Further, relative retention data of these compounds conformed with respect to squalene with those reported in literature (Metzger et al., 1985, 1986, 1990).

Fatty Acid Analysis

Lipids were extracted with chloroform - methanol (2: 1) and quantified gravimetrically. The lipid sample was dissolved in benzene and 5% methanolic hydrogen chloride and refluxed for 2 hours. Then 5% sodium chloride solution was added and the fatty acid methyl esters (FAME) were extracted with hexane. The hexane layer was washed with 2% potassium bicarbonate solution and dried over anhydrous sodium sulphate (Christie, 1982). FAME were analyzed by GC-MS (PerkinElmer, Turbomass Gold, Mass spectrometer) equipped with FID using SPB-1 [poly(dimethysiloxane)] capillary column (30 m × 0.32 mm ID 0.25 μm film thickness) with a temperature programming from 130°C to 280°C at a rate of 2°C/min. The FAME were identified by comparing their fragmentation pattern with authentic standards (Sigma) and also with NIST library.

RESULTS AND DISCUSSION

Growth and hydrocarbon profile in different B. braunii strains

The *B. braunii* strains used in this study were established in modified Chu 13 medium; however they differed in their growth and hydrocarbon content (Figure 1). Among the strains studied, LB-572 was found to be the best in terms of biomass yields followed by CFTRI -Bb-1 and 2, SAG-30.81 and N-836. Interestingly the hydrocarbon content in N-836 strain was significantly high (60% w/w) when compared to the other strains, although its biomass yield was low (Fig. 1). The strains SAG- 30.81 and LB-572 were identified as ‘A’ race based on their hydrocarbon profile (Fig. 2), however they differed in their morphology, biomass yields and hydrocarbon productivity (Fig. 1). The two strains (SAG- 30.81 and LB-572) also differed in their hydrocarbon profile as analysed by GC presented in Table 1. The strain N-836 was identified as ‘B’ race on the basis of its hydrocarbons, and found C33 and C34 botryococcenes (Fig. 3 & Table 1) as the major hydrocarbons produced. The CFTRI -Bb-1and 2 produced saturated hydrocarbons in the range of C20 to C32 (Fig. 4) with 15-18% (w/w) hydrocarbon content (Fig. 1). There were reports on *B.*

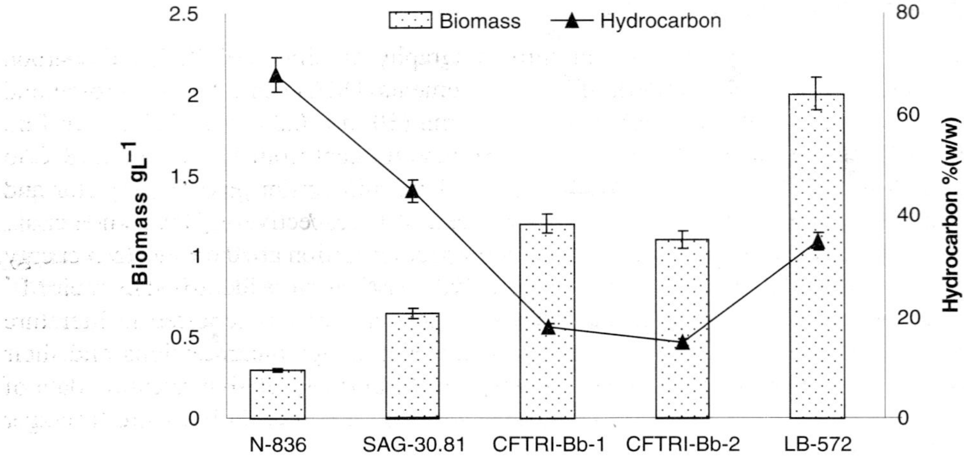

Fig. 1 Biomass and hydrocarbon yields of different strains of *B. braunii.*

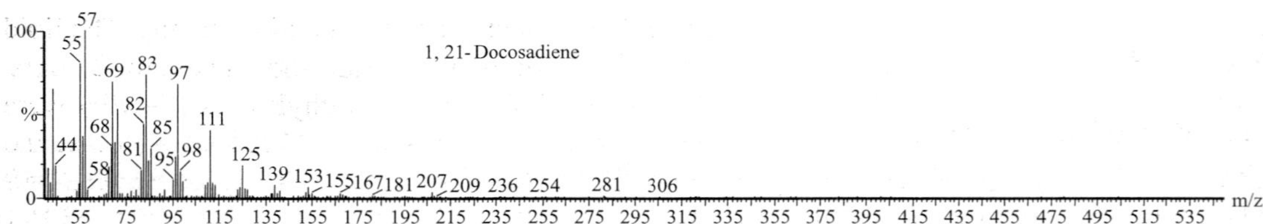

Fig. 2 Mass spectra of unsaturated hydrocarbon produced by the alga *B. braunii* (SAG 30.81 and LB-572) strains.

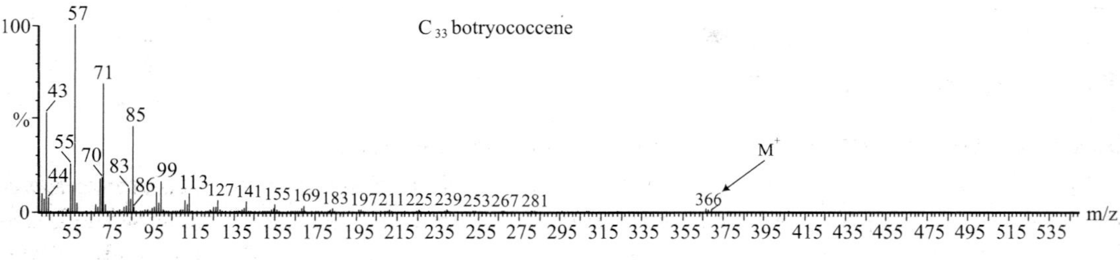

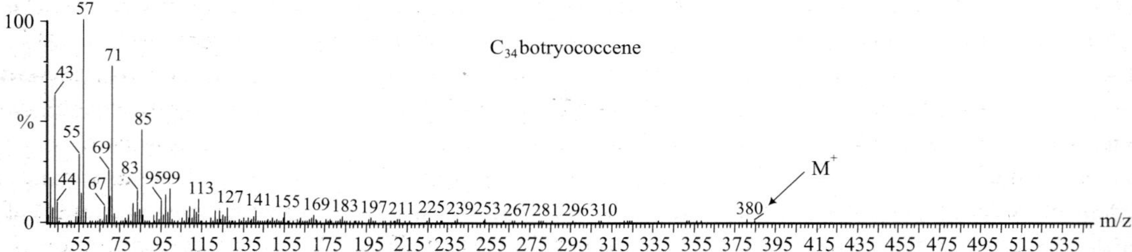

Fig. 3 Mass spectra of C_{33} botryococcene and C_{34} botryococcene hydrocarbons of *B. braunii* (N-836).

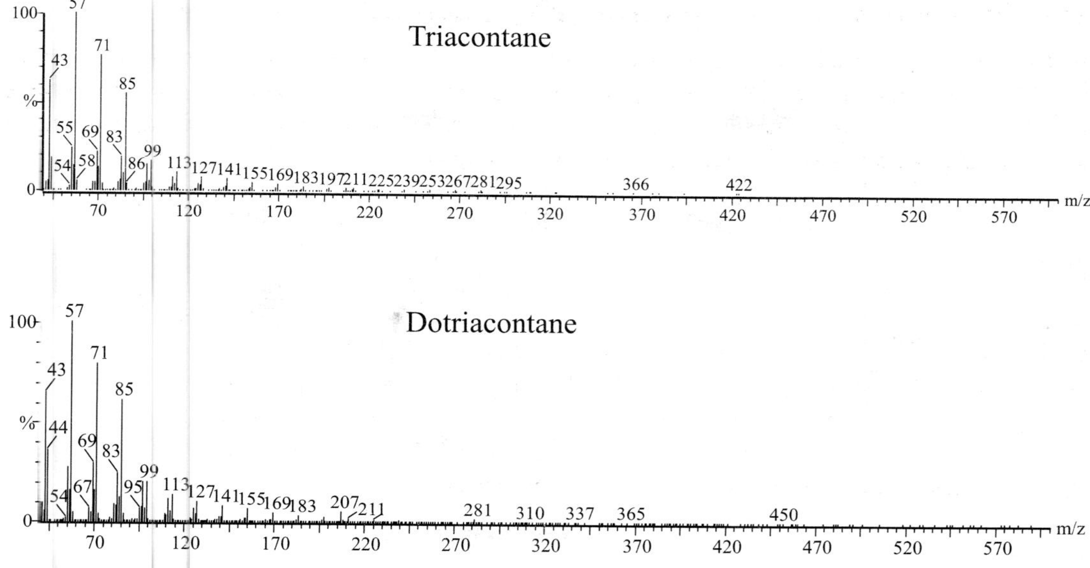

Fig. 4 Mass spectra of saturated hydrocarbons produced by the CFTRI-Bb1 & 2 strains.

braunii producing saturated hydrocarbons in the range of C21 to C31 (Yang et al., 2004). Recently Dayananda et al., 2006 also reported saturated hydrocarbons produced by a botryococcene producing strain of *B. braunii*.

Table 1 Hydrocarbon profile in *B. braunii* grown in autotrophic medium

B. braunii	< C20	C20-C24	C24-C30	>C30
SAG 30.81	3.12	14.52	2.77	79.59
LB-572	2.50	3.30	37.70	56.50
N-836*	2.20	3.35	4.45	90.00
CFTRI -Bb-1	8.94	22.99	2.67	65.4
CFTRI -Bb-2	10.25	23.66	10.50	55.59

*C_{33} and C_{34} botryococcenes constituted the major proportion of hydrocarbons (70±5%) (Identified by GCMS) while around 20±4 % are higher than C_{34} hydrocarbons which are not identified.

The hydrocarbon content varied (15 - 60% w/w) among the different strains studied. Largeau et al., 1980; Brown et al., 1969 reported the variation of hydrocarbon content in the range of 14-40% among the strains, cultural conditions and their natural habitats. Casadevall et al., 1985 have reported 27% (w/w) hydrocarbon under batch culture and 22% under continuous culture in 'A' race of *B. braunii.* Dayananda et al., 2005 reported the 50% of hydrocarbon production by the 'A' race of *B. braunii* under optimized culture conditions. Frenz et al., 1989 have reported 1.5 g/L biomass yield with 21% (w/w) of hydrocarbon production in 'B' race of *B. braunii.* Kojima and Zhang, 1999 have reported higher yields of biomass (7 g/L) with 50% (w/w) hydrocarbon production in bubble column photobioreactor, however this was reported to be debatable (Banerjee et al., 2002).

Effect of Bicarbonate on Growth Profile of *B. braunii* Strains

The strains which are having high growth rates LB-572 and indigenous strains were evaluated for the effect of bicarbonate on their growth and the data obtained is shown in Fig. 5. Of the three strains studied, indigenous strain (CFTRI-Bb-2) was able to grow at all the tested concentrations of bicarbonate while LB-572 was found to be not adoptable.

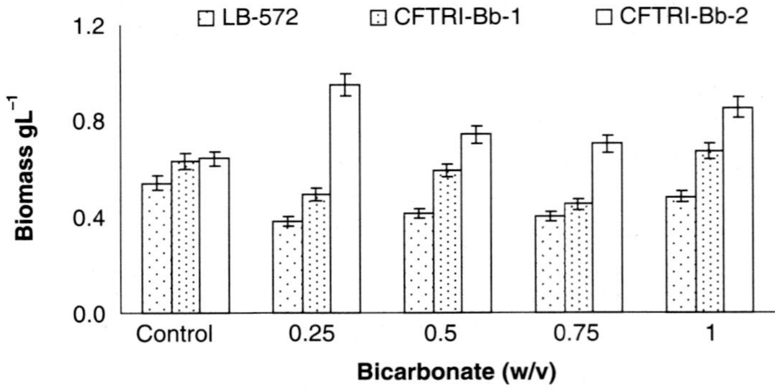

Fig. 5 Effect of biocarbonate on biomass yield and growth in *B. braunii* strains.

Fatty Acid Profile of Different *B. braunii* Strains

The fatty acid profiles of all the strains were given in Table 2. The lipid content also varied among the strains. *B. braunii* (SAG-30.81) produced 30% (w/w) of fat whereas in LB-572 it was 24% (w/w). In *B. braunii* N-836 the fat content was considerably less (15% w/w) though the hydrocarbon content was high. In CFTRI -Bb-1 and 2 strains, lipid content was high compared to their hydrocarbon content. All the strains differed in their fatty acid profile especially in the relative proportion of stearic, linoleic, palmitic and oleic acids. Irrespective of the race they belong, oleic acid was found to be the major fatty acid in all the strains of *B. braunii* studied. However, palmitic acid forms a major fatty acid in N-836 and CFTRI -Bb-1 and 2 strains. The fatty acid profile in both the indigenous strains was found to be similar. In general, the fatty acid profile changes with the physiological state of the culture. In *B. braunii* species oleic acid was reported as the precursor of the non-isoprenoid hydrocarbons produced by the race A. Oleic acid involvement in the formation of very long chain fatty acid derivatives through chain elongation was reported by Lareillard et al., 1988; Lareillard et al., 1988 reported that oleic acid as a precursor of biopolymer PRB A and PRB B present in the outer cell wall of the A and B races. Templier et al., 1984 reported changes in the concentration of oleic acid with the physiological state of the alga. Therefore the changes observed in the fatty acid profile in different species are in accordance with the earlier reports. The positive aspect of the evaluation of different indigenous species would enable to select a promising strain for both hydrocarbon and fatty acid contents and their mass cultivation in out door conditions. Recent studies from the laboratory have shown that *B. braunii* is adaptable to lower levels (17 mM to 85 mM) of salinity with an increased production of biomass, hydrocarbon, fat, carbohydrate and carotenoids (Ranga Rao et al., 2006) an added feature to explore this organism's cultivation in brackish water or seawater. Further more as shown in Fig. 5 the indigenous strains are more adoptable to bicarbonate addition which would help to cultivate these organisms in outdoor conditions with enhanced biomass yields.

Table 2 Fatty acid profile of different *B. braunii* Strains

Fatty acids	*SAG 30.81	+LB-572	Relative (%) **N-836	CFTRI -Bb-1	CFTRI -Bb-2
12:0	0.67(±0.06)	0.78 (±0.29)	0.48 (±0.28)	2.35 (±0.56)	1.98 (±0.08)
14:0	8.4 (±0.64)	2.48 (±0.39)	1.68 (±0.45)	2.69 (±0.91)	3.65 (±1.12)
15:0	0.48 (±0.04)	0.60 (±0.15)	—	—	—
16:1	2.89 (±0.32)	0.87 (±0.28)	1.79 (±0.56)	—	—
16:0	13.28 (±1.04)	10.61(±0.39)	34.89 (±7.90)	36.09 (±4.35)	41.64 (±6.89)
17:0	—	—	0.28 (±0.12)	—	—
18:3	0.66 (±0.02)	20.043(±0.74)	5.06 (±0.94)	9.03 (±2.02)	12.18 (±3.73)
18:2	8.63 (±0.48)	Trace	7.56 (±1.02)	18.28 (±3.71)	13.55 (±1.09)
18:1	56.21 (±2.16)	53.19(±5.23)	38.78 (±7.18)	26.02 (±3.41)	23.02 (±4.24)
18:0	2.2 (±0.4)	3.48 (±2.9)	9.93 (±4.56)	3.25 (±1.12)	4.52 (±3.30)

Note *19:0, 19:1 20:0, 20:1, 24:0 26:0 and 28:0 fatty acids are present in traces; +19:0, 19:1, 20:0, 20:1, 24:0 26:0 and 28:0 fatty acids are present in traces; **19:0, 19:1, 20:0, 20:1, 22:0, and 22:1 fatty acids are present in traces

ACKNOWLEDGEMENTS

Authors thank the Department of Biotechnology, Government of India, New Delhi for their financial support and Dr. V. Prakash, Director, CFTRI for his support to carry out this work.

REFERENCES

Achitouv, E., Metzger, P., Rager, M. N and Largeau, C. 2004. C31-C34 methyl branched squalene from a Bolivian strain of *Botryococcus braunii*. *Phytochemistry*. 65:3159-3165.

Banerjee, A., Sharma, R., Chisti, and Banerjee, U. C. 2002. *Botrococcus braunii*: A renewable source of hydrocarbons and other chemicals. *Critical Reviews in Biotechnology*. 22:245-279.

Brown, A. C., Knights, B. A. and Ellisie, Conway, 1969. Hydrocarbon content and its relationship to physiological state in the green alga *Botryococcus braunii*. *Phytochemistry*. 8:543-547.

Casadevall, E., Dif, D., Largeau, C., Gudin, C., Chamount, D. and Desanti, O. 1985. Studies on batch and continuous culture of *Botryococcus braunii*: hydrocarbon production in relation to physiological state cell ultra structure and phosphate nutrition. *Biotech. Bioengin.* 27: 286.

Christie, W. W. 1982. Lipid analysis, 2nd edn. Pergamon Press, New York. pp 93-96.

Dayananda, C. Sarada, R., Sila Bhattacharya and Ravishankar, G. A. 2005. Effect of media and culture conditions on growth and hydrocarbon production by *Botryococcus braunii*. *Process Biochemistry*. 40:3125-3131.

Frenz, J., Largeau, C. and Caradevall, E. 1989. Hydrocarbon recovery by extraction with a biocompatible solvent from free and immobilized cultures of *Botryococcus braunii*. *Enzyme Microbial. Tech.* 11:17-724.

Grice, K., Schouten, S., Nissenbaum, A., Charrach, J. and Sinninghe Damste, J. S. 1998. A remarkable paradox: Sulfurised fresh water algal (*Botryococcus braunii*) lipids in an ancient hyper saline auxinic ecosystem *Organic Geochemistry*. 28:195-216.

Hillen, L. W., Pollard, G., Wake, L. V. and White, 1982. Hydrocracking of the oils of *Botryococcus braunii* to transport fuels. *Biotech. Bioengin.* 24:193-205.

Kojima, and Zhang, 1999. Growth and hydrocarbon production by microalga *Botryococcus braunii* in bubble column photobioreactor. *J. Biol. Sci. Bioengin.* 87:811.

Lareillard, J., Largeau, C. and Casadevall, E. 1988. Oleic acid in the biosynthesis of the resistant biopolymers of *Botryococcus braunii. Phytochem.* 27:2095-2098.

Largeau, C., Caradevall, E., Berkaloff, C. and Dhamliencourt, P. 1980. Sites of accumulation and composition of hydrocarbons in *Botryococcus braunii. Phytochem.* 19:1043-1051.

Maxwell, J. R., Douglas, A. G., Eglinton, G. and Mc Cormick, A. 1968. The botryococcenes-hydrocarbons of novel structure from the alga *Botryococcus braunii*, Kutzing. *Phytochemistry*. 7:2157-2171.

Metzger, P., Berkaloff, C., Casadevall, E. and Coute, A. 1985. Alkadiene and botryococcene producing races of wild strains of *Botryococcus braunii. Phytochemistry*. 24:305-2312.

Metzger, P., Allard, B., Casadevall, E., Berkaloff, C. and Coute, A. 1990. Structure and chemistry of a new chemical race of *Botryococcus braunii* (Chlorophyceae) that produces Lycopadiene, a tetraterpenoid hydrocarbon. *J. Phycol.* 26:258-266.

Metzger, P. and Largeau, C. 2005. *Botryococcus braunii*: A rich source for hydrocarbons and related ether lipids. *Appl. Microbio Biotech.* 66:486-496.

Metzger, P., Templier, J., Largeau, C. and Casadevall, E. 1986. An n-alkatriene and some alkadienes from the race of the green algae *Botryococcus braunii. Phytochem.* 25: 1869-1872.

Metzger, P., Berkaloff, C., Casadevall, E. and Coute, A. 1985. Alkadiene and botryococcene producing races of wild strains of *Botryococcus braunii. Phytochem.* 24:305-2312.

Ranga Rao, A., Dayananda, C., Sarada, R., Shamala, T. R. and Ravishankar, G. A. 2006. Effect of salinity on growth of green alga *Botryococcus braunii* and its constituents — hydrocarbons, fats, carbohydrates and carotenoids. *Biores. Techno.* (In press).

Sawayama, S., Minowa, T., Dota, Y. and Yokayama, S. 1992. Growth of hydrocarbon rich microalga *Botryococcus braunii* in secondarily treated sewage. *Appli. Microbiol. Biotech.* 38:135-138.

Templier, J., Largeau, C. and Casadevall, E. 1984. Mechanism of non-isoprenoid hydrocarbon biosynthesis in *Botryococcus braunii. Phytochem.* 23:1017-1028.

Wake, L.V. and Hillen, L. W. 1980. Study of a "Bloom" of the oil rich alga *Botryococcus braunii* in the Darwin River Reservoir. *Biotechnol. Bioengin.* 22:1367-1656.

Yang, S., Wang, J., Cong, W., Cai, Z. and Ouyang, F. 2004. Effects of bisulfite and sulfite on the microalga *Botryococcus braunii. Enzyme Microbial Technol.* 35:46-50.

22

Treatment of CETP Effluent by a Seaweed Associated *Aspergillus niger* Isolate

H.V. Joshi*, Anjana K. Vala, Purva N. Bhatt and Nisha Anand
Marine Algae and Marine Environment Division, Central Salt & Marine Chemicals Research Institute, Bhavnagar-364 002.

ABSTRACT

CETP effluent receiving effluents mainly from dyes, chemicals and pharmaceutical industries (colour brown, pH 6.6, TSS 111ppm, TDS 6968 ppm, BOD 30.2 ppm, COD 301.2 ppm, total hardness 3000 ppm, sulphate 1503.53 ppm, phenolic compound 0.8 ppm, Fe 0.144 ppm, Hg 0.051 ppm, Pb 0.036 ppm, Zn 0.059 ppm) in 50% and 100% concentration was treated with seaweed (*Ulva* sp.) associated *Aspergillus niger* strain. The test isolate could decolorize both the effluent samples as evidenced by decrease in absorbance at 437 nm. There was no marked variation in percent decolorization of different concentrations of effluent. Under static condition, after 6 days incubation, 88.69% and 88.18% colour removal was observed in 50% and undiluted effluent, respectively. The test strain also exhibited biosorption ability as there was decrease in concentration of the metals present in the effluent. Dry weight of the test isolate was found to be 2.4385 g in 50% and 2.3309 g in undiluted effluent containing medium. The result suggests the test *A. niger* strain to be a promising bioremediation agent. The results further suggest that exploration of seaweeds as habitat may lead to potential microbial resources.

Keywords: *Aspergillus niger*, effluent, colour removal, % decolourization.

INTRODUCTION

Industrial development has been a boon to mankind, at the same time industries are regarded as heavy polluters of water because of their hazardous effluent discharge. A number of methods have been developed for treatment of these effluents before their release into the sea. In some cases common effluent treatment plants (CETP) also have been set up where a group of different industries release their effluent and the effluent is treated before its discharge. Despite the treatment, effluent color causes great concern in most of the cases. A suitable and cost effective method for color removal needs to be designed. Though color may not be toxic, it has a tendency to chelate metal ion and this becomes toxic to fish and other organisms (Karthikeyan, 1989). At a downstream point of discharge of colored effluents, any water supplying agencies like municipality etc., find difficulties to treat the water contaminated with color. Due to indiscriminate discharge of colored effluents, groundwater may also be contaminated (Kasthuri and Vasanthi, 2001; Trivedi 1995). Hence, it requires immediate attention. Bioremediation can be a solution to the problem, as microbes are considered omnipotent. Fungi can be screened as potential bioremediation agents due to several reasons like their high growth capacity and reach by virtue of mycelial branching and greater potential to produce number of enzymes. Effluent decolourization by *Coriolus versicolor*, *Coriolopsis polysona* and *Phanerochaete chrysosporium* have been reported (Kasthuri and Vasanthi, 2001; Kothari et al., 2001). Exploration of new habitats may lead to isolation of fungal strains with novel characteristics. Seaweeds support various epiphytic microflora. However, much attention has not been paid to them.

The aim of the present work was to examine seaweed (*Ulva* sp.) associated strain of *Aspergillus niger* for its effluent color removal potential, simultaneously, to examine the biosorption potential of the test strain and to explore possibility for its application as bioremediation agent. For the present study treated effluent from a CETP (which receives effluents mainly from dye, chemical and pharmaceutical units) was selected. The physico-chemical characteristics of the treated effluent are depicted in Table 1.

Table 1 Physicochemical characteristics of treated CETP effluent

S. No.	Parameter	Concentration*
1.	Colour	Brown
2.	pH	6.6
3.	TSS	111
4.	TDS	6968
5.	BOD	30.2
6.	COD	301.2
7.	Total hardness	3000
8.	Sulphate	1503.53
9.	Phenolic Coumpounds	0.8
10.	Iron	0.144
11.	Mercury	0.051
12.	Lead	0.036
13.	Zinc	0.059

(except pH, all values are in ppm)

MATERIALS AND METHODS

Two 1litre conical flasks containing 200 ml Potato Dextrose medium (Anonymous, 1968) were inoculated with 1ml inoculum (10^8 spores/ml) of *Aspergillus niger* strain. After two days of incubation, one flask was added with 20 ml undiluted CETP effluent and in another flask was added 20 ml 50% diluted effluent aseptically. All sets were prepared in duplicate.

At 24 h interval, 1 ml was taken out aseptically for 6 days and the residual colour was extracted with water saturated 1-butanol (2 ml). The absorbance was measured against 1-butanol. For control only 1ml broth was extracted with 1-butanol (2 ml). Decolourization was determined by monitoring the decrease in absorbance at 437 nm using UV-vis spectrophotometer (Shimadzu, 160 A) (Sani and Banerjee, 2001). Decolourization activity was calculated as follows:

Decolourization (%) = [(initial absorbance) – (observed absorbance)/initial absorbance] × 100

Dry weight of the inoculated strain was measured after removing the pellet from the medium, washing thoroughly with distilled water and keeping in oven at 80°C for 3 h.

Further, the fungal pellets grown on PD medium were transferred and exposed to 250 ml diluted and undiluted effluent, respectively, for 2 days and then tested for decolourization as described above, and for metal content using inductively coupled plasma spectroscopy (Perkin Elmer Optical Emission spectrometer Optima 2000DV).

RESULTS AND DISCUSSION

The test strain decolorized 50% diluted and undiluted CETP effluent as evidenced by decrease in absorbance at 437 nm. The percentage decolourization calculated showed that there was no marked variation in the decolorizing activity in diluted as well as crude effluent except for day one. On day one, 14% decolourization was observed for 50% diluted effluent while that for undiluted (100%) effluent was found to be 11.17%. Highest decolourization activity was observed on sixth day which was 88.69% and 88.18% for 50% and 100% effluent respectively (Fig. 1). The initial less color removal in undiluted effluent containing medium could be due to adaptation of the test strain towards the extreme conditions.

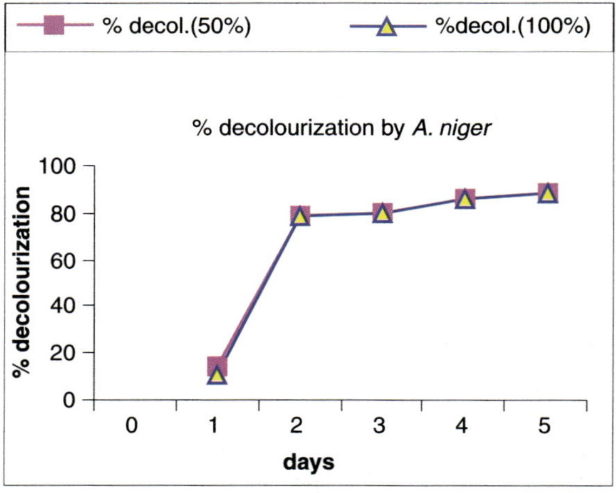

Fig. 1 % decolourization of 50% diluted and undiluted CEPT effluent by *A. niger.*

Dry weight of the fungal pellet was 2.4385 g and 2.3309 g in medium containing 50% diluted and undiluted effluent respectively, which indicates that effluent exposure did not have any significant growth inhibitory action on the test strain.

A significant increase in percentage decolourization just within a week could partly be due to rapid growth of the test strain and partly due to composition of the effluent. In case of effluent treatment with *Phanerochaete chrysosporium* and *Coriolus versicolor* 60% and 70.37% color removal has been reported after 55 days of incubation (Kasthuri and Vasanti, 2001).

The decolorization data for fungal pellets directly exposed to 50% diluted and undiluted effluent showed 24.39% and 15.28% decolorization, respectively.

The heavy metal content of the diluted effluent after exposure to fungal pellet was Fe 0.099 ppm, Hg, 0.02 ppm, Pb 0.001 ppm and Zn 0.038 ppm, whereas for undiluted effluent, it was found to be 0.096 ppm, 0.018 ppm, 0 ppm and 0.043 ppm, respectively, showing a decrease in metal concentration. These results indicate the biosorption ability of the test strain. Role of fungi in bioaccumulation and biosorption has been revealed by several workers (Bosecker, 1993; Zeljka et al., 2000).

Although fungal applications have shown good role on effluent treatment, they are still underutilized in practice this could be explained in part by a wide-spread *Apriori* assumption that fungal strains do not perform as good as bacteria (Coulibaly et al., 2003). There is a need to popularize utilization of fungal potentials for effluent treatment. This study is a step towards the same. Luxuriant growth and remarkable color removal efficiency reveals the test strain as a promising bioremediation agent.

Optimization of cultural conditions and use of immobilized biomass may further enhance the efficiency of the test strain. The present study indicated that seaweeds can be a potential resource for microbes having ecologically as well as economically important characteristics.

ACKNOWLEDGEMENTS

The authors are grateful to the Director, Dr P.K. Ghosh for providing the facilities and encouragement. They are also thankful to Prof. B. Jha, Discipline Co-Ordinator.

REFERENCES

Anonymous, 1968. Plant pathologist's pocketbook, (CAB Commonwealth Mycological Institute, Kew Surrey, England).

Bosecker, K. 1993. Biosorption of heavy metals by filamentous fungi. In : Biohydrometallurgical technologies Vol. II. edited by A.E. Torma, M.L. Apel & C.L. Brierley, (The Minerals, Metals and Materials Society, Warrendale) 1993, pp. 55-64

Coulibaly, L., Gourena G. and Agathos, N. S. 2003. Utilization of fungi for biotreatment of raw wastewaters, *African J. Biotech*., 2(12): 620-630

Karthikeyan, J., 1989. Removal of colour from industrial effluent. In : Pollution management in industries, (Environmental Publications, Karad) pp. 150-182

Kasthuri, B. R. and Vasanthi, M. 2001 Decolourization of paper and pulp mill effluent by chemical and biological methods, In : Role of microbes in the management of environmental pollution. edited by R. Tewari, K.G. Mukerji, J. K. Gupta & L. K. Gupta, (APH Publ. Corp., New Delhi) pp. 51-58

Kothari, R. K., Vyas, B. R. M. and Pathak, S. J. 2001. Decolourization of some textile azo-dyes by lignolytic enzyme system of a white-rot fungus *Coriolopsis polysona.* In : Role of microbes in the management of environmental pollution, edited by R. Tewari, K. G. Mukerji, J. K. Gupta & L. K. Gupta, (APH Publ. Corp., New Delhi), pp.1-12

Sani, R. K. and Banerjee, U. C. 2001. Biological treatment of textile and dye-stuff industrial waste, In : Role of microbes in the management of environmental pollution, edited by R. Tewari, K. G. Mukerji, J. K. Gupta & L. K. Gupta, (APH Publ. Corp., New Delhi), pp. 67-72.

Trivedi, R. K., 1995. Pollution Management in Industries. Environmental Publications, Karad.

Zeljka, F., Laszlo, S. and Felicita B. 2000. Biosorption of chromium, copper, nickel and zinc ions onto fungal pellets of *Aspergillus niger* 405 from aqueous solutions, *Food Technol. Biotech.*, 38(3): 211-216.

23

Comparative Physiological Studies amongst Cyanobacterial Isolates from Paddy Fields

Saurabh Tiwari[1], B.V. Singh[1], Dolly Wattal Dhar[1] and P.K. Singh[1]
[1]Centre for Conservation and Utilisation of Blue Green Algae (CCUBGA), Indian Agricultural Research Institute (IARI), New Delhi-110 012, India

ABSTRACT

Soil samples collected from paddy fields of J&K state of India were subjected to enrichment culture technique for the isolation of cyanobacterial flora. The isolated cyanobacterial forms were dominated by *Nostoc* (6) followed by *Cylindrospermum* (3), *Westiellopsis* (2) and *Calothrix* (1). The unialgal cultures of these isolates were raised for detailed observation on physiological/biochemical parameters related to pigment profile, total soluble proteins, N assimilatory enzymes (nitrate reductase and glutamine synthetase), nitrogen fixing potential, hydrogen production and ammonia leaking ability. There was a marked difference observed with respect to these attributes amongst the isolates.

Keywords: Cyanobacteria, pigments, proteins, N assimilatory enzymes, hydrogen.

INTRODUCTION

Cyanobacteria constitute the largest, most diverse and widely distributed group of prokaryotes that perform oxygenic photosynthesis. Many are capable of fixing atmospheric nitrogen which helps to maintain and improve productivity of rice fields (Roger et al., 1993). These organisms constitute the important component of soil especially flooded rice fields which are known to have a diverse flora of morphologically distinct forms (Roger and Kulasoorya, 1980). They are also known to withstand extremes of temperature and drought and show remarkable variation in growth, nitrogen fixation and stress compatibility (Goyal, 1997). Hydrogen exchange by nitrogen fixing, photosynthetic microorganisms has been found to either be via hydrogenase or nitrogenase (Yoch and Arnon, 1974). The heterocystous cyanobacteria have been shown to contain both so that H2 photoevolution could be coupled to either or both (Fujita and Myers, 1965; Haystead et al., 1970).

In view of this, the present investigation was undertaken to study physiological/biochemical attributes particularly N-assimilatory ability and hydrogen production amongst cyanobacterial isolates.

MATERIALS AND METHODS

Isolation and maintenance of organisms

Enrichment cultures were raised for deciphering the cyanobacterial components in the soil samples collected from paddy fields of J & K state of India (Rippka et al., 1979). For isolation of strains, one gram soil was inoculated in 50mL BG-11 medium (Stanier et al., 1971) in presence (1.5 g $NaNO_3L^{-1}$) and absence of nitrogen. Flasks were incubated for 15 days at 28±2°C temperature and 3000-5000 lux light intensity under 16/8 hr light/dark cycle. Isolation was carried out by dilution and pour plate method and identification was confirmed following the keys given by Desikachary (1959) based upon microscopic examination and these organisms were maintained under controlled laboratory conditions. The isolates used in the present study were deposited in the germplasm of CCUBGA, IARI, New Delhi (Plate 1, Table 1)

Table 1 Cyanobacterial isolates from paddy fields of Jammu & Kashmir

S.No.	Strains (CCC)*	Name
1	510	*Nostoc ellipsosporum*
2	511	*Calothrix thermalis*
3	512	*Nostoc rivulare*
4	513	*Nostoc hatei*
5	514	*Nostoc ellipsosporum*
6	515	*Westiellopsis proflica*
7	516	*Nostoc spongiforme*
8	517	*Cylindrospermum sp.*
9	518	*Cylindrospermum musicola*
10	519	*Nostoc ellipsosporum*
11	520	*Westiellopsis proflica*
12	521	*Cylindrospermum indicum*

*CYANOBATERIAL CULTURE COLLECTION

Growth studies, N assimilatory enzymes, extracellular ammonia and hydrogen evolution

Total soluble proteins were analyzed in cyanobacterial isolates at exponential phase of growth (Herbert et al., 1971). Pigment profile was tested on 15th day of incubation amongst the isolates. Chlorophyll was examined by hot-methanol extraction method (Mackiney et al., 1941). Carotenoids were extracted in acetone for estimation (Jensen, 1978) and phycobiliproteins were determined in phosphate buffer (Bennet and Bogorad, 1973). These isolates were also checked for N-assimilatory enzymes, extracellular ammonia and hydrogen production during similar time of incubation. Nitrate reductase (NR) activity (EC 1.6.6.1) was analysed by diazo coupling method

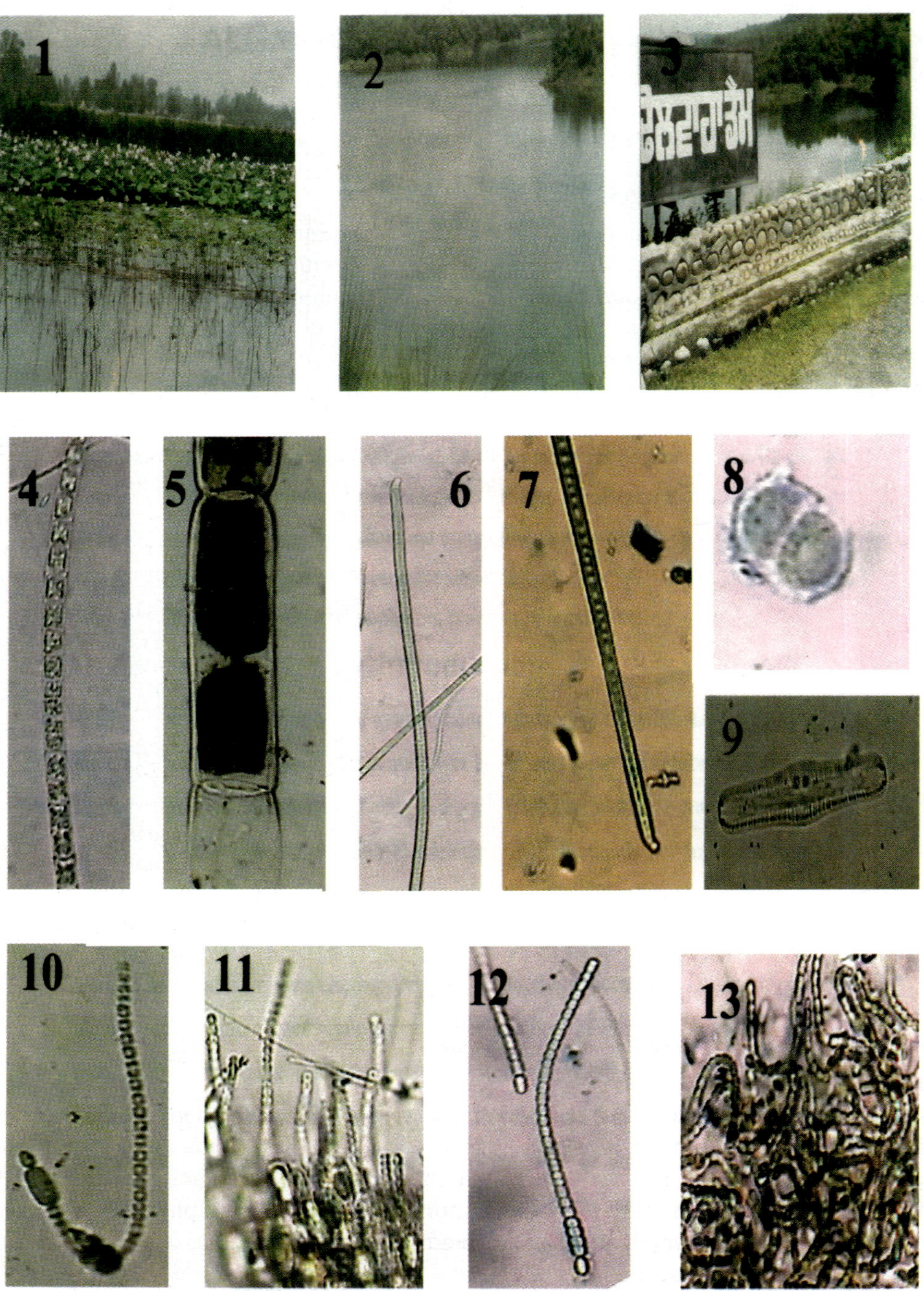

Plate 1 Photomicrographs of cyanobacterial isolates.

(Lowe and Evans, 1964) and the extracellular ammonia release was measured by employing the phenol hypochlorite method (Soloarzano, 1969). Glutamine synthetase (GS) activity (EC 6.3.1.2) was examined by the transferase assay potential (Shapiro and Stadtman, 1970; Stacey et al., 1977). Nitrogenase activity and hydrogen production were examined under aerobic as well as argon environment. Nitrogenase activity (EC 1.18.6.1) was studied as acetylene reduction assay using a gas chromatograph (Chemito Model GC 1000) fitted with Porapak N column (Hardy et al., 1973) and hydrogen production was evaluated in gas chromatograph with TCD (Kumazawa and Mitsui, 1981). The data recorded on various parameters were subjected to statistical analysis using SPSS package according to factorial CRD. Duncan's Multiple Range Test (DMRT) was used to compare the mean performance of strains for specific parameters.

RESULTS

All cyanobacterial isolates from paddy field soils of J & K state of India were heterocystous and *Nostoc* (6) represented 50% of the total isolates followed by *Cylindrospermum* (3), *Westiellopsis* (2) and *Calothrix* (1). Distinct variability was recorded amongst the cyanobacterial isolates with respect to pigments, total soluble proteins, nitrogen fixing potential, extra cellular ammonia and N-assimilatory enzymes. The chlorophyll was highest in *Nostoc ellipsosporum* which was *at par* with the chlorophyll content in *Calothrix thermalis* and *Nostoc spongiforme* followed by its content in *Westiellopsis proflica* and *Nostoc ellipsosporum*. Total carotenoids exhibited variability in the range of 9.14 mgmL^{-1} to 26.70 mgmL^{-1} respectively. Phycobiliproteins constituted more than 50% of the total pigments examined amongst the isolates and *Nostoc spongiforme* (CCC 516) showed highest phycobiliproteins which was more than seven times the lowest content exhibited by *Nostoc ellipsosporum* (CCC 514). *Nostoc ellipsosporum* (CCC 510) showed highest total soluble proteins (138.94 mgmL^{-1}) and *Nosotoc hatei* (CCC 513) exhibited the lowest soluble protein content (41.31 mgmL^{-1}). Glutamine synthetase activity was maximum in isolate *Nostoc ellipsosporum* (CCC 519) (74.16 nmole gGHmL^{-1}) which was about 75 times the lowest activity exhibited by strain *Cylindrospermum musicola* (CCC 518) (1.20 nmole gGHmL^{-1}) which also showed highest nitrate reductase activity (102.23 nmoleNO_2mL^{-1}) and NR activity was least (24.63 nmoleNO_2mL^{-1}) in *Westiellopsis prolifica* (CCC 515). Extracellular ammonia was maximum (57.58 mmoleNH_4mL^{-1}) in *Westiellopsis prolifica* (CCC 520) and minimum (10.89 mmoleNH_4mL^{-1}) in *Cylindrospermum musicola* (CCC 521). Nitrogenase activity and hydrogen production were enhanced under argon environment in comparison to aerobic situation. Isolate *Cylindrospermum indicum* (CCC 521) showed maximum nitrogenase activity (0.978 nmoleC_2H_4mg^{-1}chl hr^{-1}) while *C. musicola* (CCC 518) depicted maximum hydrogen production (3.014 nmoleH_2mg^{-1}chlhr^{-1}) which interestingly showed very low nitrogenase activity under aerobic as well as argon environment.

DISCUSSION

In the present study, only heterocystous forms of cyanobacteria could be isolated from paddy field soils of J & K state. Singh (1971) also noticed that the dominating nitrogen fixing algae from Cuttack were the heterocystous species of *Aulosira, Cylindrospermum, Nostoc, Anabaena* and *Gleotricha*. Mitra (1951) and Singh (1961) reported the occurrence of *Nostoc, Anabaena, Tolypothrix* and *Calothrix* from the Indian soils. Distinct variability was recorded amongst the cyanobacterial isolates with respect to pigments, total soluble proteins, nitrogen fixing potential, extracellular ammonia and N-assimilatory enzymes. Phycobilins comprising phycocyanin (PC), phycoerythrin (PE) and allophycocyanin (APC) were measured individually and isolates with the ability to produce enhanced phycobilins were identified. The variability observed with respect to various parameters was in accordance with the results obtained from earlier studies (Dhar et al., 2000, Mishra et al., 2001). Distinct intra- and inter-specific variability has also been reported for specific parameters in *Tolypothrix* strains (Prasanna et al. 2002). The use of nitrogen fixing cyanobacteria is currently confined mainly to their utilization as nitrogenous fertilizers for rice production in tropical countries (Whitton, 2000). Beyond their nitrogen fixing potential, their pigments and metabolite accumulation capacity stand out for their commercial value such as phycobiliproteins, carotenoids, polysaccharides and proteins as feed and food (Becker, 1988). Characterization of these organisms has also been carried out for specific parameters in relation to their utilization in value addition (Garcia- Pichel and Castenholz, 1991). Cyanobacteria are particularly attractive for the production of phycobiliproteins (Mac-Coll and Guard-Friar, 1987). These pigments can fluorescence at particular wavelengths which has found extensive use in cancer research for tagging tumor cells. These organisms depicted appreciable N assimilatory activity and extra-cellular ammonia release during incubation time. Cyanobacteria are known to assimilate nitrogen from atmospherically available nitrogen, nitrate, nitrite, ammonia, hydroxyl-amine and amino acids (Holm-Hamsen 1968). The ammonia may also be fixed up into glutamine within heterocysts and this is transported to vegetative cells where it is utilized for the synthesis of glutamate (Thomas et al., 1975; Rai et al., 1984). GS pathway is the major ammonia assimilatory route under nitrogen fixing conditions (Wolk et al., 1976; Stacey et al., 1977). Under conditions of ammonia limitations, most prokaryotes use a pathway consisting of glutamine synthetase and glutamate synthase to assimilate ammonia (MifLin and Lea, 1976). The results clearly support the contention that H_2 photoevolution by N_2-fixing, heterocystous cyanobacteria is exclusively due to the activity of the nitrogenase system. Proton reduction activity in the absence of other reducible substrates has been shown for all nitrogenase preparations including cyanobacteria (Lones, 1974). The requirement of only PS I activity and the slow decline in rate under dark anaerobic conditions indicate that like N_2 fixation, H_2 photoevolution requires both a source of a reductant (other than photosynthetic electron transport) and ATP (Lones, 1974). These organisms are known to contain two different hydrogenases. Also H_2 photoevolution is most probably an exclusive function of the heterocyst, the presence of which is a necessary prerequisite to both the activities. Studies have clearly indicated inter as well as intra-generic variations with respect to different parameters examined these can be exploited for the possible role in agriculture and specific biotechnological potential.

Table 2 Comparative total soluble proteins, N assimilatory enzymes and extracellular ammonia amongst cyanobacterial isolates

Strains (CCC*)	Total soluble proteins mg/mL	Nitrate Reductase (NR) Acticity nmole NO_2 /mL	Glutamine synthetases (GS) activity nmole γ GH'/mL	Ammonia release µmole NH_4/mL
510	138.94[a]	25.06	52.13[c]	15.56
511	63.40	32.58	35.84	24.32
512	83.72	36.43[c]	39.48	33.99

Contd.

Contd.

513	41.31	34.30	52.42[c]	34.29
514	88.26	38.33[b]	47.92	44.75
515	51.51	24.63	72.38[b]	52.43[c]
516	62.46	32.24	47.98	43.73
517	46.18	36.59[c]	31.04	30.49
518	100.60[b]	102.23[a]	1.20	15.02
519	96.14[c]	35.73	74.16[a]	56.71[b]
520	89.42	30.60	50.43	57.58[a]
521	74.47	31.59	13.79	10.89

Symbols a, b c denote LSD ranking
* CCC-Cyanobacterial Culture Collection
! GH-Glutamyl Hydroxamate

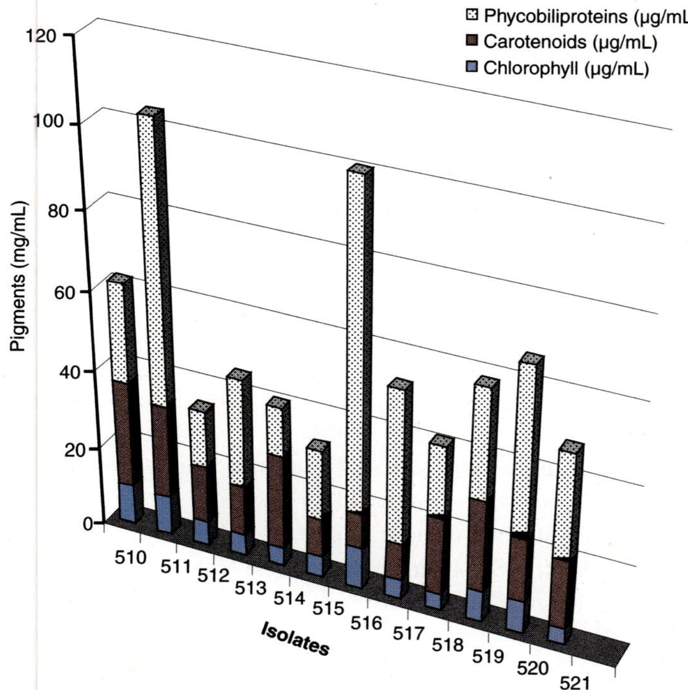

Fig. 1 Comparative pigments amongst cyanobacterial isolates.

Table 3 Comparative nitrogenase activity and hydrogen production under aerobic and argon environment amongst cyanobacterial isolates

Strains (CCC*)	Nitrogenase Activity nmole nmole C_2H_4/mgchl/hr		Hydrogen Production nmole H_2/mgchl/hr	
	Aerobic	Argon environment	Aerobic	Argon environment
510	0.074	0.123	0.223[c]	1.556[b]
511	0.377[c]	0.573[c]	–ND	0.045
512	0.376[c]	0.519[c]	0.336[b]	0.869

Contd.

Contd.

513	0.280	0.318	–ND	0.182
514	0.725[b]	0.911[b]	–ND	0.151
515	0.122	0.187	–ND	0.195
516	0.132	0.191	0.087	0.935
517	0.223	0.337	0.113	0.624
518	0.045	0.113	0.789[a]	3.014[a]
519	0.015	0.073	–ND	0.082
520	0.230	0.389	–ND	0.214
521	0.884[a]	0.978[a]	0.117	1.066[c]

CCC-Cyanobacterial Culture Collection
Symbols a, b c denote LSD ranking

ACKNOWLEDGEMENTS

The authors are grateful to Director, IARI, New Delhi and DBT, Govt. of India for the financial support.

REFERENCES

Becker, E. W. 1988. Microalgae for human and animal consumption, In:. Borowitzka MA, Borowitzka L. J. (ed.) Microalgal biotechnology, Cambridge University Press, Cambridge, pp. 222-256.

Bennett, A. and Bogorad, L. 1973. Complimentary chromatic adaptation in a filamentous blue-green algae. *J. Cell Biol.* 58: 419-435.

Desikachary, T. V. 1959. Cyanophyta, Pub. Indian Council of Agricultural Research, New Delhi.

Dhar Wattal, D., Tiwari, O. N., Prasanna R., Pabbi, S. and Singh, P. K. 2000. In: Singh P. K., Dhar Wattal Dolly, Pabbi Sunil, Prasanna Radha and Arora Anju (eds.) Biofertilizers, blue-green algae and *Azolla*. Venus Printers and Publishers, New Delhi, pp. 24-54.

Fujita, Y. and Myers, J. 1965. Hydrogenase and NADP-reduction reactions by a cell free preparartion of Anabaena cylindrical. *Arch. Biochem. Biophys.* 111:619-625.

Goyal, S. K. 1997. Algae and the soil environment. *Phykos.* 36:1-12.

Garcia, P. F. and Castenholz, W. R. 1991. Characterisation and biological implications of scytonemin, a cyanobacterial sheath pigment. *J. Phycol.* 27: 395-409.

Hardy, R. W. F., Burns, R. C. and Holsten, R. D. 1973. Applications of acetylene-ethylene assay for measurement of nitrogen fixation. *Soil Biol. Biochem.* 5: 47-81.

Herbert, D., Phipps, P. J. and Strange, R. E. 1971. Chemical analysis of microbial cells. In; Norris J. R. and Ribbons D.W. (ed.) Methods in Microbiology. Academic Press, London, pp. 209-344.

Holm, Hamsen, O. 1968. Ecology, physiology and biochemistry of blue-green algae. *Ann Rev Microbiol.* 22: 47-70.

Haystead, A., Robinson, R. and Stewart, W. D. P. 1970. Nitrogen activity in extracts of heterocystous and non-heterocystous blue-green algae. *Arch. Mikrobiol.* 74:235-243.

Jensen, A. 1978. Chlorophylls and Carotenoids. In: Hellebust J. A. and Craige J. S. (Eds.) Phycological methods: Physiological and biochemical methods. Cambridge University Press, Cambridge, pp. 59-70.

Kumazawa, S. and Mitsui, A. 1981. Characterization and optimization of hydrogen photoproduction by a saltwater blue-green alga *Oscillatoria* sp. Miami BG7. I. Enhancement through limiting the supply of nitrogen nutrients. *Int. J. Hydrogen Energy.* 339-348.

Lowe, R. H. and Evans, H. J. 1964. Preparation and same properties of a soluble nitrate reductase from *Rhizobium japonicum. Biochem. Biophys. Acta.* 85:377-389.

Lones, T. 1974 The enzyme system. In: Quispel A (ed.) The Biology of Nitrogen Fixation. North Holland Publishing Co., Amsterdam, pp. 617-638.

Miflin, B. J., Lea, P. J. 1976 The pathway of nitrogen assimilation in plants. *Phytochemistry* 15: 873-885.

MacColl, R., Guard-Friar, D. 1987. Phycobiliproteins. CRC Press, Boca Raton, Florida.

Mitra, A. K. 1951. The algal flora of certain Indian soils. *Indian J. Agric. Sci.* 21:357-373.

McKinney, G. 1941. Absorption of light by chlorophyll solutions. *J. Biol. Chem.* 140: 315-322.

Mishra, U., Pabbi, S. and Singh, P. K. 2001. Cyanobacterial diversity in terai belt of Uttar Pradesh, India I: Occurrence of some heterocystpous forms. *Phykos.* 40 (1&2): 89-94

Prasanna, R., Mohammadi, S. A., Prasanna, B. M., Singh, P. K. 2002. Intra- and interspecific variations for nitrogen fixing potential in *Tolypothrix* germplasm. *Acta. Botanica Hungarica.* 44(3-4), 383-392.

Rai, A. N., Rowell, P., Stewart, W. D. P. 1984. Evidence for an ammonium transport system in free-living and symbiotic cyanobacteria. *Arch Microbiol.* 137: 241-246.

Rippka, R., Deruelles, J., Waterbury, J. B., Herdman, M. and Stanier, R. Y. 1979. Generic assignments, strain histories and properties of pure cultures of cyanobacteria. *J. Gen Microbiol.* 111: 1-61.

Roger, P. A. and Kulasooriya, S. A. 1980. Blue-green algae and rice. IRRI, Los Banos, Philippines.

Roger, P. A., Zimmerman, W. J. and Lumpkin, T. A. 1993. Microbiological management of wetland rice fields. In: Baline Metting, F Jr., Marcel Dekker (Ed.) Soil micobiological ecology application in agricultural and ecological management, New York, Basel Hong Kong, pp. 417-455.

Singh, P. K. 1971. Blue-green algae, CRRI, Annual Report, pp. 120-127.

Singh, R. N. 1961. Role of blue-green algae in nitrogen economy of Indian agriculture, ICAR, New Delhi.

Solarzano, L. 1969. Determination of ammonia in natural waters by the phenol hypochlorite method. *Limnol Oceanogr* 14: 799-801.

Shapiro, B. M. and Stadtman, E. R. 1970. Glutamine synthatase (*E. coli*). In: Tabor H. and Tabor W. (eds.) Methods in Enzymology XVII A. Academic Press, London and New York, pp. 910-922.

Stanier, R. Y., Kunisawa, P., Mandel, M. and Cohen, B. G. 1971. Purification and properties of microcellular blue-green algae (order *Chroococcales*). *Bacteriol. Rev.* 3: 171-205.

Stacey, G., Tabita, F. R. and Baalen, V. C. 1977. Nitrogen and ammonia assimilation in a cyanobacterium. Purification of glutamine synthetase from *Anabaena* sp. Strain CA. *J. BacterioL.* 132: 596-603.

Thomas, J., Wolk, C. P., Shaffer, P. W., Austin, S. M. and Galonsky, A. 1975. The initial organic products of fixation of 15N-labelled nitrogen gas by the blue-green alga *Anabaena cylindrica. Biochem Biophys. Res. Commun.* 67: 501-507.

Wolk, C. P., Thomas, J., Shaffer, P. W., Austin, S. M. and Galonsky, A. 1976. Pathway of nitrogen metabolism after fixation of 15N-labelled nitrogen gas by the cyanobacterium, *Anabaena cylindrica. J. Biol. Chem.* 251: 5027-5234.

Whitton, B. A. 2000. Soils and rice fields. In: Whitton, B.A. and Potts M. (Ed.) Ecology of Cyanobacteria: Their Diversity in Time and Space. Kluwer Academic Publishers, pp. 235-255.

Yoch, D. C. and Arnon, D. I. 1974. Biological nitrogen fixation by photosynthetic bacteria. In: Quispel, A (ed.), The biology of nitrogen fixation, North Holland Publishing Company, Amsterdam, pp. 687-795

24

Effect of Salinity on the Growth of *Oscillatoria proboscidea*, a Freshwater Alga

R. Gunaseeli * S.K. Subramanian[1] and N. Anand[2]

*Post-Graduate Department of Microbiology, Lady Doak College, Madurai-625002 Tamil Nadu, India.

[1]Centre for Research and P.G. Department of Botany, Thiagaraja College, Madurai-625009 Tamil Nadu, India.

[2]CAS in Botany, University of Madras, Guindy, Chennai-600025, Tamil Nadu, India.

ABSTRACT

Soil salinity is an important agricultural problem. Cyanobacteria also possess outstanding capacity, second to that of bacteria, to resist adverse effects of extreme environmental conditions. The growth curve of *O. proboscidea*, isolated from Madurai region when transferred to BG_{11} medium supplemented with different concentrations of sodium chloride revealed at 1 mM and 25 mM did not affect the growth of *O. proboscidea* severely even though the amount of chlorophyll present on the 16th day was slightly less than the control. No significant difference in the length of lag or exponential growth phase was noted. Higher concentrations (50, 75 and 100 mM) reduced the chlorophyll content. Exponential growth phase was reached on the 12th day itself at 50 mM to 100 mM, and chlorophyll content showed significant reduction after the 12th day. Accumulation of proline was also noticed and the amount of proline present in the cell was concentration (Sodium chloride) dependent. Irrespective of photosynthetic rate, rate of respiration was quite high, increased by one fold. Even though the nitrogenase activity was very much reduced temporarily soon after the addition of sodium chloride, in the long term incubation recovery was noticed. In the presence of 1 mM sodium chloride, 129.16 n moles of C_2H_4/h/mg chl-*a* was fixed and the other concentrations showed decreased activity. Western blot analysis confirmed the presence of Fe protein bands in all the treatments except at 100 mM sodium chloride. Analysis of protein profile revealed the presence of 72 kD and 75 kD proteins which might be considered as stress protein. Application of fresh cultures of *O. proboscidea* to *Ixora* sp. (pot culture) and *O. sativa* variety IR 37 plants in the experimental plot definitely enhanced flowering and the yield of the treated plants compared to control. Treated *Ixora* plants were dark green with dark pink flower heads compared to control which showed yellowing and a few flower heads. Grain yield in paddy was increased by 64 per cent compared to control.

These results indicate that *O. Proboscidea*, a freshwater alga is physiologically and biochemically adapted to cope with low level osmotic stress and could be used as biofertiliser in Madurai region.

Keywords: Salinity, biofertilizer, halotolerant, protein profile, nitrogenase enzyme.

INTRODUCTION

Cyanobacterial biofertilizers constitute a perpetual source of nutrients and they do not contaminate groundwater and deplete the resources. Among the abiotic stresses that limit crop productivity, salinity and drought rank as the most detrimental. Cyanobacteria that are used as biofertilizers to partially substitute for chemical nitrogen fertilizers, respond differently to various ionic and osmotic components of salinity stress.

Over 7 million hectares of land in India is salt affected due to faulty soil and water management. Cyanobacteria have the capacity to scavenge sodium from these soils so that it could be successfully used for reclamation. Salinity stress has been shown to affect the various physiological processes. Decrease in growth rate and enhancement of respiration are the major physiological responses reported for halotolerant algae (Batterton and Van Baalen, 1971). Elevated respiration rate is responsible for maintaining the transmembrane H^+ gradient that energizes the H^+/Na^+ antiporter responsible for the removal of Na^+ from the cell during the initial salt shock and constitutes the adopting process in freshwater *Synechoccoccus* exposed to salt. Other responses include decrease in pigmentation, impairment of photosynthesis, nitrogenase activity (Apte and Thomas, 1984, 1987, 1998; Fry et al., 1986; Anand et al., 1995; Padhi et al., 1998; Apte, 1998) and protein synthesis (Blumwald and Tel-Or, 1982a, 1982b). Under salt stress, there was also an increased nitrogen demand as evidenced by high nitrogenase activity which might be due to synthesis of osmoregulatory substance like aminoacids (Measures, 1975) and quarternary ammonium compounds (Gallinski and Trueper, 1982; Mohammad et al., 1983) cyanophycin and glycogen (Madhu et al., 1998). The accumulation of intracellular proline is possibly related to the tolerance of cells to osmotic stress condition. (Csonka, 1981, 1989; Yancey et al., 1982 and Duncan and Widholm, 1987). When *Ulva fasciata was* exposed to 55 psu by adding sodium chloride to 35 psu ASW, free proline concentrations increased at 3 h and reached a plateau after 12 h. It is likely that an imbalance in ion ratio is a cue for the induction of a moderate accumulation of proline (Liu et al., 2000). In addition to its role as an osmolyte and a reservoir of carbon and nitrogen etc., it has been shown to protect plants against free radical induced damage and is also linked with the quenching of singlet oxygen (Matysik, 2002).

Protein profiles of a halotolerant *Nostoc muscorum* – R strain and the salt sensitive ARM – 221 strain showed a pronounced 72 kD protein band during early salt stress. (Shobhana Goel et al., 1997).

Based on these, the present study was carried out to find out the tolerance limit of *O. Proboscidea*, a freshwater alga isolated from Madurai region to salinity stress.

MATERIALS AND METHODS

O. Proboscidea filaments were isolated from the rice field soil following the method of Allen (1968) and maintained in BG_{11} medium under fluorescent illumination (3000 lux) with 12h L/D cycle. 1 ml of exponentially growing culture was inoculated into 100 ml BG_{11} medium supplemented with 1, 25-50, 75 and 100 mM sodium chloride. Growth was estimated by determining the chlorophyll-*a* (Mac Kinney, 1941) content every 4th day up to 24 days. Proline (Sadasivam et al., 1992) and Nitrogenase activity was assayed by the acetylene reduction technique (Stewart et al., 1967). Photosynthetic oxygen evolution and respiratory oxygen consumption in the intact filaments were measured (Allen and Holmes, 1986) at 25° C using Clark type (YSI model 53) oxygen electrode at a light intensity of 100 μ E/m^2/s. The illumination was provided by a 150 W incandescent bulb. Sodium dodecyl sulphate (SDS) polyacrylamide gel electrophoresis (PAGE) was carried out by the modified method of Laemmli (1970) to find out the protein profile. Western blot analysis was done to detect the dinitrogenase reductase of *Oscillatoria proboscidea* by the method of Zehr et al., (1993). The primary antibody for the analysis was a gift from Prof. P. Ludden, University of Wisconsin, USA.

RESULTS AND DISCUSSION

The growth curve of *O. proboscidea* in terms of chlorophyll-*a* revealed that it could tolerate up to 25 mM (16.5 μg/ml in 1 mM and 15.71 μg/ml in 25 mM) and approached control growth rates. No significant differences in the length of lag or exponential growth phase, but at 50 mM and 100 mM, exponential growth phase was reached on the 12th day itself and chlorophyll content showed significant reduction after the 12th day. At 50 mM, the cultures remained green up to the 12th day with a maximum amount of 8.87 μg/ml chl-*a* followed by gradual decline. At 75 mM and 100 mM NaCl, cultures showed yellowing and the maximum amount of chl-*a* recorded were 4.5 μg/ml in 75 mM and 1.8 μg/ml in 100 mM on the 12th day (Fig. 1).

Analysis of amino acid content of the cells revealed the accumulation of proline in response to salt stress. Proline level was more in salt stressed cells compared to control and was influenced by the concentrations of sodium chloride. At 1 mM and 25 mM concentrations steady increase in

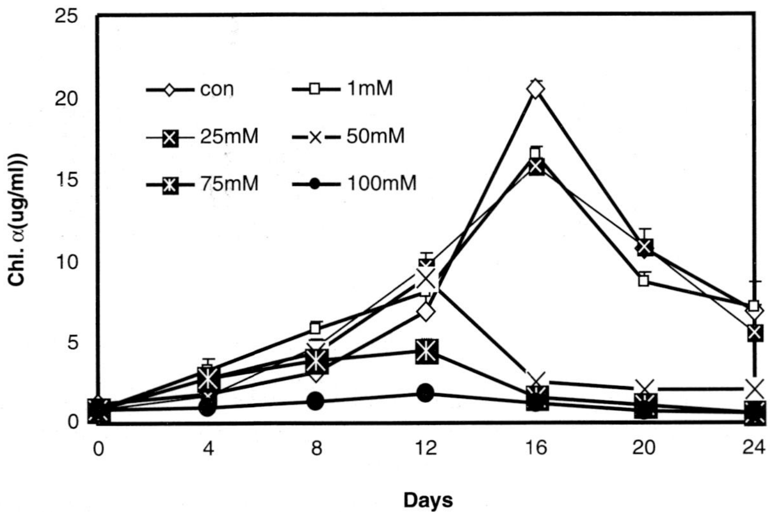

Fig. 1 Effect of different concentrations of sodium chloride on the growth of *O. proboscidea.*

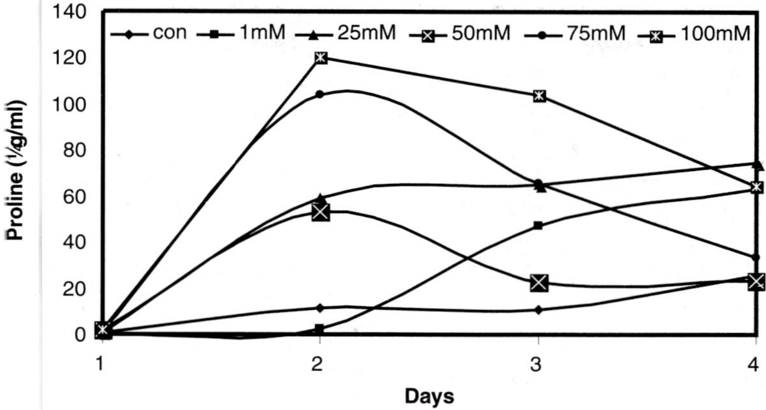

Fig. 2 Effect of different concentrations of sodium chloride on Proline content of *O. Proboscidea.*

proline level was noticed till the 24th day, but 50, 75 and 100 mM recorded increase in proline level upto the 8th day followed by a gradual decrease (Fig. 2). Proline appears to be a suitable osmoregulator because of its high solubility in water and its low inhibition of enzyme activities even at higher concentrations (Stewart and Lee, 1974). In *Spirulina platensis*, proline is the only or the main amino acid involved in halotolerance (Riccardi et al., 1981a, 1981b, 1983).

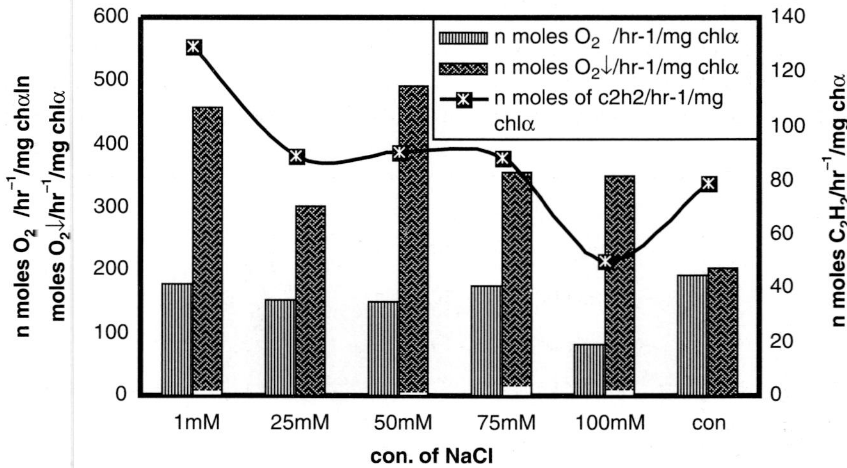

Fig. 3 Long term effect of sodium chloride on photosynthetic/respiratory/nitrogenase activity of *O. Proboscidea.*

Irrespective of photosynthetic rate, rate of respiration was quite high, increased by one fold. (Fig. 3) Fry et al. (1986) also observed similar increase in respiration by tenfold during the first 24h in *Synechococcus* 6311 under saline conditions.

Salinity significantly reduced the activity of nitrogenase in the early stages or soon after inoculation but on continued growth, the cells got adapted and the nitrogenase activity was

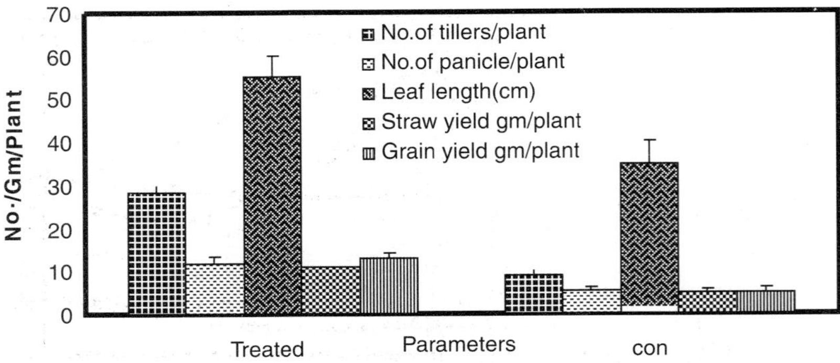

Fig. 4 Effect of addition of *O. Proboscidea* on the growth of *Oryza sativa* var IR. 37.

enhanced. In the presence of 1 mM sodium chloride, 129.16 n moles of C_2H_4/h/mg chl-*a* was fixed while at other concentrations nitrogenase activity was decreased (88 to 90 n moles of C_2H_4/h/mg chl-*a*) In the presence of 100 mM the activity was reduced to 55 n moles of C_2H_4/h/mg chl-*a*. (Fig. 4).

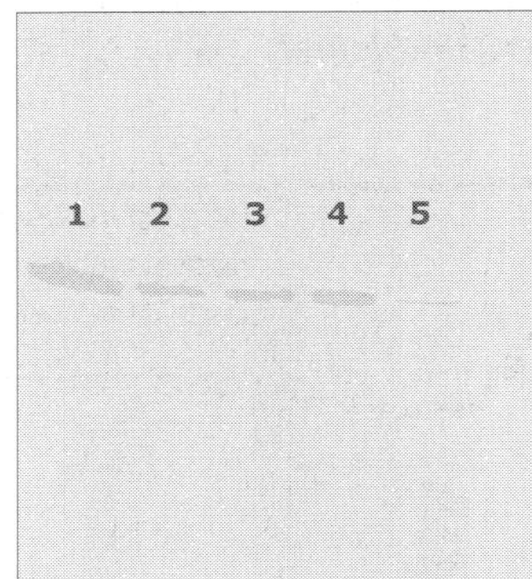

Fig. 5 Western blot for Fe protein of nitrogenase in *O. Proboscidea* grown in the presence of soldium chloride.

Apte and Thomas (1984) found that sodium was essential for nitrogenase activity rather than for the synthesis in cyanobacteria. Even though the nitrogenase activity was very much reduced temporarily soon after the addition of sodium chloride, in the long term incubation recovery was noticed in *Anabaena* sp. and also observed increased nitrogen demand during salt stress. Nitrogenase activity was further confirmed by the presence of Fe protein bands in all the treatments except at 100 mM sodium chloride through Western blot analysis (Fig. 5).

Analysis of protein profile revealed the presence of 72 kD and 75 kD proteins which might be considered as stress protein because similar band was noticed by Shobhana et al. (1997) in the protein profile of halotolerant *Nostoc muscorum* - R strain (Fig. 6).

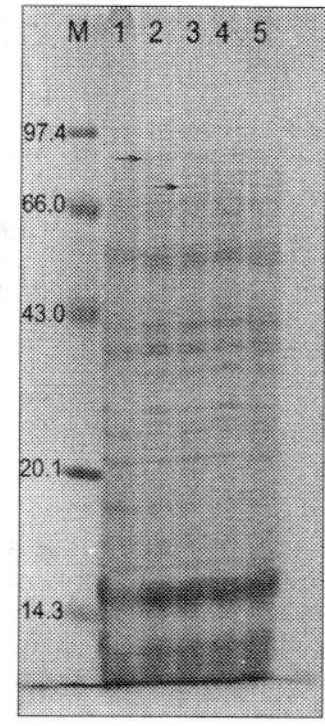

1. mM NaCl 2. 25 mM NaCl
3. 50 mM NaCl 4. 75 mM NaCl
5. 100 mM NaCl

Fig. 6 Polypeptide pattern of *O. Proboscidea* grown in the presence of sodium chloride.

So metabolic disruption is transient, in the long term, growth and activity are not completely suppressed at low level of salinity.

Application of *O. proboscidea* to *Oryza sativa*, and *Ixora* sp. showed positive results. Treated *Ixora* sp. plants were dark green healthy with many flower heads, while the untreated plants showed small and pale yellow leaves and a few flower heads. Application of *O. proboscidea* to *O. sativa* variety IR 37 also showed significant increase in all the parameters studied. Percentage of increase obtained over control was 68.2% (No. of tillers/plant) 54.7% (No. of panicle) 36% (Leaf length) 55.9% (straw yield) and 64% reference to grain yield (Fig. 4).

These results indicate that *O. proboscidea*, a freshwater alga, physiologically and biochemically adapted to cope with low level osmotic stress and could be used as biofertiliser in Madurai region.

ACKOWLEDGEMENTS

We thank the Director, CAS in Botany for providing the facilities and one of us R. Gunaseeli thanks the UGC for the award of FIP fellowship.

REFERENCES

Allen, J. F. and Holmes, N. G. 1986. The liquid phase oxygen electrode. *In*: Photosynthesis energy transduction a practical approach. (Ed.) M. F. Hipkins and N. R. Baker, IRL Press, pp. 108-116.

Allen, M. M. 1968, Simple conditions for growth of unicellular blue-green algae on plates. *J. Phycol.* 4: 1-4.

Anand, N. and Shanthakumar, Hopper, R. S. 1995. Distribution of blue green algae in rice fields of Kerala state, India. *Phykos.* 34(1 and 2): 55-64.

Apte, S. K. and Thomas, J. 1984. Effect of sodium on nitrogen fixation in *Anabaena torulosa and Plectonema boryanum. J. Gen. Microbiol.* 130: 1161-1168.

Apte, S. K. Reddy, B. R. and Thomas, J. 1987. Relationship between sodium influx and salt tolerance of nitrogen fixing cyanobacteria, *Appl. Environ. Microbiol.* 53: 1934-1939.

Apte, S. K. 1998. Expression and possible role of stress responsive proteins in *Anabaena.*, *J. Bio. Sci.* 23 (4): 399-406.

Batterton, J. C. J. R. and Van, Baalen, C. 1971. Growth responses of blue-green algae to sodium chloride concentration. *Arch. Microbiol.* 76:151-165.

Blumwald, E. and Tel, Or. E. 1982a. Osmoregulation and cell composition in salt-adaptation of *Nostoc muscorum*. *Arch. Microbiol.*132: 168-172.

Blumwald, E. and Tel, Or. E. 1982b. Structural aspects of the adaptation of *Nostoc muscorum* to salt. *Arch. Microbiol.* 132: 163-167.

Csonka, L. N. 1981. Proline over-production results in enhanced osmotolerance in *Salmonella typhimurium. Mol. Gen. Genet.* 182: 82-86.

Csonka, L. N. 1989. Physiological and genetic responses of bacteria to osmotic stress. *Microbiol. Rev.* 53: 121-147.

Duncan, D. R. and Widholm, J. M. 1987. Proline accumulation and its implication in cold tolerance of regenerable maize callus. *Plant Physiol.* 83: 703-708.

Fry, I. V. Huflejt, M., Erber, W. A., Peschek, G. A. and Packer, L. 1986. The role of respiration during adaptation of the freshwater cyanobacterium *Synechococcus* sp. 6311 to salinity. *Arch. Biochem. Biophy.* 244: 686-691.

Galinski, E. A. and Trueper, H. G. 1982. Betaine a compatible solute in the extremely halophillic phototrophic bacterium *Ectothiorhodospira halochloris*. *FEMS Microbiol. Lett.* 13: 357-360.

Laemmlie, U. K. 1970. Cleavage of structural proteins during the assembly of the head of bacteriophage T4. *Nature.* 227: 680-685.

Larher, F., Jolivet, Y., Briens, M. and Goas, M. 1982. Osmoregulation in higher plant halopytes: Organic nitrogen accumulation in glycine, betaine and proline during the growth of Aster tripholium and Suaeda macrocarpa under saline conditions. *Plant Sci. Lett.* 24: 201-210.

Liu, C. H., Shis, M. C. and Lee, T. M. 2000. Free proline levels in ULVA (CHLOROPHYTA) in response to hypersalinity: Elevated NaCl in sea water versus concentrated sea water. *J. Phycol.* 36: 118-119.

Mackay, M. A. Norton, R. S. and Borowitzka, L. J. 1983. Marine blue-green algae have a unique osmo-regulatory system. *Mar Biol.* 73: 301-307.

Mackinney, G. 1941. Absorption of light by chlorophyll solution. *J. Biol. Chem.* 140: 315-322.

Madhu, P. S. Behn, Carolyn, A. and Geoffrey, S. D. 1998. Cyanophycin and glycogen synthesis in a cyanobacterial *Scytonema* species in response to salt stress. FEMS *Microbiol. Letters.* 160(1): 11-15.

Matysik, J. Allia Bhalu, B. and Mohanty, P. 2002. Molecular mechanisms of quenching of reactive oxygen species by proline under stress in plants. *Curr. Science.* 82(5): 525-532.

Measures, J. C. 1975. Role of amino acids in osmoregulation of non-halophilic bacteria. *Nature.* 257: 398-400.

Mohammad, F. A. A., Reed, R. H. and Stewart, W. D. P. 1983. The halophilic cyanobacterium *Synechocystis* DUN 52 and its osmotic responses. *FEMS Microbiol. Letters.* 16: 287-290.

Padhi, H. Rath, H. and Adhikary, S. P. 1997-1998. Tolerance of nitrogen fixing cyanobacteria to NaCl. *Biol Plant.* 4(2): 261-268.

Riccardi, G. Sora, S. and Ciferri, O. 1981a. Production of amino acids by analog resistant mutants of the cyanobacterium *Spirulina platensis*. *J. Bact.* 147: 1002-1007.

Riccardi, G. Sanangelantoni, A. M. Carbonerra, D. Savi, A. and Ciferri, O. 1981b. Characterization of mutants of *Spirulina platensis* resistant to amino acid analogues. *FEMS Microbiol.* 12: 333-336.

Riccardi, G. Cella, R. Camerino, G. and Ciferri, O. 1983. Resistance to Azetidine-2-carboxylic acid and sodiumchloride tolerance in carrot cell cultures and *Spirulina platensis. Plant Cell Physiol.* 24(6): 1073-1078.

Sadasivam, S. and Manickam, A. 1992. Biochemical methods for agricultural sciences. Wiley Eastern Limited, New Delhi, pp. 42-43

Shobhana, G. Mamta, G. and Kaushik, B. D. 1997. Nitrogen fixation and protein profiles of halotolerant *Nostoc muscorum*: R strain isolated from rice fields and ARM. 221 strain. Indian. *J. Expt. Biol.* 35: 746-750.

Stewart, W. D. P., Fitzgerald, G. P. and Burris, R. H. 1967. *In situ* studies on nitrogen fixation using the acetylene reduction technique. *Proc. Natl. Acad. Sci.*, USA. 58: 2071-2078.

Stewart, G. and Lee, J. A. 1974. The role of proline accumulation in halophytes. *Planta.* 120: 279-289.

Yancey, P. H., Clark, M. E., Hand, S. C., Bowlus, R. D. and Somero, G. N. 1982. Living with water stress: evolution of osmolyte systems. *Science* (Wash. D.C.). 217: 1214-1222.

Zehr, J. P., Wyman, M., Miller, U., Duguay, L. and Capone, D.G. 1993. Modification of the Fe protein of nitrogenase in natural populations of *Trichodesmium thiebautti. Appl. Envir. Microbiol.* 59: 669-676.

25

Influence of Salinity Changes on the Morphology of *Cladophora vagabunda* (Linn.) Hoek

G. Rani
Department of Plant Biology & Plant Biotechnology, S.D.N.B. Vaishnav College for Women, Chromepet, Chennai-44, India.

ABSTRACT

The marine species *Cladophora vagabunda* are intertidal algae and are therefore often exposed to changing salinity; they show morphological variations in the pattern of branching, which is responsible for the difficulty in the delineation of the species. Such variations in branching pattern, cell dimensions in response to salinity and the rate of growth are observed in the present study.

Keywords: *C. vagabunda*, Chlorophyceae, delineation of species, intertidal, morphological variation, salinity.

INTRODUCTION

Marine species of *Cladophora* occur in the intertidal zone. The intertidal zone is potentially a stressful environment with nutritional limitation, high light, high and low temperature, high tide, desiccation and osmotic stress. Delineation particularly of marine species of *Cladophora* has always been a problem due to variation in thallus organization influenced by varying environmental conditions. Therefore, phenotypic plasticity is a notable feature among the members of Cladophoraceae.

Since the present taxonomic work on *Cladophora vagabunda* has revealed the fact that they show morphological variations among the intertidal and rock pool forms exposed to varying salinity, salinity tolerance in this species has been emphasized. Ultrastructural studies were carried out in *C. vagabunda* grown in different salinities to observe the change in the structure of the cell wall and cell organelles.

MATERIALS AND METHODS

Cladophora vagabunda was collected from Kovalam, Mahabalipuram, Rameswaram, the east coast of Tamilnadu, India. The algae collected were washed thoroughly in sterilized seawater. The morphological features of the collected materials were observed under stereo-dissection binocular microscope and identified with the help of 'Phycologia Generalis' (Kuetzing, 1843), Monograph of Hoek (1963) 'Revision of the European species of *Cladophora*'. The healthy uncontaminated filaments of marine species of *Cladophora* were brought under culture (Provasoli 1958).

A systematic study of the characteristic morphological features of the field material as well as material brought under culture of marine *Cladophora vagabunda* (L.) has been carried out. Cell dimensions of both field and culture materials were subjected to Single ONEWAY ANOVA TEST and Q-TEST (Snedecor & Cochran, 1967).

Field materials and live cultures were photographed using Nikon-Labophot-2 Photomicroscope and Nomarsky (DIF) and polarized light microscopic photographs were taken with Nikon Microphot FXA.

Different salinities of Provasoli's enriched seawater medium were prepared following the method of Reed (1989). The salinity in parts per thousand (S‰) for each concentration of seawater was estimated using ATAGO Hand Refractometer S/Mill-E. Electron microscopic studies were carried out using Philips EM200 (Moss and Jones, 1977).

Fresh weight and dry weight of the filaments were taken in Sartorius electronic balance every 7th day from the day of inoculation upto the 35th day (Reed, 1989).

RESULTS

In the present study *C. vagabunda* and *C. laetevirens*, which appear similar under field conditions (Figs. 1 & 2), showed distinctly different branching pattern under culture conditions thus helping in differentiating the two species. It is also interesting to note that *C. vagabunda* collected from rocks in the intertidal zone in Kovalam, which are often bathed by the tides and those collected from rock pools that have higher salinity (40-45‰) differed morphologically in their pattern of branching (Figs. 3 & 4).

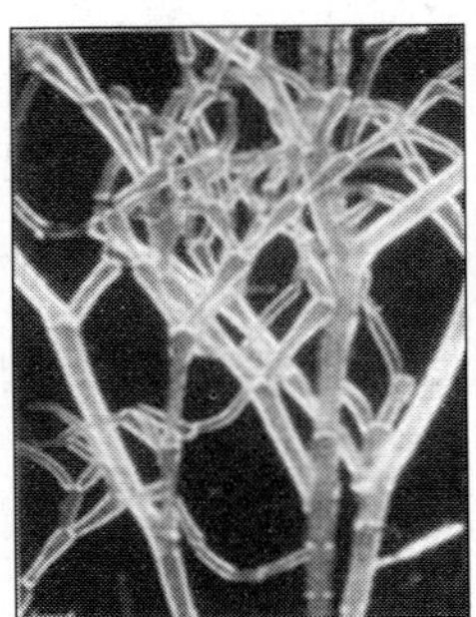

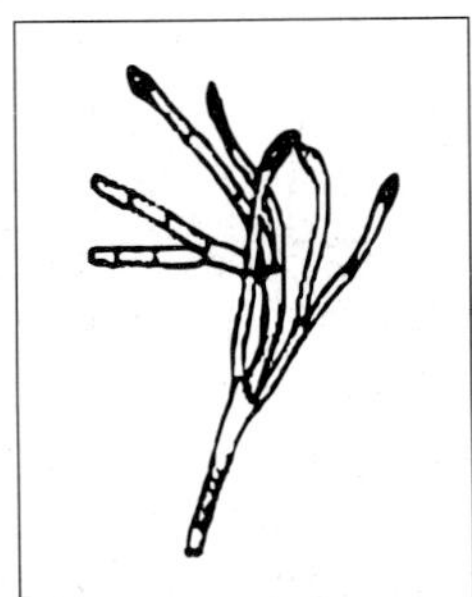

Figs. 1, 2, 3, 4

Description of *Cladophora vagabunda* (Linn.) Hoek

Thallus light green to dark green, soft, epizoic or epilithic; branched tufts 6-12 cm high attached to the substratum by matted rhizoids (Fig. 5); fasciculate terminal branches; unilateral, ultimate branches (tertiary branches) falcate or straight, 1-2 celled; growth acropetal in the upper portion and intercalary in mid and lower portions; main axis sparsely branched; pseudodichotomously or pseudotrichotomouly branched, 3-5 lateral branches (primary branches) at each node; insertion of lateral branches below the septum, inclined cross wall in young branches, becoming horizontal cross wall in older branches resulting in pseudodichotomy; angle of ramification acute. Apical cells cylindrical, falcate, tip rounded; cells of the main axis clavate; terminal cells of the laterals develop into reproductive structures or zooidangia; (Fig. 7) zooids are released through apical or subapical pores. Zooids germinate into young plants.

Cladophora laetevirens (Dillwyn) Kuetzing

Thallus yellow green, rough, epilithic; branched tufts 3-8 cm high; attached to the substratum by branched rhizoids forming a disc (Fig. 6); fasciculate terminal branches; every cell of the terminal branch giving rise to 1 or 2 laterals without subtending cross walls initially; ultimate branches falcate or straight, opposite; growth acropetal; main axis sparsely branched & pseudotrichotomous, with 3 lateral branches per node; insertion of lateral branches above the cross wall; cross wall in young branches inclined, becoming horizontal in older branches and in the main axis; angle of ramification acute in young branches, wide in old branches; apical cells cylindrical, falcate, rounded tip with small protuberance.

Morphological Variation in Varying Salinity

C. vagabunda shows a range of salinity tolerance between 10‰ and 65‰. Morphological variation in the pattern of branching in response to salinity level is an interesting feature observed. Stunted growth with long lateral branches at hyposalinity (10‰) and longer vegetative cells with sparse branching in hypersalinity (65‰) have been observed. Algae at higher salinity (55‰ – 65‰) showed swollen tips at lateral branches (Figs. 8 & 9).

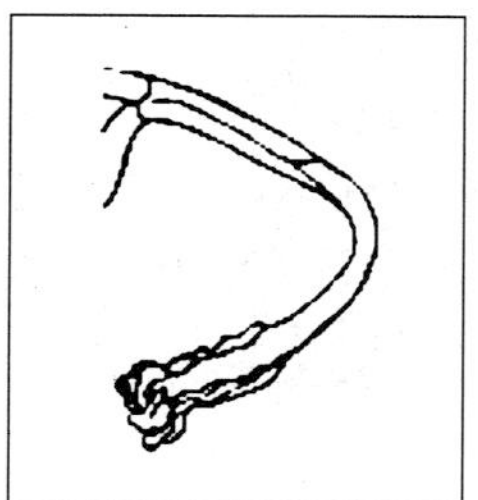
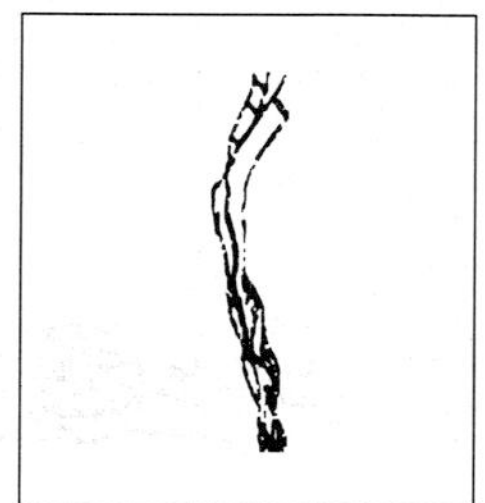
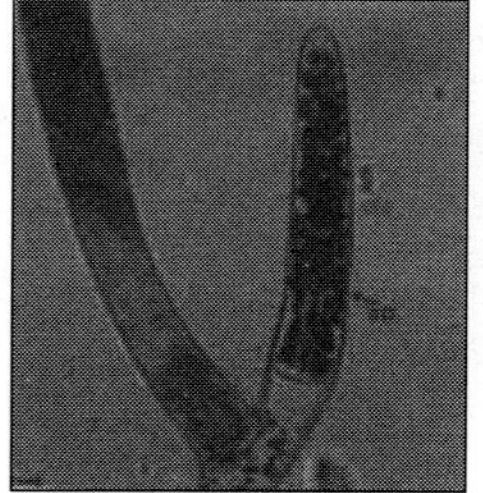
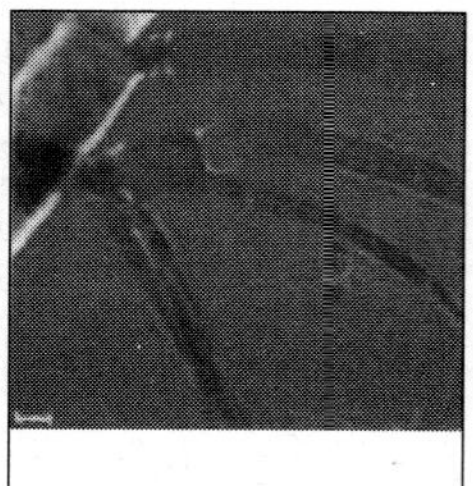

Figs. 5, 6, 7, 8

Morphometric Studies

Survey of length and width of apical cells, cells of main axis, basal cell, lateral branch cell, ultimate branch cell has been taken up. Data obtained has been subjected to ANOVA ONE WAY and Q-TEST to test the significance at 5% level.

Electron Microscopic Studies

The cell wall is thick and lamellated with many layers. The lamellae contain parallel, densely packed microfibrils. The arrangement of microfibrils is seen clearly in the longitudinal section of the cell wall. The cell organelles observed are the interphase nucleus with nucleolus. The nucleolus is distinguished into an outer granular layer (g) and an inner fibrillar layer and a central core region (cc). Outside the nucleus, centrioles have been observed (c) (Fig. 10). Chloroplast with round to oblong starch grains (st) with a clear zone (z) between the grains and the stroma are seen (Fig. 11).

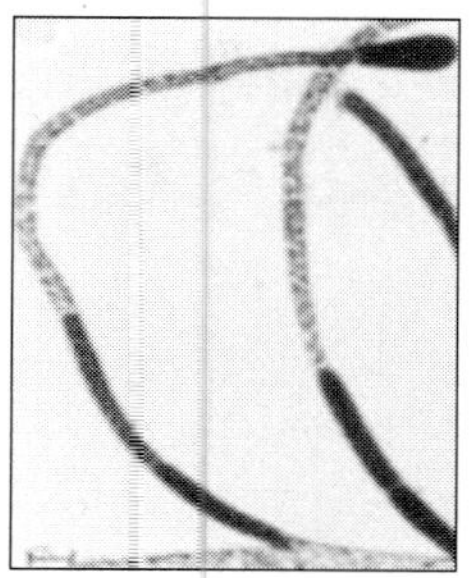
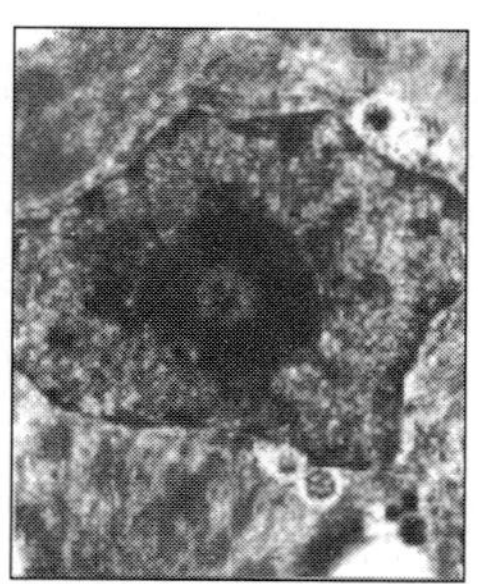
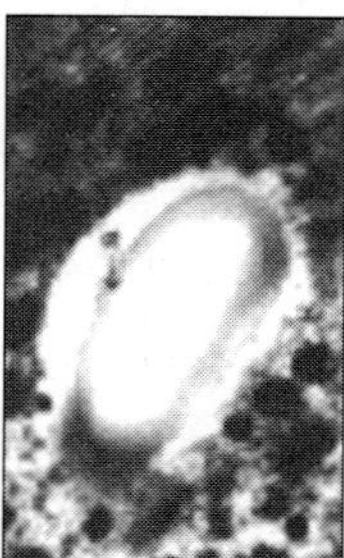

Figs. 9, 10, 11

Fine structure of *Cladophora vagabunda* grown under hyposalinity (10‰) and hypersalinity (65‰) showed unusual structure of the cell wall. The cell wall of the alga subjected to hyposalinity lack distinction of the cell wall layers into microfibrillar and amorphous layers (Fig. 13) compared to the control (Fig. 12). Cell wall appeared to be stretched at hypersalinity (65‰) (Fig. 14). Due to salinity stress even the cellular organelles were found disturbed. Alteration in the shape of the nucleus and chloroplast was observed at hypersalinity (Fig. 15).

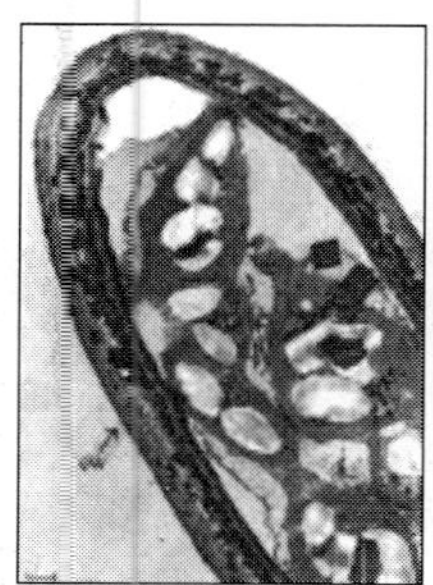
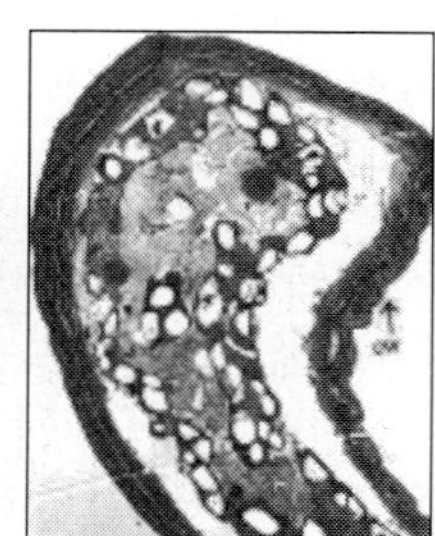
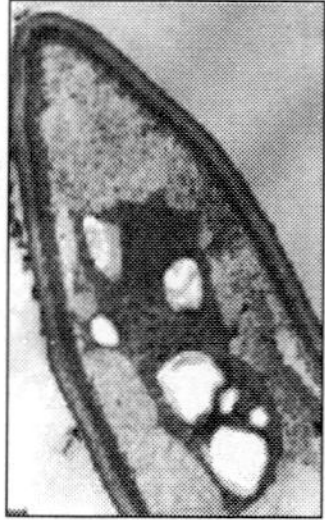
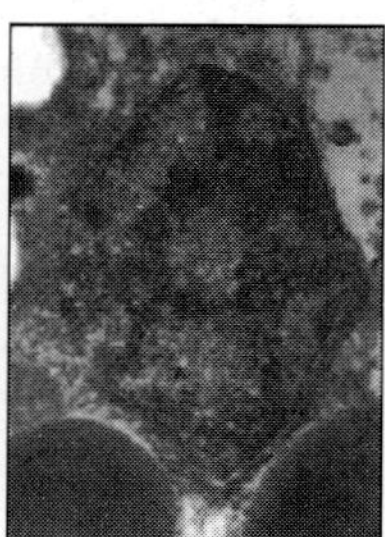

Figs. 12, 13, 14, 15

Growth Rate

Growth of *C. vagabunda* (as increase in fresh weight) at different salinities (hyposalinity 10‰, hypersalinity 65‰ and seawater equivalent media 34‰) over a period of 35 days is plotted (Fig. 16). Plants grown in 34‰ salinity showed sustained growth with more than double the initial fresh weight (1120 mg). Comparatively the lag period in 17‰ and 22‰ was only up to 7th day. However, the increase in fresh weight at 17‰ was 785 mg (±0.1) and at 22‰ it was 820 mg (±0.05).

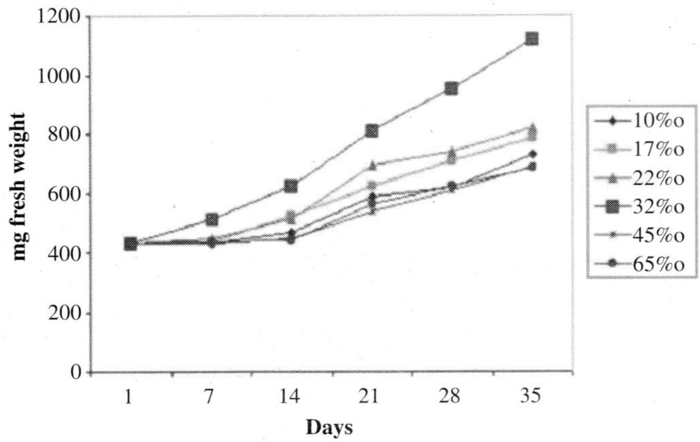

Fig. 16 Effect of salinity on growth (as fresh weight-mg) of *C. vagabunda*.

In hypersalinity of 45‰ and 65‰ the increase in tissue fresh weight was very minimal upto 14 days and only during latter half there was an increase in fresh weight. The alga grown in salinity below 10‰ and above 65‰ lost their pigmentation.

DISCUSSION

Morphometric studies employing statistical analyses (ANOVA test) have revealed that cell dimensions alone cannot be considered an adequate criterion for the delineation of the species of *Cladophora*. On the contrary, it should be considered along with other valid and relatively more stable morphological characters, viz., pattern of branching of the main axis as well as the rhizoidal system.

C. laetevirens resembles *C. vagabunda* in having fasciculate branches and falcate apical cells, but it differs from it in showing greater diameter of apical cells, lesser diameter of main axis cell and insertion of lateral branches above the septum, protruded tip of apical cell, absence of cross wall of ultimate branches and by disc-like rhizoids. Similarly the isolates of *C. vagabunda* collected from Kovalam; Mahabalipuram; Rameswaram and Rock pool, show distinct variations in branching pattern and cell dimensions in field condition but the pattern of branching is almost similar in all the above-mentioned isolates under controlled culture conditions.

The growth at hypo- and hypersalinity is lesser compared to the control. Statistical analyses with ANOVA and Q-test showed significant difference at 5% in different salinities.

Based on the present experimental studies it can be inferred that salinity has great influence on the branching pattern of *C. vagabunda*. Branching is profuse and the branches are long under conditions of hyposalinity (10‰) whereas branching is sparse and branches are secund, i.e.

branches on one side only, at hypersalinity (65-68‰). This experimental work clearly indicates that salinity influences the morphological features of *Cladophora* to a great extent. Hence, culture studies help to a large extent in the evaluation and delineation of species of *Cladophora* as pointed out by Hoek (1982).

ACKNOWLEDGEMENTS

My sincere thanks to my Professor Dr. S.Bharathan, Professor, Pachaiyappas College, Chennai for his guidance and constant encouragement.

REFERENCES

Hoek, C.V. 1963. Revision of the European species of *Cladophora*. Proefschrift … Rijksuniversiteit te Leiden … Leiden: E.J. Brill.

Hoek, C. V. 1982. A Taxonomic revision of the American species of *Cladophora* (Chlorophyceae) in the North Atlantic Ocean and their Geographic Distribution, pp. 1-236. North Holland Pub. Co., Amsterdam.

Kuetzing, F.T. 1843. *Phycologia generalis*——Leipzig. XXXII.

Moss, S.T. and Gareth Jones, E.B. 1977 Ascospore appendages of marine ascomycetes: *Halosphaeria mediosetigera. Trans. Br. Mycol. Soc.* 69:(2) 313-344.

Provasoli, L. 1968. Media and prospects for the cultivation of marine algae. In: Culture and collection of Algae. Proc. U.S-Japan conf. Hakane. Sept. 1966. *Jap. Soc. Plant Physiol.* 63-75.

Reed, R.H. 1989. Osmotic adjustment and organic solute accumulation in *Chaetomorpha capillaries. Brit. Physiol. Society.* 16: 21-37.

Snedecor, G.W and Cochran, W.G. 1967. Statistical methods. 6th edition, IOWA State University Press, pp. 330-342.

26

Ecological Study of Ujani Dam Backwater at Siddhteka Tal, Karjat, Ahmednagar, Maharashtra, India

S.D. Pingle[1] and A. B. Gore[2]

[1]Head, Dept. of Botany, P.V.P. College Pravaranagar, Dist. Ahmednagar, Maharashtra, Pin 413713

[2]Dept. of Botany, Shri Chhatrapati Shivaji Mahavidyalaya, Shrigonda, Dist. Ahmednagar, Maharashtra, Pin 413701.

ABSTRACT

An attempt has been made to assess the water quality of Ujani Dam backwater at Siddhateka and nearby stations. Collecting water samples from five sampling stations carried out the study. The samples were collected monthly from June 2001 to March 2002. The samples were analysed for parameters such as temperature, pH, dissolved oxygen, electric conductivity, TDS, BOD, Free CO_2, total alkalinity, chlorides, hardness, calcium, magnesium and phosphate.

At the same time macrophytes collected in aquatic and marginal area tolerating alkaline nature of water are recorded.

Keyword: Bhima river.

INTRODUCTION

The Ujani Dam is constructed on Bhima river near Pune-Solapur Road. Its backwater spread over about 40 square kilometers reaching to Siddhateka. In the present study water quantity at different selected stations as S_1-Pedgoan, S_2-Jalalpur, S_3-Shirapur, S_4-Siddhateka, S_5-Bhambora has been studied by physicochemical methods.

MATERIALS AND METHODS

In the present study five different sampling stations were chosen at backwater area. The sampling was done monthly from October 2000 to March 2001. Water samples were collected the from surface. The sampling time was uniformly maintained from 9.00 to 11.00 a.m. every month.

APHA (1985) and Trivedy and Goel (1986) procedures were used for water analysis. The parameters such as temperature pH, electrical conductivity, dissolve oxygen, free carbon dioxide were analysed at the site and the samples were brought to laboratory for further analysis.

The macrophytes found in water and bank of Bhima river are collected and identified. The list of the same is given below.

RESULTS AND DISCUSSION

The results of physicochemical parameters of average of five sampling stations are given in the Tables 1 and 2.

Temperature of selected sampling stations was from 21.06 to 28.56°C. The pH values ranged from 8.44 to 8.86. The decreased pH values were observed in December 2001. Electrical conductivity values ranged from 144 to 164.2 mmhos/cm. Total dissolved solid values ranged between 94.8 and 165.8 mg/lit. The values of dissolved oxygen ranged from 10.92 to 17.24 mg/lit. The values of free CO_2 ranged from 0.02 to 1.46 mg/lit. Free CO_2 was absent at many stations. The levels of total alkalinity ranged from 125 to 142.4 mg/lit.

Chloride level fluctuate from 110 to 168.2 mg/lit. The value of total hardness ranged from 105.6 to 173.6 mg/lit. The calcium ranged from 12.54 to 33.98 mg/lit. The level of magnesium fluctuated from 14.04 to 24.28 mg/lit. The phosphate values ranged from 0.048 to 0.142 mg/lit.

Following macrophytes are reported at the study area in water and the banks of Bhima river.

1. *Abutilon indicum* sweet
2. *Astorcantha longifolia,* Ness
3. *Boerhavia diffusa,* L
4. *Caesulia axillaries,* Roxb
5. *Ceratophyllum demersum,* L
6. *Chara wallichi,* Braun
7. *Corchorus oiltorius,* L
8. *Cynodon dactylon,* Pers
9. *Cyperus corymbosus,* L
10. *Cyperus rotundus,* L
11. *Eclipta alba,* Haask
12. *Eichhornia crassipes*(Mars) 50/ms
13. *Eriocaulon setaceum,* L
14. *Hydrilla verticillata,* Perst.
15. *Ipomoea fistulosa,* Mlarl
16. *Lemna polyrhiza,* L
17. *Limnophyton abtusifolium,* Miq,
18. *Marsilea ageyptia*
19. *Nitella acuminata,* A. Braum
20. *Ottelia alismoides,* Pers
21. *Parthenium hysterophorus,* L
22. *Pistia stratiotes,* L
23. *Polygonum glabrum,* wild
24. *Potamogeton perfolixatus,* L
25. *Prosopis juliflora,* Dc
26. *Scirpus articulatus,* L
27. *Scirpus corymbosus,* Heyne
28. *Sida rhombifolia,* L
29. *Tephrosia purpurea,* Pers
30. *Tridax procumbens,* L
31. *Triumfetta rotundifolia,* L
32. *Typha angustata,* Bary & chaub
33. *Vallisneria spiralis,* L

Table 1 Physicochemical features of Ujani Dam backwater at different sampling stations

Sr. No.	Name of Physicochemical Parameter	June-2001						July-2001					
		S1	S2	S3	S4	S5	Mean	S1	S2	S3	S4	S5	Mean
1	Temperature (0°C)	22.50	22.60	22.60	22.50	22.60	22.56	22.80	22.70	22.70	22.60	22.70	22.7
2	pH	8.4	8.5	8.5	8.5	8.4	8.46	8.6	8.6	8.6	8.5	8.5	8.56
3	Dissolved oxygen mg/lit	10.8	11.1	11.2	11.3	11.3	11.14	11.4	11.3	11.13	11.4	11.4	11.326
4	Elec. cond. mho/cm	152.00	155.00	151.00	148.00	147.00	150.6	149.00	150.00	148.00	147.00	149.00	148.6
5	T.D.S. mg/lit	167.00	168.00	161.00	167.00	165.00	165.6	160.00	167.00	167.00	167.00	168.00	165.8
6	BOD mg/lit	8.40	8.30	8.30	8.30	8.30	8.33	8.40	8.40	8.40	8.40	8.40	8.4
7	Free CO_2 mg/lit	0.00	0.30	0.10	0.00	0.10	0.13	0.00	0.10	0.10	0.10	0.10	0.075
8	Total alkalinity mg/lit	132	132	140	144	130	135.6	139	131	134	134	140	135.6
9	Chlorides mg/lit	113.50	112.10	111.70	113.40	113.40	112.82	110.40	109.40	108.60	107.40	111.40	109.44
10	Total hardness mg/lit	128.40	127.40	130.00	119.50	118.50	124.76	140.40	139.30	129.40	129.00	119.00	131.42
11	Calcium mg/lit	12.80	12.60	12.40	12.40	12.50	12.54	13.50	13.50	13.40	13.30	13.30	13.425
12	Magnesium mg/lit	19.50	19.40	19.80	18.10	18.50	18.88	18.50	18.20	18.20	18.50	18.50	18.425
13	Phosphate mg/lit	0.07	0.07	0.08	0.08	0.08	0.076	0.08	0.06	0.06	0.06	0.06	0.064

Table1 (Contd.) Physicochemical features of Ujani Dam backwater at different sampling stations

Sr. No.	Name of Physicochemical Parameter	August-2001						September-2001					
		S1	S2	S3	S4	S5	Mean	S1	S2	S3	S4	S5	Mean
1	Temperature (0°C)	24.50	24.60	24.60	24.60	24.40	24.54	27.50	26.90	27.50	27.40	27.40	27.34
2	pH	8.6	8.5	8.6	8.5	8.5	8.54	8.6	8.6	8.6	8.5	8.7	8.6
3	Dissolved oxygen mg/lit	11.4	11.5	11.3	11.4	11.2	11.36	10.9	10.8	10.8	11.0	11.1	10.92
4	Elec. cond. mho/cm	147	148	147	156	139	147.4	148	145	143	143	143	144.4
5	T.D.S. mg/lit	150.00	149.00	140.00	150.00	149.00	147.6	140.00	145.00	149.00	145.00	138.00	143.4
6	BOD mg/lit	9.20	9.10	9.10	9.10	9.10	9.12	9.50	9.60	9.70	9.70	9.80	9.66
7	Free CO_2 mg/lit	0.00	0.10	0.10	0.00	0.00	0.04	0.00	0.10	0.00	0.00	0.00	0.02
8	Total Alkalinity mg/lit	125.00	121.00	123.00	128.00	128.00	125	142.00	143.00	143.00	142.00	142.00	142.4
9	Chlorides mg/lit	115.00	116.00	116.00	116.00	116.00	115.8	112.00	112.00	112.00	110.00	110.00	111.2
10	Total hardness mg/lit	122.50	124.30	123.50	122.50	122.10	122.98	138.30	138.40	138.40	135.10	136.50	137.34
11	Calcium mg/lit	13.40	13.30	13.20	13.10	13.50	13.3	15.50	15.00	15.60	16.10	15.30	15.48
12	Magnesium mg/lit	18.60	18.40	18.50	17.50	17.50	18.1	19.10	18.90	18.80	18.80	18.80	18.88
13	Phosphate mg/lit	0.06	0.07	0.07	0.06	0.06	0.064	0.08	0.07	0.07	0.08	0.08	0.076

Table 1 (Contd.) Physicochemical features of Ujani Dam backwater at different sampling station

Sr. No.	Name of Physicochemical Parameter	October-2001						November-2001					
		S1	S2	S3	S4	S5	Mean	S1	S2	S3	S4	S5	Mean
1	Temperature (0°C)	29.55	29.10	20.30	20.40	29.50	25.77	28.40	29.30	29.00	29.10	27.00	28.56
2	pH	8.7	8.8	8.7	8.5	8.8	8.7	8.7	8.5	8.5	8.7	8.6	8.6
3	Dissolved oxygen mg/lit	12.8	12.6	12.7	12.1	12.8	12.6	12.9	13.1	11.9	9.0	10.2	11.42
4	Elec. cond. mho/cm	140	152	145	147	141	145	149	148	154	142	151	148.8
5	T.D.S. mg/lit	142	151	142	105	109	129.8	119	118	119	118	117	118.2
6	BOD mg/lit	14.3	13.9	14.5	14.5	13.8	14.2	8.9	8.8	8.9	8.7	8.1	8.68
7	Free CO_2 mg/lit	0.00	0.20	0.10	0.00	0.10	0.08	0.00	0.30	0.10	0.10	0.10	0.12
8	Total Alkalinity mg/lit	140.00	105.00	160.00	167.00	130.00	140.4	139.00	138.00	139.00	145.00	146.00	141.4
9	Chlorides mg/lit	101.00	110.00	115.00	109.00	115.00	110	125.00	121.00	129.00	141.00	132.00	129.6
10	Total hardness mg/lit	122.00	111.50	122.40	113.60	135.00	120.9	138.00	142.00	150.00	158.00	128.00	143.2
11	Calcium mg/lit	14.20	14.30	14.50	14.10	15.00	14.42	18.50	21.30	20.40	19.70	18.50	19.68
12	Magnesium mg/lit	22.25	23.50	24.30	22.15	23.10	23.06	24.80	25.20	25.10	23.10	23.20	24.28
13	Phosphate mg/lit	0.04	0.07	0.06	0.06	0.07	0.06	0.18	0.12	0.13	0.16	0.12	0.142

Table 1 (Contd.) Physicochemical features of Ujani Dam backwater at different sampling station

Sr. No.	Name of Physicochemical Parameter	December-2002						January-2002					
		S1	S2	S3	S4	S5	Mean	S1	S2	S3	S4	S5	Mean
1	Temperature (0°C)	21.40	21.30	21.40	21.50	21.40	21.4	20.10	20.10	20.30	20.30	20.40	20.24
2	pH	8.30	8.40	8.60	8.40	8.50	8.44	8.80	8.70	8.80	8.70	8.70	8.74
3	Dissolved oxygen mg/lit	15.10	14.80	14.70	14.80	14.30	14.74	17.10	17.20	17.00	17.40	17.50	17.24
4	Elec. cond. mho/cm	148	150	142	150	130	144	151	150	157	149	148	151
5	T.D.S. mg/lit	98	98	92	90	96	94.8	113	112	117	119	129	118
6	BOD mg/lit	11.2	11.8	11.4	11.5	12.0	11.58	6.5	6.1	6.12	6.3	6.4	6.284
7	Free CO_2 mg/lit	0.00	0.00	0.30	0.00	0.10	0.08	0.00	0.00	0.10	0.00	0.10	0.04
8	Total Alkalinity mg/lit	144.00	143.00	141.00	139.00	148.00	143	145.00	140.00	143.00	142.00	140.00	142
9	Chlorides mg/lit	168.00	173.00	179.00	176.00	145.00	168.2	170.00	169.00	168.00	165.00	165.	168
10	Total hardness mg/lit	175.00	164.00	180.00	179.00	170.00	173.6	160.00	161.30	160.5	160.4	160.8	160.6
11	Calcium mg/lit	34.10	33.90	33.90	34.00	34.00	33.98	32.50	32.30	32.10	32.20	32.80	32.38
12	Magnesium mg/lit	22.40	22.40	22.30	22.10	22.10	22.26	18.40	18.30	18.50	18.60	18.10	18.38
13	Phosphate mg/lit	0.05	0.04	0.05	0.05	0.05	0.048	0.04	0.05	0.06	0.06	0.05	0.052

Table 1 (Contd.) Physicochemical features of Ujani Dam backwater at different sampling stations

Sr. No.	Name of Physicochemical Parameter	February-2002						March-2002					
		S1	S2	S3	S4	S5	Mean	S1	S2	S3	S4	S5	Mean
1	Temperature (0°C)	21.10	21.00	21.10	21.00	21.10	21.06	23.50	24.00	24.00	24.00	24.00	23.9
2	pH	8.50	8.60	8.50	8.50	8.60	8.54	8.90	8.90	8.80	8.80	8.90	8.86
3	Dissolved oxygen mg/lit	11.30	11.50	11.60	11.40	11.50	11.46	13.10	13.40	13.50	13.60	13.40	13.4
4	Elec. cond. mho/cm	152	151	156	156	156	154.2	160	165	170	162	164	164.2
5	T.D.S. mg/lit	119	122	123	122	121	121.4	138	142	143	133	138	138.8
6	BOD mg/lit	8.1	7.8	7.7	7.9	8.1	7.92	6.1	6.2	6.1	6.1	6.2	6.14
7	Free CO_2 mg/lit	0.00	0.80	0.90	0.90	1.30	0.78	1.10	1.60	1.80	1.50	1.30	1.46
8	Total Alkalinity mg/lit	139.00	138.00	139.00	139.00	135.00	138	138.00	139.00	141.00	143.00	140.00	140.2
9	Chlorides mg/lit	170.00	173.00	160.00	159.00	176.00	167.6	176.00	177.00	175.00	160.00	150.00	167.6
10	Total hardness mg/lit	101.50	103.40	105.00	108.10	110.00	105.6	111.00	112.40	113.60	111.40	112.60	112.2
11	Calcium mg/lit	21.30	21.40	20.60	20.40	20.40	20.82	26.4	18.50	18.60	17.60	17.40	19.7
12	Magnesium mg/lit	17.40	17.30	17.50	17.50	17.60	17.46	14.10	14.30	15.10	13.32	13.40	14.044
13	Phosphate mg/lit	0.08	0.07	0.07	0.06	0.06	0.068	0.07	0.06	0.07	0.07	0.06	0.066

Table 2 Mean values of the Physicochemical parameters at all station of study area

Year 2001-2002	Temperature (0°C)	pH	Dissolved oxygen mg/lit.	Elec. cond. mho/cm	T.D.S. mg/lit.	BOD mg/lit.	Free CO_2 mg/lit.	Total Alkalinity mg/lit.	Chlorides mg/lit.	Total hardness mg/lit.	Calcium mg/lit.	Magnesium mg/lit.	Phosphate mg/lit.
Jun-01	22.56	8.46	11.14	150.6	165.6	8.33	0.13	135.6	112.82	124.76	12.54	18.88	0.76
Jul-01	22.7	8.56	11.33	148.6	165.8	8.4	0.075	135.6	109.44	131.42	13.43	18.43	0.064
Aug-01	24.54	8.54	11.36	147.4	147.6	9.12	0.04	125	115.8	122.98	13.3	18.1	0.064
Sept-01	27.34	8.6	10.92	144.4	143.4	9.66	0.02	142.4	111.2	137.34	15.48	18.88	0.076
Oct-01	25.77	8.7	12.6	145	129.8	14.2	0.08	140.4	110	120.9	14.41	23.06	0.06
Nov-01	28.56	8.6	11.42	148.8	118.2	8.68	0.12	141.4	129.6	143.2	19.68	24.28	0.142
Dec-01	21.40	8.44	14.74	144	94.8	11.58	0.08	143	168.2	173.6	33.98	22.26	0.048
Jan-02	20.24	8.74	17.24	151	118	6.28	0.04	142	168	160.6	32.38	18.38	0.052
Feb-02	21.06	8.54	11.46	154.2	121.4	7.92	0.78	138	167.6	105.6	20.82	17.46	0.068
Mar-02	23.9	8.86	13.4	164.2	138.8	6.14	1.46	140.2	167.6	112.2	19.7	14.04	0.066

CONCLUSION

Physicochemical analysis of Ujani Dam backwater at Siddhateka and nearby stations reveals that all the parameters except pH fall within the permissible limits. The pH values is high at many sampling stations. The water is alkaline and macrophytes those tolerating this condition can grow in water and at the bank of river.

ACKNOWLEDGEMENTS

The author is thankful to P.A. Lawande, Principal, S.C.S College, Shrigonda, Dr. S. D. Pingle, Head P.G. Dept. of Botany P.V.P. College, Pravaranagar for providing facilities. Thanks are due to Dr. D. R. Shirke and Dr. S. V. Thite for their help during study.

REFERENCES

APHA, 1985. Standard methods for the examinations of water and waste water.

Trivedi, R. K. and Goel, P. K. 1986. Chemical and biological methods for water pollution studies. Environmental Publication Karad.

27

Studies on Genus *Vaucheria* of Jammu Region

Namerta Sharma and V.K. Anand
Department of Botany, University of Jammu, Jammu-180 006, India

ABSTRACT

Genus *Vaucheria* has long been a subject of controversy among the workers for its systematic placement. Venkataraman (1961) placed *Vaucheria* under Vaucheriaceae of order Vaucheriales of class Xanthophyceae and reported 9 species of *Vaucheria* from India whereas Saxena (1962) reported 13 species from India.

During the course of the present investigation on the occurrence of Genus *Vaucheria* from Jammu region, more than 17 species have been collected from different habitats (soil and water streams, ponds, ditches, etc.). Out of 17 species (10 species and 2 varieties) were identified, i.e., *Vaucheria amphibia* Randhawa, *V. arechavaletae* Magnus and Wille, *V. debaryana* Woronin, *V. geminata* De Condolle, *V. hamata* Walz, *V. pachyderma* Walz, *V. sessilis* (Vauch) De Condolle, *V. sessilis* (Vauch) De Condolle f. *orthocarpa* (Reinsch) Hansgirg, *V. sessilis* (Vauch) De Condolle f. *repens* (Hassall) Hansgirg, *V. taylorii* Blum, *V. terrestris* Lyngbye em Walz and *V. walzi* Rothert.

Studies revealed that 3 species, i.e., *V. amphibia, V. sessilis* and *V. walzi* restricted to the aquatic habitats (lotic only) while rest to the terrestrial habitats usually forming loose mats particularly during the severe winter period.

In the present communication, species diversity, distribution and occurrence of Genus *Vaucheria* within the region Jammu and habitat ecology have been discussed.

INTRODUCTION

Genus *Vaucheria* has long been a subject of controversy among the workers for its systematic placement. Bohlin (1901) placed *Vaucheria* under Vaucheriaceae of heterokontae constituting a separate order Vaucheriales. Pascher (1914) and Smith (1938) kept it along with Bacillariophyceae and Chrysophyceae, division/phyllum of Chrysophyta, while Papenfuss (1955) placed it under Chrysophycota. On the basis of reserve food material (starch) under continuous or long light conditions and development of typical Chlorophycean characters, Fritsch (1935) kept it under siphonales of Chlorophyceae. Later, Fritsch's view was also supported by Iyengar (1925) who gave the affinities with Chlorophyceae, i.e., presence of chl-*a*, pyrenoids in certain marine species, development of starch as reserve food materials under prolonged illumination, formation of zoospores with two flagella and oogamous type of sexual reproduction.

Genus *Vaucheria* does show dissimilarities with Chlorophyceae being siphonaceous and coenocytic thallus, cell wall rich in hemicellulose, presence of continuous central vacuole, pigment bearing bodies less developed with the presence of carotenoid (β-carotene) pigments, different chlorophylls; carotenoids proportion, absence of siphonein and siphonoxanthein, pyrenoids absent, peculiar arrangement of chromatophores and nuclei (chromatophores being peripheral and nuclei towards the inner side), oils as their reserve food material, compound and specialised zoospores called synzoospores, antherozoids with two different types of flagella and haplontic life cycle.

Considering the affinities and dissimilarities of *Vaucheria* from Chlorophyceae, modern phycologists like Smith (1955), Chapman (1962) and Round (1965) placed *Vaucheria* under heterosiphonales. This view of Smith has also been supported by Morris (1967), Prescott (1969) and Trainor (1978), who believed that numerous basic characters of *Vaucheria* show similarities with Xanthophyceae, i.e., absence of chl-*b*, presence of chl-*e*, large quantities of carotenoids (Xanthophylls predominates), storage of food reserve in the form of oil, absence of characteristic xanthophylls (siphonein and siphonoxanthein) of siphonales, unequal length of paired flagella of compound zoospores.

All the mentioned characters together with siphonous and coenocytic nature of thallus bring *Vaucheria* more close to Xanthophyceae than Chlorophyceae. Formation of septa either at the time of injury or during the formation of reproductive bodies indicate that alga is multicellular. In certain species of *Vaucheria* (*V. pseudohamata*) septa are often present. Such a filament which contains a continuous cytoplasm and a large number of nuclei is called coenocytic.

Vaucheriaceae has not attracted the phycologists in the past, so needed to be worked out extensively from various parts of the globe. Many stray reports on Vaucheriaceae have appeared from various parts of India, but no consolidated work has ever been done on Vaucheriaceae of J&K State (region Jammu) in particular.

Keeping in mind the paucity of work done on Vaucheriaceae, the present research problem has been undertaken to work out the Vaucheriaceae of Jammu region and hence a survey has been made from December, 2002 to April, 2004 with the following objectives in mind:

- Survey and enumeration of Vaucheriaceae of Jammu region of J&K state and identification of the species.
- Morphological variations in various species.
- Occurrence and distribution of different species in different habitats.
- Ecological status of the habitats (soil/water).

Comparing all the proposed classifications, i.e., Fritsch (1935), Smith (1935), Chapman (1962), Round (1965), Prescott (1969), Pandey and Trivedi, (1995) Dillard's classification is more applicable and has been adopted for present studies.

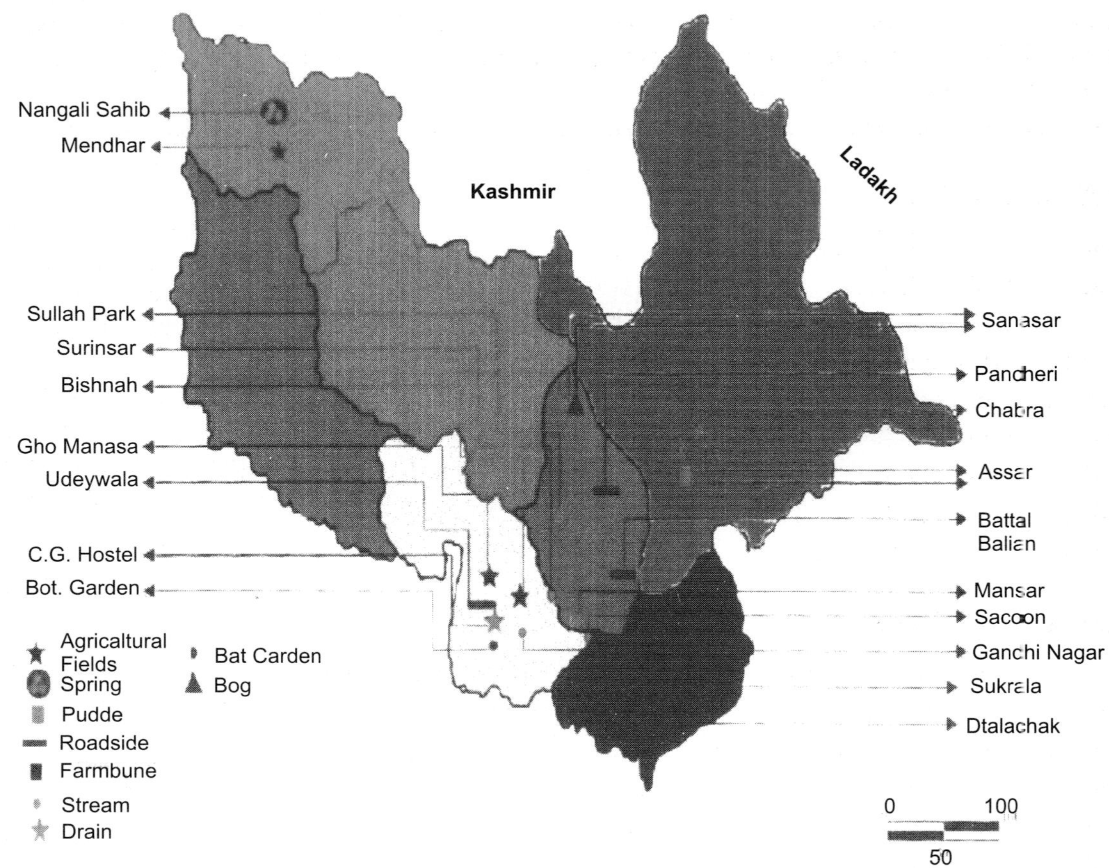

Plate 1 Location and topography of the collection spots in province Jammu.

12 species of *Vaucheria* (10 species and 2 varieties), i.e., *Vaucheria amphibia* Randhawa, *V. arechavaletae* Magnus and Wille, *V. debaryana* Woronin, *V. geminata* De Condolle, *V. hamata* Walz, *V. pachyderma* Walz, *V. sessilis* (Vauch) De Condolle, *V. sessilis* (Vauch) De Condolle *f. orthocarpa* (Reinsch) Hansgirg, *V. sessilis* (Vauch) De Condolle *f. repens* (Hassall) Hansgirg, *V. taylorii* Blum, *V. teresstris* Lyngbye em Walz and *V. walzi* Rothert have been collected.

MATERIALS AND METHODS

Study Area and Collection Sites

Jammu, the winter capital of J&K state is situated at a longitude of 74° to 76°15′ East and latitude of 32°15′ to 30°-30′ North at 304.8 – 3658.5 metres above the mean sea level. Jammu province has six districts, viz. Jammu, Udhampur, Doda, Kathua, Rajouri and Poonch. All the districts have been bestowed with numerous lotic and lentic water bodies which are situated at different altitudes. These harbour a great deal of aquatic algal diversity. The Jammu province also harbours a good deal of terrestrial algal diversity. Collections of *Vaucheria* species have been made from different habitats particularly from those areas, which could be easily approached, for a period from Dec. 2002 to April 2004 (Plate 1).

The details regarding the spots visited from different districts are:

JAMMU DISTRICT

Spot 1 : Moist soil of a field at Gho-manasa

Spot 2 : Moist soil of an agricultural field behind a school at Bishnah

Spot 3 : Wall of a cemented water tank in the Botanical Garden of Jammu University

Spot 4 : Moist soil in the Botanical Garden of Jammu University

Spot 5 : Moist soil near a well in Surinsar

Spot 6 : A grass lawn at Trikuta Nagar

Spot 7 : Small drain in Chanderbhaga Girls' Hostel in Jammu University Campus

Spot 8 : Slow moving small seasonal stream at Udeywala.

Spot 9 : Slow moving water of the lake Surinsar

UDHAMPUR DISTRICT

Spot 12 : Moist soil of small ditches on the shore line of lake Mansar

Spot 13 : A thick green mat in a small slow moving stream at Sacoon.

Spot 14 : Moist soil under the shade of a tree at Pancheri

Spot 15 : Moist mud of the bog in Sanasar

Spot 16 : Moist soil along the pathway at Battal Balian

Spot 17 : Small stream near the bog at Sanasar

KATHUA DISTRICT

Spot 18; Moist soil of a farmhouse at Dyala Chak

Spot 19: Moist mud in the fields at Sukrala bus stand

POONCH DISTRICT

Spot 20: Freshwater spring at Nangali Sahib, Poonch

Spot 21: Moist soil of an agricultural field at Mendhar

DODA DISTRICT

Spot 22: Moist soil of a puddle at Assar

Spot 23: Small cold, clear and slow running stream at Baggar, Assar

Spot 24: Cold, clear water stream at Chabra, Dist. Doda

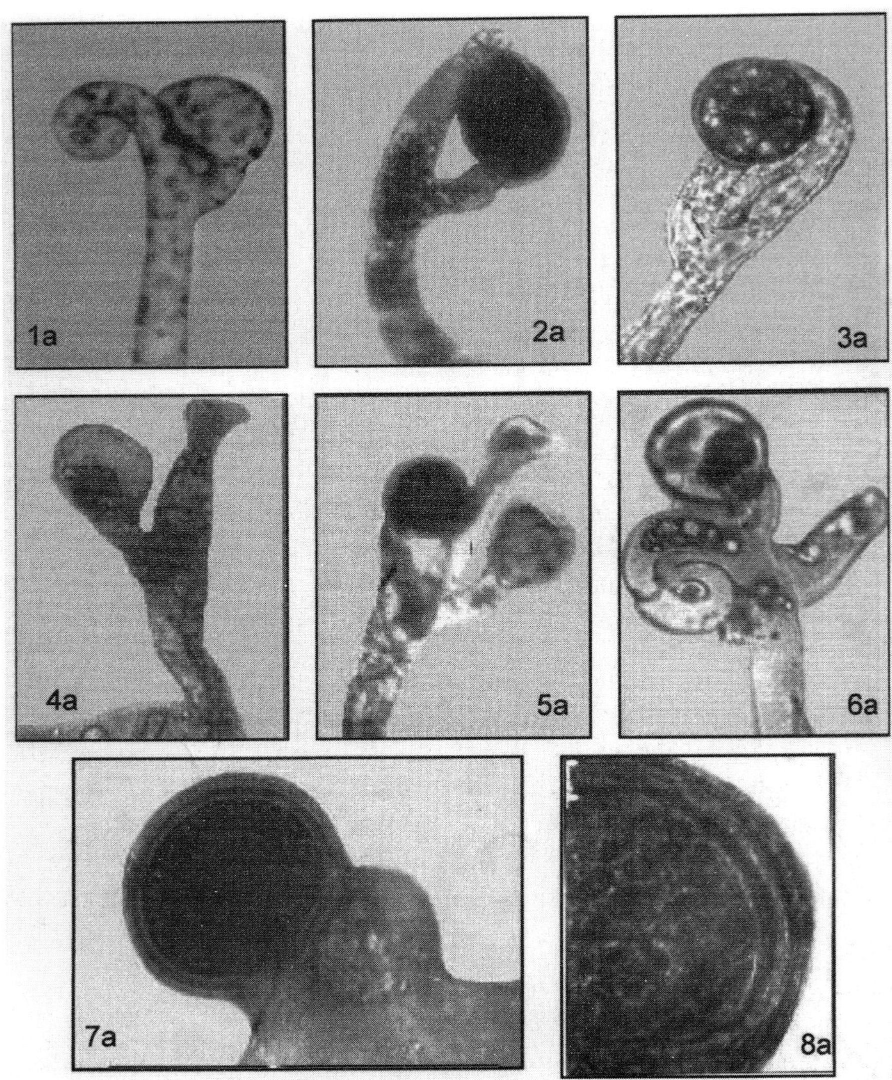

PLATE 2

Microphotographs (showing sex organs) [10 × 40X]

1a: *V. amphibia* Randhawa

2a,3a: *V. arechavalatae* Magnus and Wille

4a, 5a: *V. debaryana* Woronin

6a: *V. hamata* Walz

7a: *V. pachyderma* Walz

8a: *V. pachyderma* Walz (multilayered oospore wall)[10 × 100X]

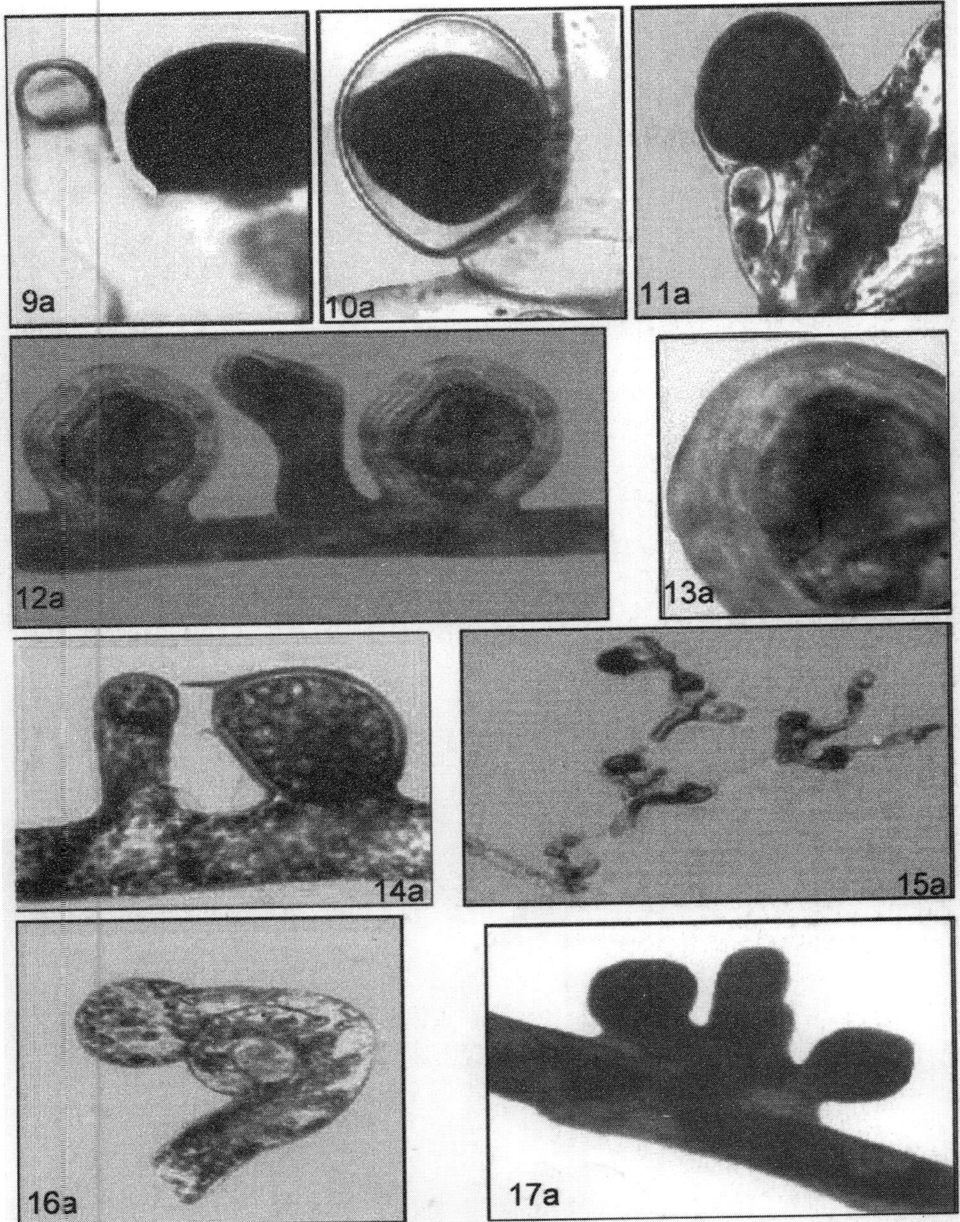

PLATE 3

9a: *V. geminata* De Condolle
10a: *V. geminata* De Condolle (oospore)
11a: *V. sessilis* (Vauch) De Condolle
12a: *V. sessilis* f. *orthocarpa* [Reinsch] Heering
13a: *V. sessilis* f. *orthocarpa* (multilayered oospore wall) [10 × 100X]
14a: *V. sessilis* f. *repens* [Hassall] Hansgirg
15a: *V. taylorii* Blum
16a: *V. terrestris* Lyngbye em Walz
17a: *V. walzi* Rothert.

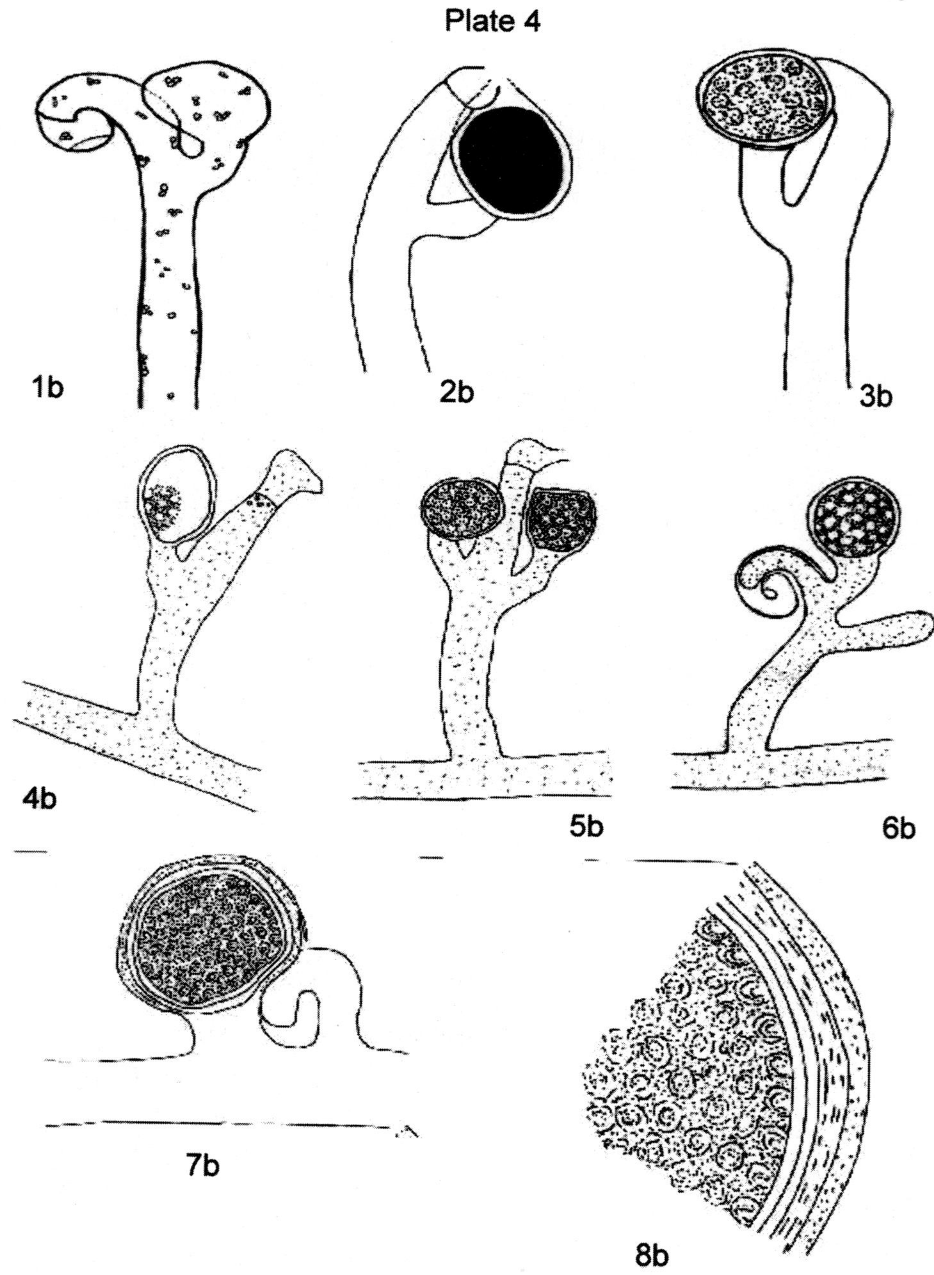

PLATE 4

Line drawings (showing sex organs) [10 × 40X]

1b: *V. amphibia* Randhawa

2b, 3b: *V. arechavalatae* Magnus and Wille

4b, 5b: *V. debaryana* Woronin

6b: *V. hamata* Walz

7b: *V. pachyderma* Walz

8b: *V. pachyderma* Walz (multilayered oospore wall)[10 × 100X]

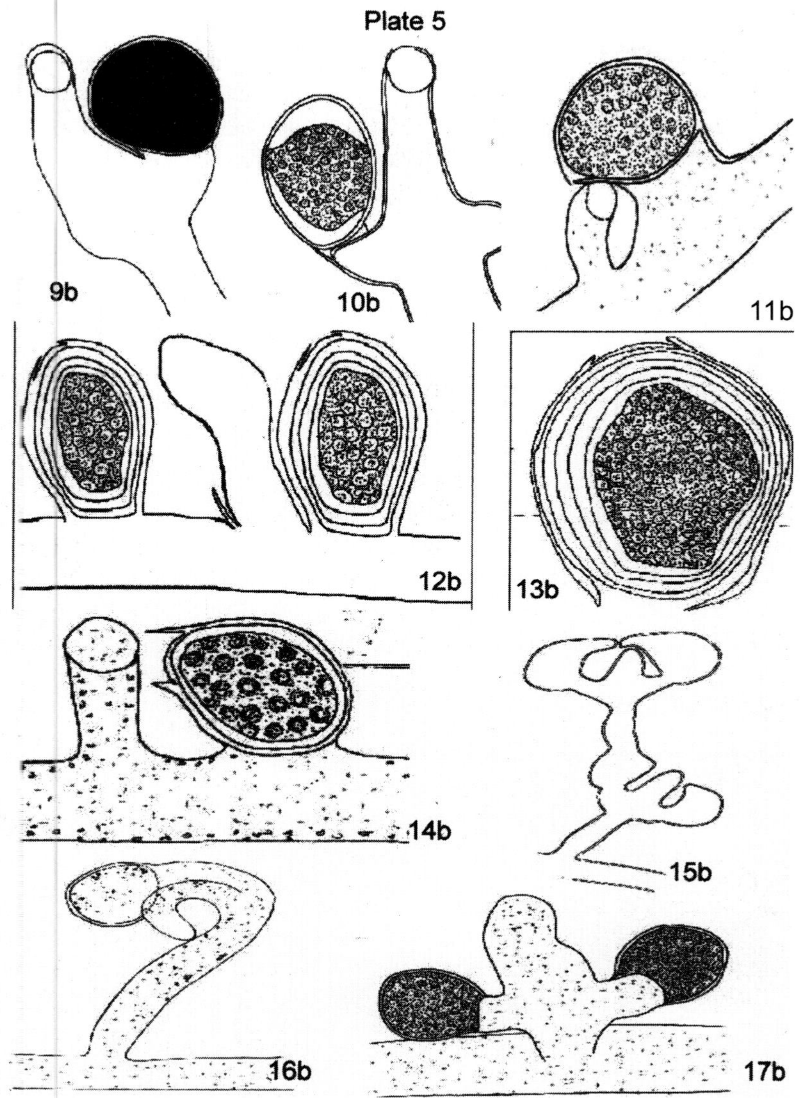

PLATE 5

9b: *V. geminata* De Condolle

10b: *V. geminata* De Condolle (oospore)

11b: *V. sessilis* (Vauch) De Condolle

12b: *V. sessilis* f. *orthocarpa* [Reinsch] Heering

13b: *V. sessilis* f. *orthocarpa* (multilayered oospore wall) [10 × 100X]

14b: *V. sessilis* f. *repens* [Hassall] Hansgirg

15b: *V. taylorii* Blum

16b: *V. terrestris* Lyngbye em Walz

17b: *V. walzi* Rothert.

Morphological Studies

Freshwater species of *Vaucheria* are normally observed to grow along the moist shores of ponds, streams, drainage ditches, marshes and swamps, while terrestrial species occupy the moist and wet soil habitats like the roadside catchments, fields or even the overexposed moist soil during the rainy season especially the winter period when temperatures are very low.

Collection of Material

More than 50 habitats representing drainage ditches, wetlands, drying ponds, puddles, fields, etc. had been visited from Dec. 2002 to April 2004. Collections of crude samples had been made either as fuzzy green algal mats or just imparting bright green to dark green colour to the substrate soil. Sediment quadrats with siphons were cut with a scalpel along with mud or sand 2-3 cm thick, placed in labelled, locked plastic bag and transported to laboratory for further observations.

Laboratory Culture

In the laboratory, samples were first observed under a compound microscope for the presence or absence of gametangia. Vegetative samples were transferred to culture vessels (Petri dishes or flasks). Removed the vascular plants, sticks and leaves from the surface, added Bold's Basal culture medium (c.f. Bischoff and Bold, 1963) to each culture dish (100 × 20) approximately to the 1/3rd thickness of crude sample. Kept the cultures in the culture room.

Light Microscopic Studies

After 5-10 days, microscopical observations on each crude culture were made. After 2-3 weeks, the incubated cultures had been observed by pulling some filaments from several locations within the sample and made a wet mount on the glass slide.

Camera Lucida Drawings

Line drawings of different species of *Vaucheria*, their sex organs, gametophores, etc., were drawn with the help of camera Lucida (prism type, Japan make). Measurement of sex organs and various other parts of *Vaucheria* thallus were recorded with the help of ocular micrometer by standardizing it against the stage micrometer.

Microphotography

Microphotographs of vegetative and reproductive structures of different species of *Vaucheria* were taken using a microphotographic camera.

Identification of Material

Identification of various species of *Vaucheria* were made using relevant standard literature, i.e., Venkataraman, 1961 and Saxena, 1962. Some of the research papers were also used for identification:

Hoppaugh, K.W., 1930; Randhawa, M.S. 1942; Tiffany and Britton, 1952; Sarma, P. 1973 and 1974; Entwisle, T.J. 1987 and 1988; Schneider and Land 1999, 2000; Schneider et al., 2001.

Preservation

Preservation of material was done in 4% formalin and specimens containing fixed samples and bearing their respective numbers are labelled, and have been deposited in the herbarium, Department of Botany, University of Jammu, India.

OBSERVATIONS

Ecological Studies

Biotic: Observations on the occurrence, habit, habitat and distribution of various species of *Vaucheria* in different habitats had been recorded. Observations were also made on associated algae with the different species of *Vaucheria* in different habitats and had been represented in the form of table.

Abiotic: Soil samples were cut with the help of scalpel, collected and dried in an oven at 60°C for 4-6 hours and stored in an airtight container for further analysis. Soil samples and water samples were studied with respect to various parameters as given in the Table 1.

Taxonomic Observation

Vaucheria amphibia Randhawa

(Venkataraman, 1961. p. 73-74, Figs. 49 a-c; Saxena, 1962. p. 35-37, Figs. 11, 12, 13]

Plates 2, 4, Figs. 1a, 1b

Vegetative Filament: Monoecious, coenocytic, siphonous, aseptate, unbranched, 40-60 μ broad.

Antheridium: Single, circinate, terminating the branch, 40-46 μ broad.

Oogonium: Single, globose to pear-shaped, pushed to one side with an antheridium, shortly stalked, beak towards antheridium, 140 μ long and 122 μ broad.

Vaucheria arechavalatae Magnus and Wille

(Venkataraman. 1961. p. 94-95, Figs. 171a-e.)

Plates 2, 4, Figs. 2a, 3a, 2b, 3b.

Vegetative Filament: Monoecious, unbranched, 16-18 μ wide, coenocytic, siphonous, yellowish green.

Antheridium: Single, cylindrical, 12-16 μ wide and 20-25 μ long, terminal on a fruiting branch, bent towards the oogonium, opening by a terminal pore.

Oogonium: Single, ovoid, 38 μ in diameter, on a short pedicel below the antheridium, opening by an apical pore.

Oospore: Fills the oogonium, 35-40 μ wide and 35-45 μ long, sometimes having a dark central spot, membrane smooth, two layered.

Vaucheria debaryana Woronin

(Hoppaugh, 1930. 17: 343-344, Pl XXVII, Figs. 17 and 18; Venkataraman, 1961, p. 90-91, Fig. 63)

Plates 2, 4, Figs. 4a, 4b, 5b

Vegetative Filament: Monoecious, unbranched, 16-30 μ broad, siphonous, coenocytic and yellowish green.

Antheridium: Single, 12-16 μ broad, straight or hooked, above the oogonium or between two oogonia, broader than the stalk, open by one or two lateral pores.

Oogonium: One to two, 40-50 μ long and 40-45 μ broad, oval or spherical with a short terminal beak, shortly stalked, situated under the antheridium.

Oospore: 38-52 μ broad and 40-50 μ long, wall three layered and smooth.

V. geminata De Condolle

[Hoppaugh, 1930. 17:338-339, Pl. XXV, Figs. 8 and 9; Tiffany and Britton, 1952. p. 213, Pl. 37, Fig. 385; Venkataraman, 1961. p. 83, Fig. 60a; Saxena, 1962. p. 39, Fig. 15; Sarma, 1974. p. 88, Pl. 2,7, Figs. 7, 8, 10, 24 and 25;]

Plates 3, 5, Figs. 9a, 10a, 9b, 10b.

Vegetative Filament: Thallus monoecious, branched, 58-66 μ wide, coenocytic, siphonous, aseptate and yellowish green.

Gametophore: Fruiting branch - 58-74.7 μ wide and of varying length (66-664.7 μ) bearing one or two oogonia and a single antheridium.

Antheridium: Single, cylindrical, 58-74 μ in diameter, terminating the branch, present besides a single oogonium or between two oogonia, usually on a stipe longer than that of oogonia.

Oogonium: 1-2, ovoid to slightly reniform, 99.6-132.8 μ broad and 124-149.4 μ long, usually in pairs, born laterally near the end of a fruiting branch on short pedicels below the antheridium.

V. hamata Walz

(Hoppaugh, 1930. 17:341, Pl. XXVI, Figs. 12 and 13; Venkataraman, 1961. p. 85, Fig. 62; Saxena, 1962. p. 32 and 33, Fig. 9; Sarma, 1973. p. 168, Figs. 1-3, 9, 11 and 13)

Pls. 2, 4, Figs. 6a, 6b.

Vegetative Filament: Monoecious, siphonous, yellowish green, branched, aseptate, 24-30 μ broad.

Antheridium: Single, terminal or lateral, circinate and coiled, 16-20 μ in diameter.

Oogonium: Single, ovoid to pear-shaped, born on long, upwardly curved pedicel at the end of a lateral fruiting branch from the main axis, oogonium and antheridium bent in the same direction, 50-55 μ wide, 60-70 μ long.

Oospore: Fills the oogonium, three layered, smooth, 52-56 μ wide and 56-72 μ long.

V. pachyderma Walz

(Hoppaugh, 1930. 17:334-335, Pl XXIV, Figs. 6 and 7; Venkataraman, 1961. p. 66, Fig. 44; Sarma, P. 1973. p. 168, Figs. 1 and 5; Sarma, 1974. p. 87, Pl. 1, Fig. 6, Pl. 3, Fig. 17)

Plates 2, 4, Figs. 7a, 8a, 7b, 8b.

Vegetative Filament: Thallus monoecious, branched, 40-58 μ in diameter, siphonous, coenocytic, yellowish green.

Antheridium: Single, 20-26 μ broad, sessile, saccate, develops besides the oogonium, has a terminal pore.

Oogonium: Single, sessile, globose, 110-120 μ long and 90-100 μ broad, beak distinct, parallel to the filament.

Oospore: Fills the oogonium, 96-118 μ long and 84-98 μ broad, wall thick, multilayered with distinct markings on its upper layer.

***V. sessilis (Vauch)* De Condolle**

[Hoppaugh, 1930. 17: 335, Pl. XXIV, Fig. 1; Venkataraman, 1961. p. 68-69, Fig. 46a; Saxena, 1962. p. 27-28, Fig. 7].

Plates 3, 5, Figs. 11a, 11b.

Vegetative Filament: Monoecious, unbranched, 40-58 μ in diameter, coenocyctic, aseptate, slender.

Antheridium: Solitary, 25-28 μ in diameter, usually between two oogonia, on a short, more or less straight stalk, opening by a terminal pore.

Oogonia: 1-2, sessile, 60-82 μ in width and 80-92 μ in length, ovate or oblong ovate, slightly oblique with a short beak towards antheridium.

Oospore: Completely fills the oogonium, 82 μ wide and 92 μ long, dark spotted, membrance smooth, three layered.

***V. sessilis* f. *orthocarpa* (Reinsch) Heering**

[Hoppaugh, 1930. 17: 336, Pl. XXIV, Fig. 2; Randhawa, 1942. 21:263, Figs. 1-4; Venkataraman, 1961. p. 69, Figs. 46d-e; Saxena, 1962. p. 29-31, Fig. 8].

Plates 3, 5, Figs. 12a, 13a, 12b, 13b.

Vegetative Filament: Monoecious, unbranched, 26-50 μ wide, siphonous, coenocytic, aseptate, yellowish green.

Antheridium: Single, straight or circular, 15-30 μ wide, present either besides an oogonium or between two oogonia with a short straight pedicel.

Oogonia: 1-2, sessile, oblong-ovate, 70-85 μ long and 60-72 μ wide, beak vertical, short and broad.

Oospore: Fills the oogonia, 74-90 μ long, 50-72 μ wide, dark spotted, membrane several layered (more than 3)

***V. sessilis* f. *repens* (Hassall) Hansgirg**

(Hoppaugh, 1930, 17:336, Pl. XXIV, Fig. 2; Tiffany and Britton, 1952. p. 212, Pl. 37, Fig. 383; Venkataraman, 1961. p. 70, Fig. 46b; Sarma, 1974. 20: 87-88, Figs. 2, 4, 5, 15 and 16]

Plates 3, 5, Figs. 14a, 14b.

Vegetative Filament: Siphononus, coenocytic 20-30 μ.

Antheridium: Single on a straight pedicel, develops near the oogonium 20 μ in diameter

Oogonium: Single sessile, oblong, beak horizontal to main filament, 74-78 μ long 64 μ broad.

Table 1 Occurrence of different species of *Vaucheria* sp. under different physicochemical characters of the habitat

Spot no.	%age of moisture	Soil texture	Organic carbon (%)	Organic matter (%)	CO''_3(%)	HCO'_3(%)	Ca^{++}(%)	Mg^{++}(%)	Cl^-(%)	pH
01	7.48	Sandy Loam	0.93	1.6	A	0.0793	0.005	0.004	0.0062	8.53
02	5.0	Silt Loam	0.99	1.7	A	0.0732	0.0035	0.0021	0.0028	8.1
04	13.7	Sandy Loam	0.799	1.37	A	0.0158	0.0033	0.0018	0.0034	5.32
05	23.1	Sandy Loam	0.52	0.89	A	0.0175	0.0031	0.0021	0.008	6.34
06	14.8	Silt Loam	0.811	1.39	A	0.0305	0.0033	0.0009	0.0059	5.79
07	10.4	Sandy Loam	0.82	1.41	A	0.0162	0.0032	0.0015	0.003	6.41
12	41.2	Sandy Loam	0.468	0.806	A	0.0183	0.0032	0.002	0.0098	7.2
14	19.6	Loam	0.731	1.26	A	0.0192	0.0105	0.0023	0.0072	8.2
16	13.0	Sandy Loam	0.819	1.411	A	0.0183	0.0038	0.001	0.0082	5.8
18	11.06	Loam	0.702	1.21	A	0.0366	0.0033	0.0003	0.0023	5.4
19	27.4	Loam	0.413	0.71	A	0.0168	0.0026	0.0007	0.0035	6.2
22	19.6	Loam	0.791	1.36	A	0.0183	0.0006	0.0001	0.016	7.8
25	14.0	Sandy/Silt	0.801	1.38	A	0.0427	0.0049	0.0017	0.0038	8.1

Table 2 Physicochemical analysis of various water habitats (spots)

Spot No.	DO_2	Free CO_2	Water Temp. (°C)	Air Temp. (°C)	CO_3^- (mg/l)	HCO_3^- (mg/l)	Ca^{++} (mg/l)	Mg^{++} (mg/l)	Cl^- (mg/l)	pH
03	10.2	A	15	–	A	329.4	33.6	8.748	15.9	8.0
08	9	A	16.3	–	A	366	37	11.08	16.9	7.0
09	8.04	2.91	25	25.5	A	744.1	12.4	27.5	25.3	9.3
11	1.7	A	18	–	A	549	45.4	17.15	28.9	8.6
23	12.24	A	11.5	–	A	274.5	71.8	7.41	10.49	7.9
24	8.31	14	16	17	A	536	17.6	4.2	20	8.1

Table 3 Qualitative composition of associated algal flora on different species of *Vaucheria* collected during Dec. 2002 to April 2004

S. No.	Name of species	Chlorophyceae						Cyanophyceae			Bacillariophyceae								
		Cosmarium	*Volvyox*	*Cladophora*	*Ulothrix*	*Spirogyra*	*Stigeoclonium*	*Oscillatoria*	*Merismopedia*	*Anabaena*	*Navicula*	*Nitzschia*	*Ropalodia*	*Synadra*	*Surirella*	*Pinnularia*	*Cymbella*	*Amphora*	*Gomphonema*
1.	*V. amphibia*	–	–	–	–	+	–	+	–	–	+	+	+	–	–	–	–	–	–
2.	*V. arechavaletae*	–	–	–	–	–	–	+	–	–	+	–	–	+	+	–	–	–	–
3.	*V. debaryana*	+	+	–	–	+	–	+	–	–	+	–	–	+	–	+	–	–	–
4.	*V. geminata*	–	–	–	–	–	–	+	–	–	+	+	–	+	–	–	–	–	–
5.	*V. hamata*	–	–	–	–	–	–	+	–	–	+	+	–	+	–	–	–	–	–
6.	*V. pachyderma*	+	–	+	–	+	–	+	–	–	+	–	–	+	–	–	+	+	–
7.	*V. sessilis*	–	–	–	–	–	–	+	–	–	+	+	+	+	+	–	–	–	–
8.	*V. sessilis* f. *orthocarpa*	–	–	–	–	+	+	–	–	+	+	–	–	+	–	+	–	–	–
9.	*V. sessilis* f. *repens*	–	–	–	–	–	–	–	–	–	+	+	–	+	–	–	–	–	+
10.	*V. taylorii*	–	–	–	+	–	–	–	+	–	+	–	+	+	–	+	–	–	–
11.	*V. terrestris*	–	–	–	–	–	–	+	–	–	+	+	–	+	–	–	–	–	–
12.	*V. walzi*	–	–	–	+	–	–	+	–	–	+	+	+	+	+	+	–	–	–

+ = Present

– = Absent

Oospore: Completely fills the oogonium, having dark spots and triple membrane 50 μ broad and 58 μ long.

V. *taylorii* Blum

[Entwisle, 1988. 27:183-200, Figs. 16-20; Schneider, 1999. 101 : 245-247, Figs. 26]

Plates 3, 5, Figs.15a, 15b.

Vegetation Filament: Monoecious, branched, aseptate, 28-38 μ wide, yellowish green, siphonous and coenocyctic.

Antheridium: Single, straight or circinate, 18 μ and 38 μ long, besides a single oogonium or between two oogonia.

Oogonia: 1-2, ovate-oblong, 44-56 μ wide and 50-80 μ long, born at erect angles on pegs from the fruiting branches with their tips pointing towards the antheridum.

Oospore: Fill the oogonium, 45-55 μ wide and 55-85 μ long, wall layers smooth.

Reproductive Branches: The reproductive branches are swollen centrally where oogonial pegs are formed.

V. terrestris Lyngbye em Walz

(Hoppaugh, 1930. 17:341, Pl. XXV, Figs. 10 and 11; Venkataraman, 1961. p. 76, Fig. 53; Saxena, 1962. p. 32 and 33, Fig. 9; Sarma, 1974. 20:88, Pl. 2, 5, 7, Figs. 9, 11, 12, 14, 20 and 25]

Plates 3, 5, Figs.16a, 16b.

Vegetative Filament: Monoecious, slender, 20-54 μ wide, coenocytic, aseptate and yellowish green.

Antheridium: Single, curved or circinate 20 μ wide, terminating the lateral fruiting branch, but pushed aside by the oogonium, thereby appearing to be lateral.

Oogonium: Single, sessile or with a short stalk on the pedicel of antheridium, 65-68 μ wide and 66-70 μ long, subglobose to ovoid, bent in the same plane as the antheridium.

Oosporre: Fills the oogonium, 66 μ wide and 84 μ long, globose with many brown spots and a three layered wall.

Fruiting Branch: 66 μ in diameter and of varying length (176-192 μ) bearing a single oogonium and a lateral antheridium.

V. walzi Rothert

(Venkataraman, 1961, p. 79-80, Fig. 14)

Plates 3, 5 Figs. 17a, 17b.

Vegetative Filament: Thallus monoecious, branched, 58-84 μ broad, siphonous, coenocytic and thick.

Antheridium: Single, stalked, 16-26 μ in diameter more or less coiled, terminating the fruiting branch.

Oogonium: Oogonia 2-3, ovoid, 60-80 μ in diameter and 55-65 μ long, 1-4 on pedicels, usually inclined downward towards the filament, occasionally horizontal or turned upward.

Oospore: 70-80 μ long and 55-65 μ wide, membrane three layered, mature oospore with brown or black spots.

Physicochemical parameters

Tables 1 and 2 reveal the physicochemical characteristics of various habitats from where collections have been made. It has been observed that specimens of *Vaucheria* growing in a particular habitat (aquatic, terrestrial, epilithic) are greatly being influenced by the factors like temperature, pH, carbonates, bicarbonates, calcium, magnesium, chlorides, etc.

Diversity has been observed with respect to algal association with different species of *Vaucheria* growing at different habitats (Table 3).

DISCUSSION

Taxonomy

Vaucheriaceae has been placed under the order Vaucheriales of Xanthophyceae. Venkataraman, 1961). Venkataraman (1961) reported 9 species of *Vaucheria* while Saxena (1962) reported 13 species of *Vaucheria* from India. During the course of present investigations, more than 18 species have been collected from different habitats of province Jammu. Out of the 18 species, 12 have been identified, viz. *V. hamata* Walz, *V. geminata* De Condolle, *V. terrestris* Lyngbye em Walz, *V. walzi* Rothert, *V. arechavaletae* Magnus and Wille, *V. debaryana* Woronin, *V. sesilis* (Vauch) De Condolle, *V. sessilis* (Vauch) De Condolle f. *repens* (Hassalle) Hansgirg, *V. sessilis* (Vauch) De Condolle f. *orthocarpa* (Reinsch) Heering, *V. pachyderma Walz, V. taylorii* Blum and *V. amphibia* Randhawa. Six species could not be identified as they varied greatly in their morphological (phenotypic) characters.

Present studies have been restricted to the taxonomy and ecology of *Vaucheria* of Jammu. Species of *Vaucheria* have been observed to grow on terrestrial as well as in aquatic (lentic and lotic) habitats. Species of *Vaucheria* have also been observed to grow as epiliths on stone submerged in the stream water and also on the wall of a tank, particularly the side from which water oozes out. Out of the 12 species, two species, i.e., *V. sessilis and V. walzi* occupy exclusively the flowing waters indicating their specific habitat preference. Similarly, 10 species, i.e., *V. hamata, V. terrestris, V. geminata, V. debaryana, V. arechavaletae, V. sessilis* f. *repens, V. sessilis* f. *orthocarpa, V. pachyderma, V. taylorii* and *V. amphibia* have been collected from the habitats like moist shady places of agricultural fields, roadside catchments, under the shades of trees etc. 3 species, i.e., *V. geminata, V. sessilis* f. *repens* and *V. pachyderma* preferred to grow both in terrestrial and aquatic habitats. 2 species were observed to grow as epiliths, i.e., *V. walzi* has been observed to grow on stone submerged in water of slow running stream while other (unidentified species) grew on the wall of a water tank.

Venkataraman (1961) categorized Vaucheriaceae into 9 sections under Vaucheriales, i.e., Androphorae, Piloboloidae, Woroninia, Tubuligerae, Globiferae, Corniculatae, Anomalae, Pseudoanomalae and Contortae. Out of the 9 sections, only four section, i.e., Globiferae, Corniculatae, Pseudoanomalae and Anomalae had the represented members of Vaucheriaceae growing in the Jammu province. 12 species of *Vaucheria* belonging to above-mentioned four sections have been collected from different habitats. Section Corniculatae has been studied in detail since nine species of *Vaucheria* belong to this section only. Corniculatae is further divided into two subsections, viz. subsection Sessilis and subsection Racemosae. Accordingly, out of 9 species, 3 belong to subsection Sessilis while 6 belongs to subsection Racemosae.

I Section Globiferae

V. pachyderma is the only representative in this section.

V. pachyderma Walz *V. pachyderma* Walz had been collected from Sukrala (Kathua Dist.). Air and water temperatures were very low with moist and cold climate. The present specimens were much smaller than those reported by Venkataraman (1961) who reported it from Europe and USA. Earlier the species had been reported from Europe, USA, the Netherlands, Bangladesh, China and Japan which speak about the type of habitat it required. Hoppaugh (1930) made three collections: one as dense mat on damp soil at Stanford University, shallow pool and other from a stream. Present collected specimens matched in their dimensions of vegetative filaments (40-58 μ) and oogonial dimensions (110-120 × 90-100 μ) to those described by Hoppaugh (1930) [Vegetative filament: 50-70 μ and oogonia: 90-100 × 98-120 μ]. Present collected material had been observed to grow on damp soil in an agricultural field in Sukrala (Dist. Kathua). It had been reported for the first time from India.

The variations in the vegetative and reproductive structural dimensions were due to differences in prevailing environmental conditions of the habitats.

II Section Corniculatae

It has two subsections.

II. i Subsection Sessilis: This section is represented by three species, i.e., *V. sessilis, V. sessilis* f. *repens* and *V. sessilis* f. *orthocarpa*.

II. ii Subsection Racemosae: In this sections there are six species, i.e., *V. amphibia, V. taylorii, V. terrestris, V. hamata, V. walzi and V. geminata*.

2. *V. sessilis* (Vauch) De Condolle

Material had been collected from the shore of slow moving lake Surinsar. The material collected showed marked variations in terms of position of beak and dimensions of oospore but matched well to that of *V. sessilis* in features like dimensions of oogonia, antheridium and oospore. So, the material collected from Surinsar deserved to be raised to the status of new forma.

3. *V. sessilis* (Vauch) De Condolle f. *orthocarpa* (Reinsch) Heering

Material had been collected from Sanasar (Udhampur Dist.) and was smaller in its dimensions as compared to those described by Hoppaugh (1930), Randhawa (1942), Venkataraman (1961) and Saxena (1962). It resembled to *V. sessilis* f. *clavata* in all its characters. The only difference being the number of wall layers of the oospore. In *V. sessilis* f. *clavata*, three wall layers had been reported while in *V. sessilis* f. *orthocarpa* seven wall layers of oospore had been reported by Hoppaugh, 1930; Venkataraman, 1961 and Saxena, 1962. In the present collected specimen not exactly seven wall layers had been observed but as more than three wall layers had been recorded, so present collected specimen came close to *V. sessilis* f. *orthocarpa*.

4. *V. sessilis* (Vauch) De Condolle f. *repens* (Hassall) Hansging

The material had been collected from damp soil along the shore of a drain in Girls' Hostel (Jammu Dist.). Vegetative filaments and oospores of the specimens differed slightly in their dimensions from those described by Hoppaugh (1930) and Sarma and Chapman (1975) but matched well with the description given by Tiffany and Britton (1952) and Venkataraman (1961). The main characters which distinguished the species *V. sessilis* f. *repens* (Hassall) Hangirg from *V. sessilis* (Vauch) De Condolle are:

(a) Narrower dimensions of vegetative filament

(b) Parallel beak of oogonia

5. *V. amphibia* Randhawa

The material had been collected from a small, slow moving, clear water stream at Assar (Dist. Doda). It differed from its nearest related species *V. terrestris* in its peculiar habitat and certain other characters which differed from those of *V. terrestris* and conform to the description of *V. amphibia* are:

(a) Spherical shape of oogonia and oospore

(b) Terminal position of oogonia

(c) Less curved nature of antheridia

Similar observations had also been reported by Venkataraman, 1961 and Saxena, 1962.

6. *V. taylorii* Blum

The species had been collected from the moist soil of small ditch along the shore of the lake Mansar. Number of oogonia, position of oogonia, position of antheridium, swollen pedicels of the presently studied algal material, agreed with that of *V. taylorii* described by Schneider, 1999. But at the same time, it showed some resemblance to *V. prona* in certain characteristic features, viz. dimensions of oogonia and proliferation of reproductive branch. Schneider (1999) distinguished *V. prona* from other species of *Vaucheria* on the basis of its peculiar character, i.e., pendent oogonial pegs that had not been observed in the present collected specimens. Thus, the present studied material is very close to *V. taylorii*.

7. *V. terrestris* Lyngbye em Walz

Present study material had been collected from. Sanasar (Udhampur Dist.) in the form of a mat of thick green filaments. Present materials of *V. terrestris* exhibited wide range of variations in width of filament and dimensions of sex organs. Comparative studies on the dimensions as reported by various authors, viz. Hoppaugh (1930), Venkataraman (1961), Saxena (1962) and Sarma (1973) had not been observed in the present material. Instead, a combination of measurements from these had been observed. Specimens of *V. terrestris* were observed to be much smaller in size than the one described by Hoppaugh (1930), Venkataraman (1961) and Saxena (1962) but slightly larger than those described by Sarma (1974). The present collected specimens coincided with the characteristic shape of species concerned. Sarma (1974) emphasized that for taxonomic studies "size" should be given less importance as variation in size was just an ecological feature. While shape of sex organs (antheridium and oogonium) was very characteristic for a particular species and is a genetic expression which must be given much importance.

Venkataraman (1961) and Saxena (1962) reported four wall layers of oospores while in the species collected from Sanasar only three wall layers had been observed. Several workers (C.f. Hoppaugh, 1930) had emphasized that oogonial wall adheres to mature zygote, but when it detached from pedicel, it gelatinizes and disappears. During the course of present study, this gelatinization could not be observed in the specimen collected from Sanasar.

The main characters which distinguished the species are:

(a) Presence of a solitary oogonium which was usually so shortly stalked that it appeared practically sessile.

(b) Parallel bending of antheridium and oogonium.

8. *V. hamata* Walz

The material had been collected from moist soil from grassy lawns at Dyalachak (Dist. Kathua). Present specimen resembled in its description more with *V. terrestris* as far as the position of antheridium and oogonium was concerned (terminal) but it had its own distinguishing features which differed it from *V. terrestris*, i.e., long upwardly curved pedicel of antheridium and oogonium. As far as dimensions of vegetative and reproductive organs were concerned, present specimens coincided with the description given by Venkataraman (1961) and Sarma (1973), but much smaller than those described by Hoppaugh, 1930. Only three walls of oospores had been observed in present collected specimens like that of Hoppaugh (1930), oogonial wall adheres to zygote when it falls and does not gelatinize as in *V. terrestris*. Otherwise, its description conforms to the description of *V. hamata.*

9. *V. walzi* Rothert

V. walzi Rothert collected from Jammu region did not exhibit prominent variations from those described by Venkataraman (1961) and Saxena (1962) except for small variations which could be due to different habitat conditions, i.e., present specimen had been collected from a stone submerged in a stream water while Saxena (1962) collected it from an irrigation canal (U.P.).

10. *V. geminata* De Condolle

Material had been collected from Mendhar (Poonch Dist). The present specimens coincided with the description of species *V. geminata* [(Hoppaugh (1930), Tiffany and Britton (1952), Venkataraman (1961), Saxena (1962) and Sarma (1974)] in terms of:

(a) antheridium born on a stripe longer than that of oogonium

(b) oogonium born laterally near the end of a fruiting branch on short pedicels below the antheridium.

The present collected specimens certainly showed some distinct variations in terms of larger dimensions of oogonium, antheridium and vegetative filament. So, needed to be raised to the level of a new variety under *V. geminata*. Hence, could be named as *V. geminata* var. *major* De Condolle.

III Section Anomale Hansgirg

The section is represented by *V. debaryana* only.

11. V. debaryana Woronin

Material had been collected from damp soil of an agricultural field at Bishnah. No striking variations had been observed from the one collected by Hoppaugh, 1930. Hoppaugh (1930) collected it from a damp sandy ground along the bank of a stream and on the bank of a river. Woronin (C.f. Hoppaugh, 1930) and observations of later workers showed that *V. debaryana* when grown in water deposited calcium carbonates as crystals or as a continuous sheath on the filament. But physicochemical analysis of the habitat soil showed complete absence of carbonates. So, it seemed that the habitat variability contributed variations with respect to certain morphological characters.

IV Section Pseudoanomalae Jao and Ley

This section has only one representative, i.e., *V. arechavaletae*

12. V. arechavaletae Magnus and Wille

The present specimen had been collected from the damp, dark soil of an agricultural field at Ghomanasa (Jammu Dist.). It showed small variations in terms of smaller dimensions of vegetative filament and that of antheridium from those described by Venkataraman, (1961) but otherwise very close to *V. arechavalatae* in terms of following characters:

(a) Cylindrical or slightly curved antheridium
(b) Oogonium born on a lateral below the antheridium
(c) Antheridium slightly bent towards the oogonium.

Ecological Observations

Due to altitudinal variations, different districts of Jammu province experience climatic variability and are responsible for the distribution of different species of *Vaucheria*. Of all the species, *V. sessilis, V. debaryana* and *V. arechavaletae* restricted to the planes of Jammu only. *V. sessilis* f. *repens, V. hamata* and *V. pachyderma* occurred both in plains as well as in high mountainous regions of Jammu province, indicating their adaptability to both low and high altitudinal climatic conditions. While species like *V. geminata, V. sessilis* f. orthocarpa, *V. terrestris, V. taylorii, V. walzi* and *V. amphibia* restricted to high altitudes of the province only, indicating their preference for high altitudinal conditions. Two species had been observed to occur in Kathua district, four in Jammu, five each in Udhampur and Doda districts while one species in Poonch district (which had not thoroughly been surveyed) respectively.

In general, it had been observed that though the species of *Vaucheria* occurred at both low and high altitudes, yet they preferred to grow at high altitudes where temperature was low coupled with cool and moist habitat conditions.

Seven species, viz. *V. hamata, V. terrestris, V. arechavaletae, V. debaryana, V. sessilis* f. *orthocarpa, V. taylorii* and *V. amphibia* were restricted to terrestrial habitats whereas three species *V. geminata, V. sessilis* f. *repens* and *V. pachyderma* occupied both terrestrial and aquatic habitats. *V. sessilis* occupied purely the aquatic habitat while *V. walzi* occurred as epilith in aquatic habitat, thereby, corresponding to 58.31%, 25%, 8.31% of species composition to respective habitats.

Species of *Vaucheria* preferred to occupy characteristic moist habitats, with moisture contents ranging from 10–30%. *V. debaryana* preferred soils with low moisture contents (5.1%) indicating its remarkable ability to overcome water stress conditions and grew successfully when conditions improved. Schneider et al. (2001); Schneider and Devit (2002) also pointed out that some species of *Vaucheria,* viz *V. undulata, V. prona,* and *V. frigida* had inbuilt ability to overcome the stress conditions, thus, conforming to ecological strategy. In general, species occurring in plain areas of districts possessed low moisture contents while those at higher altitudinal areas had high percentage of moisture contents.

Bear (1965) stated that moisture contents and organic matter of the soil showed positive correlation. But during the course of present studies, no such correlation had been observed. At spot 2 (Bishnah, Jammu Dist.) moisture contents had been observed to be lowest, i.e., 5.0% while organic matter was highest, i.e., 1.7%. Similarly, at spot 12 (Mansar, Udhampur Dist.) moisture contents had been highest, i.e., 41% while organic matter content to be the lowest, i e., 0.5%. According to Wilde et al., 1972, organic matter of the soil varies with respect to climatic factors, nature of the soil and specific requirement of plants. During the course of present studies it had been observed that, organic matter contents perhaps depended upon soil texture. Species *Vaucheria* preferred to grow at spot 2 (Bishnah, Jammu Dist.) having silt loamy soil, exhibiting low bulk density, consequently possessed high organic matter. On the contrary, species of *Vaucheria* also grew at spot 12 (Mansar, Udhampur Dist.) being sand loamy, having high bulk density with low organic matter. Brady (1974) also observed similar correlation between bulk density and organic matter. Odum (1971) had also reported that the amount of organic matter of soil was dependent upon soil texture. Thus, soil texture, bulk density and organic matter contents play a povital role in controlling the distribution of species of *Vaucheria*.

Present studies also revealed a positive correlations between soil organic matter and soil organic carbon which remained variable throughout the study period.

SUMMARY AND CONCLUSION

Eco-morphological studies on Vaucheriaceae of Jammu province (J&K state) were conducted during Dec. 2002 to April 2004. In all 24 spots representing different habitats (soil, water-streams, lakes, ponds, ditches, etc.) were visited, covered a range of 200 to 1010 metres of altitude.

The amplitude of variations with respect to morphological characters of various species of *Vaucheria* was quite significant. The variations have been discussed under the description of each species.

Soil samples and water samples collected from different habitats were analyzed with respect to various physicochemical parameters, i.e., soil texture, moisture contents, organic matter, temperature, pH, DO_2, free CO_2, CO_3^{--}, HCO_3^-, Cl^-, Ca^{++}, Mg^{++}, etc.

Since collections were made irregularly from different habitats during the period of one year, no correlation could be derived with respect to the occurrence of Vaucheriaceae and water and soil quality.

A list of associated algal flora has also been tabulated.

ACKNOWLEDGEMENTS

One of the authors (Namerta Sharma) is thankful to H.O.D. for the help and also to her friends for the cooperation.

REFERENCES

A.P.H.A. 1985. Standard Methods.

Bear, F. E. 1965. Chemistry of the soil (2nd ed. 1968). Oxford and IBH Publishing Co., New Delhi.

Bischoff, H. W. and Bold, H. C., 1963. Phycological studies in some algal from enchanted rock and related algal species. The Univ. of Texas Pub. No. : 6318..

Bohlin, K. 1901. Utkast till de groan algernas och arkengoni aternas fylogeni. (1): pp. 43. folding chart. Thesis. Lund. Publ. by authors.

Brady, N. C., 1974. The nature and properties of soil., 8th edition, New York, MacMillian Publishers.

Chapman, V. J. 1962. The Algae. Macmillan, London.

Entwistle, T.J. 1987. An evolution of taxonomic characters in the subsection Sessilis, Section Corniculatae of *Vaucheria. Phycologia* 26 (3) : 297-321.

Entswistle, T. J. 1988. An evolution of taxonomic characters in the *Vaucheria prona* complex. *Phycologia* 27 (2) : 183-200.

Fritsch, F. E. 1935. The structure of reproduction of algae. Vol. I. Camb. Univ. Press.

Hoppaugh, K.W. 1930. A taxonomic study of species of the genus *Vaucheria* collected in California. *Amr. J. Bt.* 17, (5): 329-347.

Iyengar, M. O. P. 1925. Notes on any two species of *Botrydium* from India. *J. Indian. Bot. Soc.* 4 : 193-201.

Morris, I. 1967. An introduction to the algae. Hutchinson, London

Odum, E. P., 1971. Fundamental of ecology. W. B. Saunders Co. Philadelphia, New York.

Pandey, S. N. and Trivedi, P. S., 1995. A textbook of algae. Vikas Publishing House (P) Ltd.

Pascher, A. 1914. Ber. Deutsch. *Bot. Ges.* 32: 136-160.

Papenfuss, G. F. 1955. Classification of algae. In a century of progress in the natural sciences. *Proc. Calif. Acad. Sci.* 115-224. San Francisco.

Prescott, G. W. 1969. The algae: A review. Thomas Nelson and Son, London

Randhawa, M. S. 1942. Further observations on Vuacheriaceae from Northern India. *J. Indian Bot. Soc.*, 21 (5&6) (1982), pp. 263-266.

Round, F. E. 1965. The biology of the algae. Edward Arnold Pub. London.

Sarma, P. 1973. Addition to the freshwater algae of New Zealand-I. Tane 19 : 167-173.

Sarma, P. 1974. Addition to the freshwater algae of New Zealand-II. Tane 20 : 86-100.

Sarma, P. and Chapman, V. J. 1975. Light and SEM study of *Vaucheria pachyderma* Walz collected from Auckland, New Zealand. Nova Hedwigia. XXVI.

Saxena, P. N. 1962. Algae of India-2. Vuacheriaceae. Bull. National Botanic Garden, Lucknow. No. 70.

Scheneider and Lane, 1999. The freshwater species of *Vaucheria* from Connecticut. Rhodora. Vol. 101, No. 907. 234-263.

Scheneider and Lane, 2000. Two species of Vaucheria new for New England, V. Lii and V. Racemosa. Northeastern Naturalist. 7 (1) : 25-32.

Scheneider, C.W., Dunphy, M. E, Devit, D. C. and Lane, C. E. 2001. The survival of *Vaucheria propagules* in desiccated New England R1 parian sediments. Rhodora, Vol. 103, No. 916, pp. 416-426

Scheneider and Mc Devit 2002. The survival of *Vaucheria propogules* in New England Riparian Sediments after reported freeze/thaw cycle. *Rhodora* 104, (918): 161-169.

Smith, G.M. 1938, 1955. Cryptogamic Botany. Vol. 1, McGraw-Hill, New York

Tiffany, L.H. and Britton, M.E. 1952. The algae of Illinois. The University of Chicago Press, Chicago.

Trainor, F.R. 1978. Introductory phycology. John Willy and Sons, Inc.

Venkataraman, G.S. 1961. Vaucheriaceae. I.C.A.R., New Delhi.

Wilde et al., 1972. Soil and plant analysis. Oxford and IBH Publishing Co.

28

Biodiversity of Freshwater Algae in Some Districts of Tamil Nadu, India

Mahendra Perumal, G. Ganesan and N. Anand
Centre for Advanced Studies in Botany
University of Madras, Chennai-600 025, India

ABSTRACT

Collections were made from freshwater bodies in Tiruvallur, Kanchipuram, and Tiruvnnamalai districts of Tamil Nadu. A total number of 43 genera belonging to Cyanophyceae, Chlorophyceae and Bacillariophyceae have been recorded. Chlorophyceae was found to be dominant. The genera were *Gloeocapas, Chroococcus, Gloeothece, Merismopedia, Oscillatoria, Lyngbya, Microcoleus, Spirulina, Arthrospira, Nostoc, Calothrix, Westiellopsis, Pandorina, Eudorina, Volvax, Characium, Hydrodictyon, Pediastrum, Coelastrum, Scenedesmus, Ankistrodesmus, Kirchneriella, Schizomeris, Pithophora, Cladophora, Oedogonium, Bulbocheate, Zygnema, Spirogyra, Closterium, Staurastrum, Euastrum, Micrasterias, Cosmarium, Cyclotella, Synedra, Fragilaria, Nitzschia, Hantzchia, Cymbella, Navicula, Pinnularia and Phacus.* The illustrations are presented.

Keywords: Cyanophyceae, Chlorophyceae, Tamil Nadu.

INTRODUCTION

Biodiversity refers to the variability among the different life forms or the number of species or races, within the species and also includes the habitat diversity. The freshwater ecosystem is generally divided into lotic and lentic type. The lotic waters include the springs, the streams, canals, the waterfalls and rivers and rivulets. The lentic system includes the pools, puddles, ponds, reservoirs, lake and the agricultural fields like paddy fields (Anand et al., 1995). The freshwater algae are also differentiated into various types as planktons (free floating), benthons (attached to sediments), epiphytic algae (attached to larger algae and hydrophytes) epilithic algae (on stones and rock of reservoirs and lakes) and epipelic algae (attached to sand and mud). The biodiversity of freshwater algae, which appear, disappear and reappear according to the seasons.

MATERIALS AND METHODS

Random sampling techniques have been applied in the algal collection procedure. Sample collections were made between October 2004 and October 2005, from freshwater bodies in Tiruvallur, Kanchipuram and Tiruvnnamalai of Tamil Nadu. The different type of algal forms were collected from lotic and lentic environment. The soil is productive and climates are best suited for different classes of algae, viz. Cyanophyceae, Chlorophyceae, Bacillariophyceae and Euglenophyceae.

Collection of desmids and diatoms were made after Williamson (1999) method, by using a bulb pipette attached to 25 cm of plastic tube which was used to suck up the detritus and grit from selected aquatic habitats, after an initial examination of the living sample, the coarser material was removed by filtration through a mesh net.

Table 1 Places of collection of freshwater algae in different districts of Tamil Nadu

District	Collection Area	No. of Samples
Tiruvallur	Gummidipundi, Ponneri, Uttukottai, Ambathur, Poonamallee, Tiruttani, Senthamangalam and Pallipattu.	38
Kanchipuram	Sriperumbudur, Chengalpattu, Tirukkalukkundram, Madurantakam, Tambaram, Uttiramerur and Cheyur.	34
Tiruvnnamalai	Polur, Chengam, Arani, Tiruvettipuram and Dushinattakollai.	42
	Total	**114**

The algal samples were preserved in 4% formalin (aqueous solution of formaldehyde), Bacillariophycean forms were studied after cleaning the frustules (Prasad and Singh, 1996). Cyanophycean forms were stained by Methylene blue whereas Chlorophycean forms were stained by iodine. Glycerine was used for mounting the material. The entire organisms have been photographed using a microscope and SANYO Digital Colour CCD camera.

MORPHO-TAXONOMIC OBSERVATIONS

Genera belonging to class Cyanophyceae have been arranged according to Fritsch (1945), Desikachary (1959), and Anand (1998). Arrangement of Chlrophycean genera is according to Fritsch (1935), Prasad and Srivastava (1992). Bacilllariophycean taxa are systematized according to Prasad and Srivastava (1992).

Class: Cyanophyceae

Member of Prokaryotes and also referred to as Cyanobacteria. Vegetative cells may be unicellular or form multicellular colonies of regular or irregular forms expanded sheets and branched or unbranched filaments. Flagella are absent. The gelatinous sheath of certain filamentous genera may enclose one or more trichomes. The sheath may be colourless, transparent and so thin they are observed with difficulty or they may be thick, frequently stratified, firm, coloured and readily evident due to prokaryotic nature, a definite nucleus never observed. Motality is characteristic of many filamentous genera, movement consists of forward and backward gliding or may be an oscillating or waving type.

Order–Chroococcales
Family–Chroococcaceae

***Aphanocapsa grevillei* (Hass.) Rabenh Fig. 1**

Thallus light blue-green algae, cell spherical 2-3.5 µm in diameter closely arranged in pairs or in groups of four

***Aphanocapsa pulchra* (Kutz.) Rabenh. Fig. 2**

Thallus gelatinous, homogenous, blue-green algae, free, cell spherical 4.7-6.0 µm, in diameter loosely arranged, single or twos in pale blue-green individual sheath of cell indistinct.

***Aphanocapsa roeseana* de Bary Fig. 3.**

Thallus brownish-green, irregular shaped, gelatinous mucilage, cells spherical or sub spherical 3-7 µm in diameter.

***Aphanothece bullosa* (Menegh) Rabenh. Fig. 4.**

Thallus yellowish blue-green algae, more or less spherical or irregularly shaped, cells long broad, cylindrical, 2-4 cells grouped together in a common mucilaginous envelope.

***Aphanothece clathrata* West, W. and G. S. Fig. 5.**

Thallus microscopic, small, when planktonic or large when benthic, often irregular, calathrate, with colourless homogeneous mucilage, cells long 3.5-4.5 µm, pale blue green.

***Aphanothece gelatinosa* (Henn.) Lemmermann Fig. 6.**

Cells ovate, nearly spherical, compactly arranged in globular, blue green to brownish gelatinous attached masses, cell 3.8-4 µm in diameter.

***Chroococcus disperses* var. minor G. M. Smith Fig. 7.**

A variety differing from the typical in having smaller cells, 1.7-2.5 µm in diameter, arranged at some distance from one another.

***Chroococcus limneticus* var. elegans G. M. Smith Fig. 8.**

Free floating colony cells, cells 18-22 µm. cells irregularly scattered in a wide, gelatinous colonial envelope.

***Chroococcus turgidus* (Kuetz) Naegeli Fig. 9.**

A free-floating colony of 2-4 ovoid or hemispherical cells inclosed by a very wide, colonial sheath, cells bright blue-green algae.

***Gloeocapsa gelatinosa* Kuetz. Fig. 10.**

Colonies single or many forming an expanded mass, cells without sheath, 2.5 µm broad.

***Gloeocapsa nigrescens* Naeg. Fig. 11**

Colonies single or many, cells up to 6.0 µm.

***Gloethece rupestris* (Lyngb.) Bornet Fig. 12.**

Sheath and colonies similar to *Gloeocapsa*, mucilage envelope colourless, cells ellipsoidal 12.3 µm broad.

Biodiversity of freshwater algae in Tamil Nadu.

Districts of Tamil Nadu

***Merismopedia elegans* A. Braun Fig. 13.**

Colonies small or big, 16-4000 celled, cell spherical or oblong, more or less closely arranged, 5.5-6.5 μm broad and 8 μm long, bluish green.

Order-Nostocales

Family-Oscillatoriaceae

***Arthrospira platensis* (Nordst) Gam. Fig. 14.**

Filaments short, spiral broad, showing distinct cross wall of cells.

***Lyngbya spiralis* Geitler Fig. 15.**

Filaments spirally coiled, 6.2 μm broad, with thin or very thick sheath.

***Oscillatoria subbrevis* Schmidle. Fig. 16.**

Trichomes single or forming a flat without sheath, motile, trichomes 5.5 μm broad.

***Spirulina major* Kuetz. ex. Gomont Fig. 17.**

Trichomes 1.2 μm-1.8 μm in diameter, forming more or less somewhat irregular spirals.

***Spirulina meneghiniana* Zanard. ex. Gomont Fig. 18**

Trichomes 1.0-1.5 μm in diameter, forming more or less regular spirals.

***Spirulina subsalsa* Gomont. Fig. 19.**

Trichomes 1-1.5 μm in diameter, forming somewhat irregular very close or rarely regular spirals, the turns between two spirals either contiguous.

***Microcoleus acutissimus* Gardner Fig. 20.**

Filaments with many trichomes in bundles inside the firm or gelatinizing sheath, trichomes aggregated, end of trichomes attenuated, Trichomes up to 3.0. μm.

Family: Nostocaceae

***Anabaena ambigua* Rao, C. B. Fig. 21.**

End cells obtusely rounded, heterocysts 4.5-5.5 μm in diameter spores ellipsoidal one on each side of the heterocyst.

***Anabaena azollae* Strasburger Fig. 22.**

Trichomes straight or coiled, filamentous epiphytic, cells subglobose to ellipsoid, heterocyst ovate 6-9.5 μm in diameter.

***Nostoc muscorum* Ag. Fig. 23.**

Thallus mucilaginous gelatinous of various shapes, some time brownish in colour, cells spherical, trichomes 5.0 μm broad.

Family- Rivulariaceae

***Calothrix brevissima* West, G. S. Fig. 24.**

Filaments epiphytic, Trichome very short, not gradually attenuated, length 53-94 μm, sheath firm, colourless.

***Calothrix parietina* Thuret ex. Born. and Flah. Fig. 25.**

Filaments single, thallus brownish to dark black in colour, Trichome 5-10 μm broad.

Order-Stigonematales

Family-Stigonemataceae

***Westiellopsis prolifica* Janet Fig. 26.**

Freshwater Algae of *Cyanophyceae*

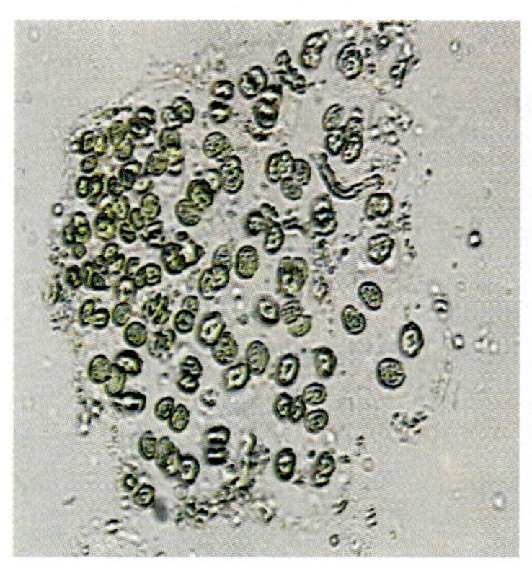

1. *Aphanocapsa grevillei*

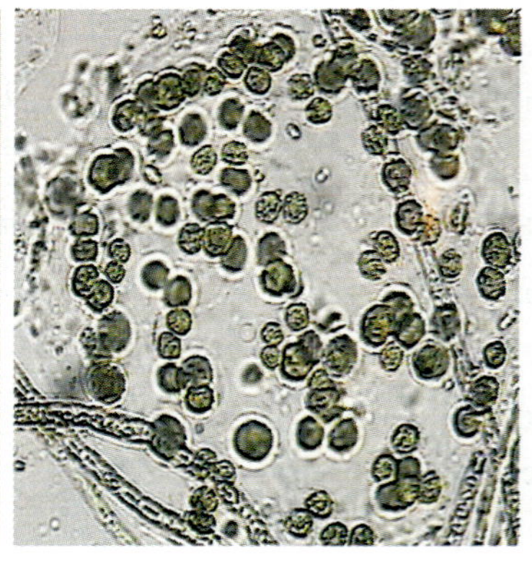

2. *Aphanocapsa pulchra*

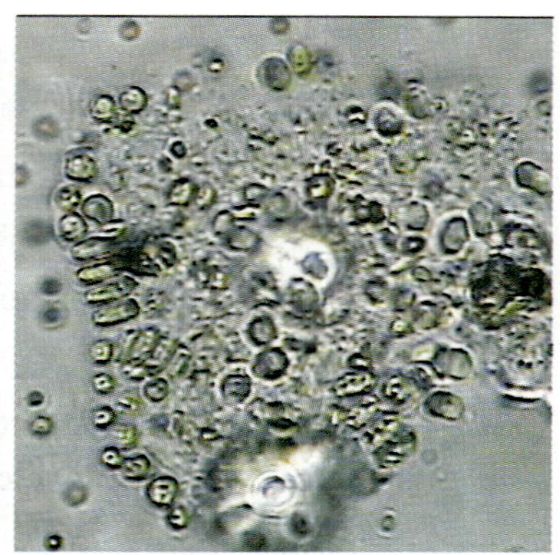

3. *Aphanocapsa roeseana*

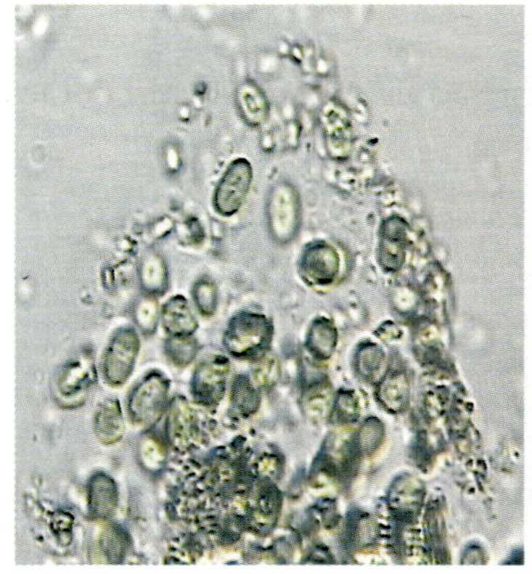

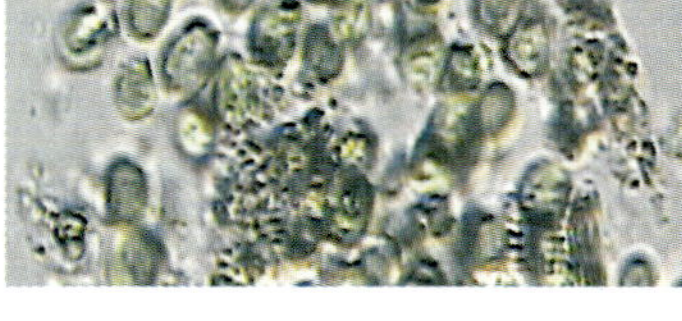

4. *Aphanothece bullosa*

5. *Aphanothece clathrata*

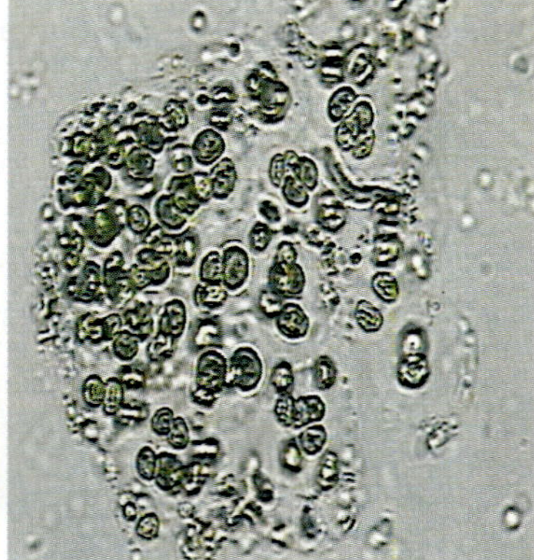

6. *Aphanothece gelatinosa*

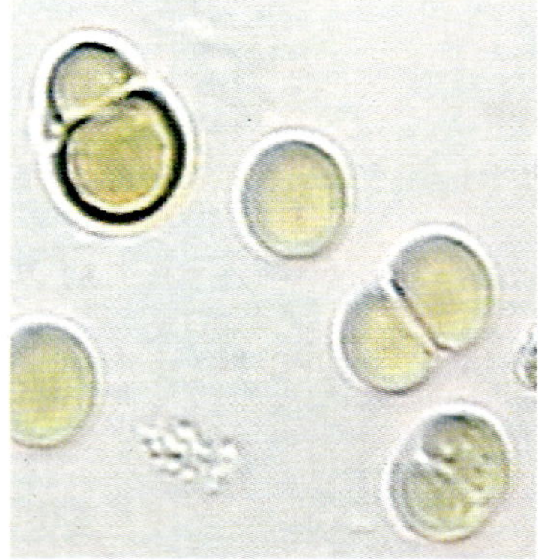

7. *Chroococcus disperses*

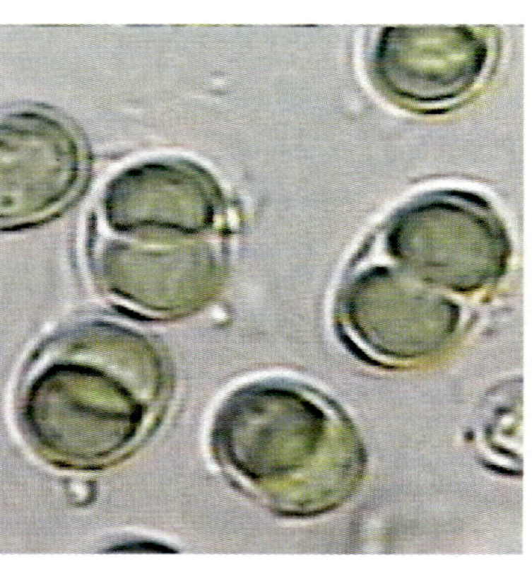

8. *Chroococcus limneticus*

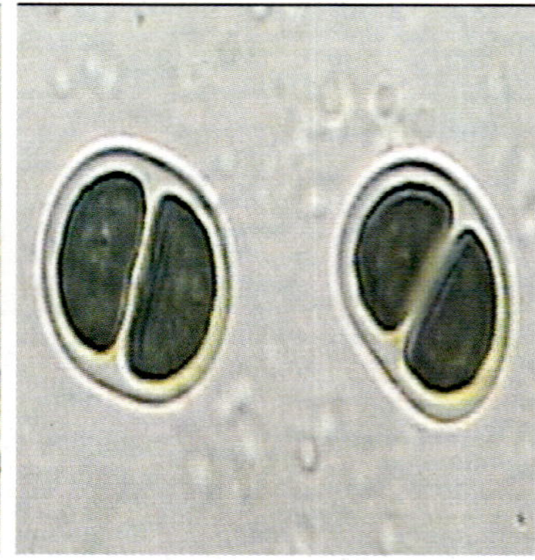

9. *Chroococcus turgidus*

Freshwater algae of *Cyanophyceae*

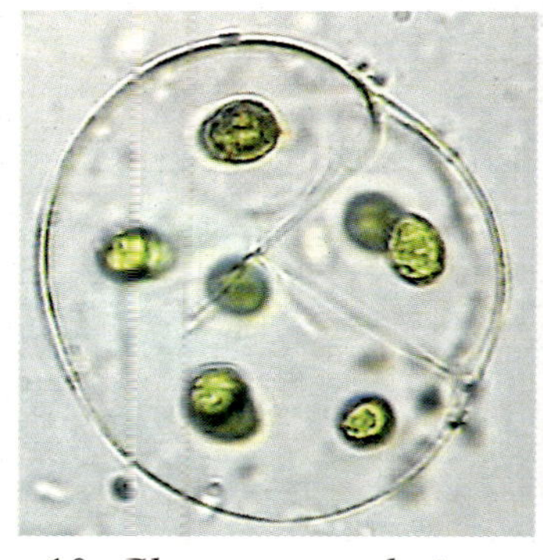

10. *Gloeocapsa gelatinosa*

11. *Gloeocapsa nigrescens*

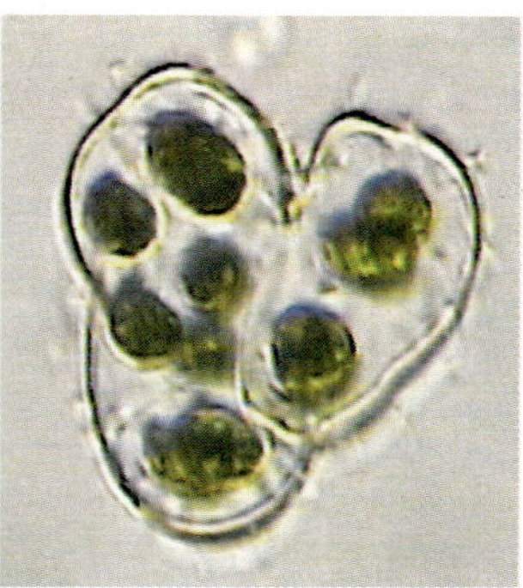

12. *Gloethece rupestris*

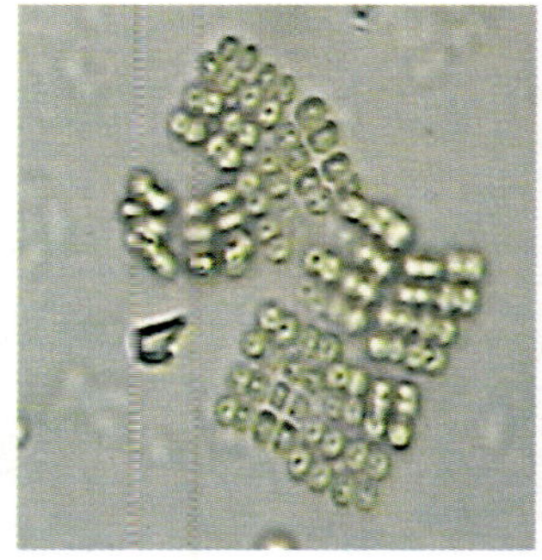

13. *Merismopedia elegans*

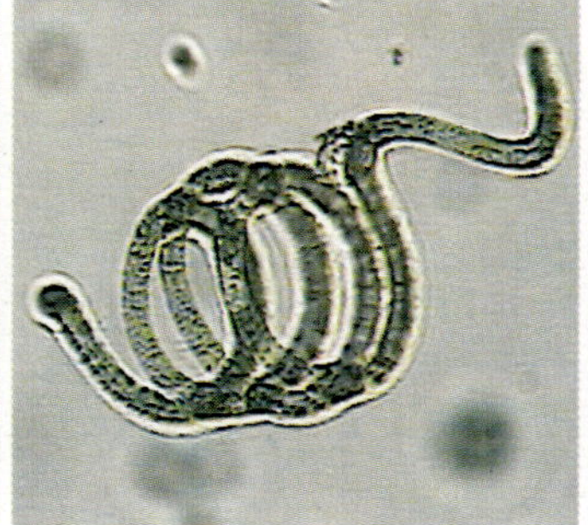

14. *Arthrospira platensis*

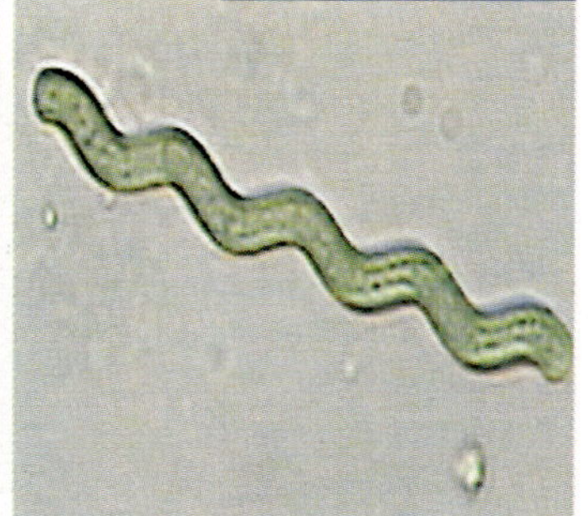

15. *Lyngbya spiralis*

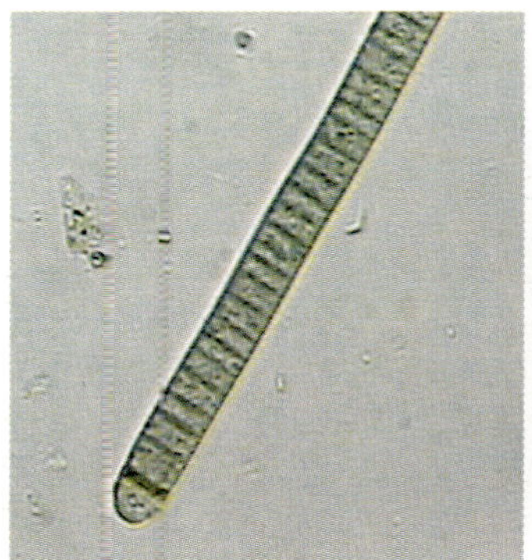

16. *Oscillatoria subbrevis*

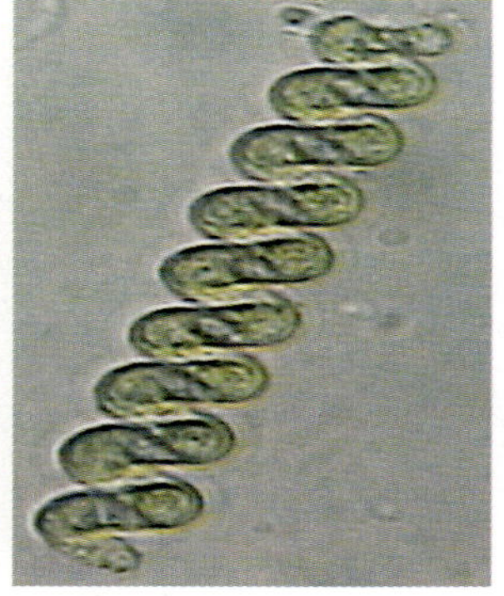

17. *Spirulina major*

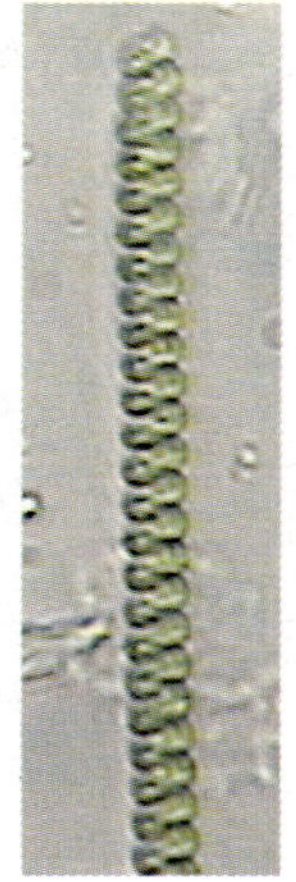

18. *Spirulina subsalsa*

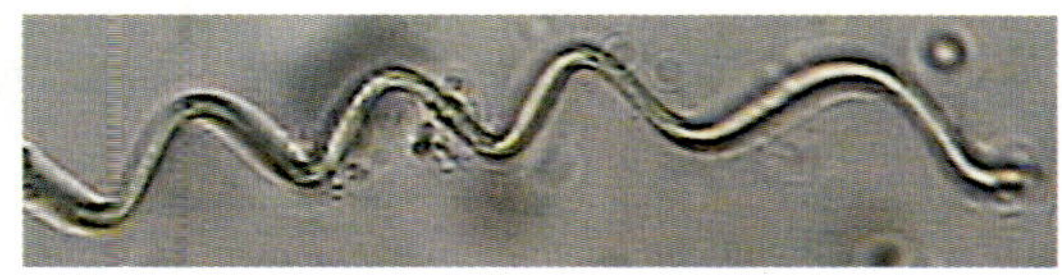

19. *Spirulina meneghiniana*

Freshwater algae of *Cyanophyceae*

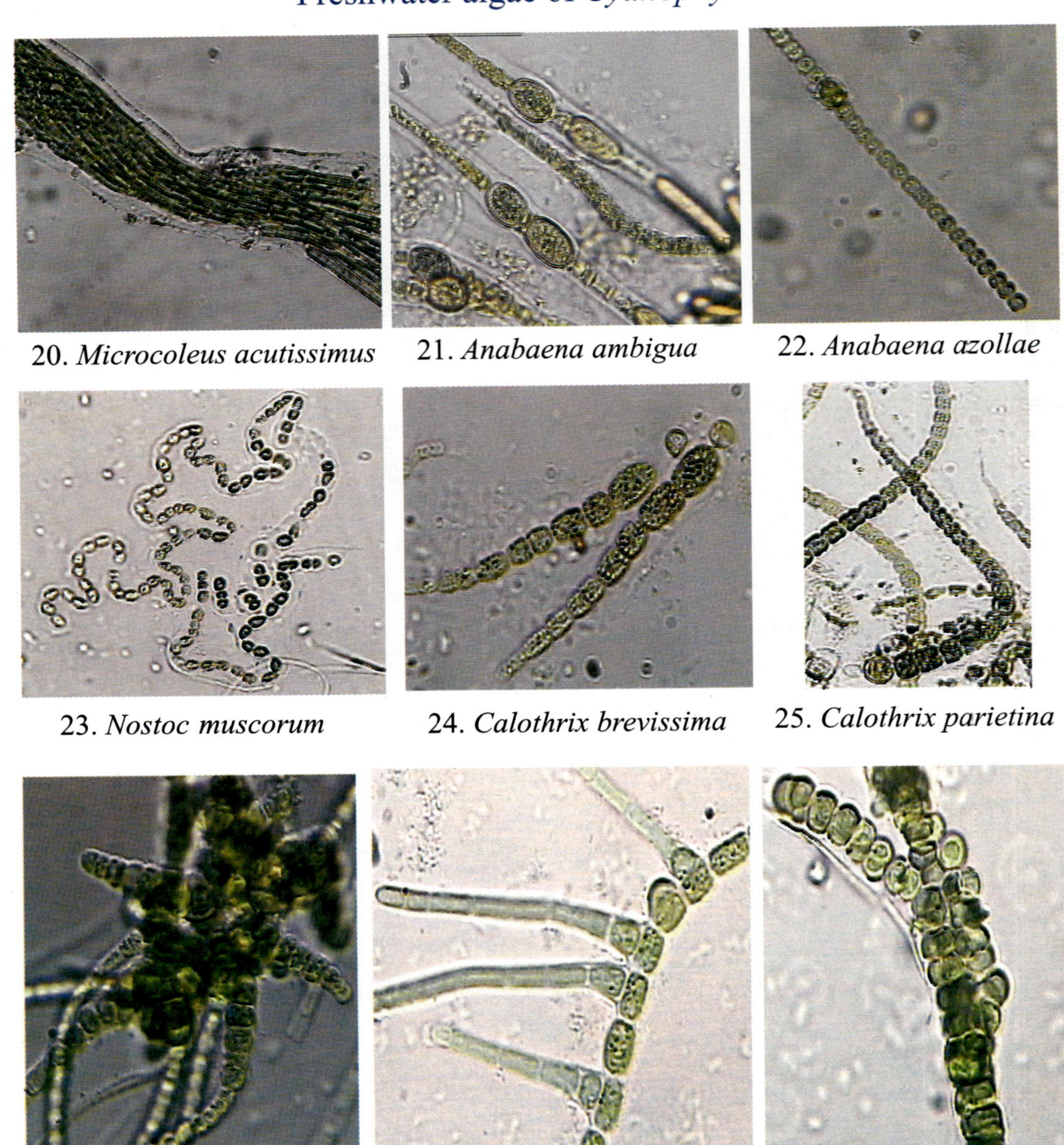

20. *Microcoleus acutissimus* 21. *Anabaena ambigua* 22. *Anabaena azollae*

23. *Nostoc muscorum* 24. *Calothrix brevissima* 25. *Calothrix parietina*

26. *Westiellopsis prolifica* 27. *Hapalosiphon welwitschii* 28. *Stigonema turfaceum*

Filaments loosely entangled, lateral branches clearly seen, cells of main filaments short, barrel shaped, cells of main filament 5-9 μm broad.

Hapalosiphon welwitschii **West and West Fig. 27.**

Trichomes consisting of one row of cells, irregularly branched, trichome light blue-green algae, main filament 4.4.5 μm.

Stigonema turfaceum **(Berkeley) Cooke Fig. 28.**

Filaments much-branched, long and curved, lamellate and yellow brown in colour, filaments 27-37 μm.

Class: Chlorophyceae

Chlorophyceae (green algae) may be unicellular or multicellular thalli, branched or unbranched filamentous and hallow or solid cylinders. Vegetative cells are uni- or multinucleate, motile or immobile, free floating or sessile and usually contain one or more chloroplasts that are highly variable among different species. Chloroplasts are single to numerous in each cell, axial or parietal in position entire or perforate and may be cup-shaped discoid, stellate laminate or a variety of other shape. Motile cell whether vegetative or reproductive bear anterior flagella usually of equal length, 2 in number, they may be 2-8 or many.

Order- Volvocales

Family- Volvocaceae

Pandorina morum **(Muell) Bory Fig. 29.**

Colonies usually distinctly ovate, cells pyriform, crowed, diameter12-17 μm long.

Eudorina elegans **Ehr. Fig. 30.**

Colony spherical or ovate with 16-32 ovoid cells evenly disposed within a gelatinous envelope.

Volvox aureus **Ehr. Fig. 31.**

Colony of 1300-3200 cells, which are not enclosed by individual sheaths.

Order- Chlorococcales

Family- Characiaceae

Characium rabenhorstii **DeToni Fig. 32.**

Cells elongate–fusiform or lanceolate, narrowed anteriorly but with a bluntly rounded apex, with a brown basal attaching disc, cells 16-18 μm long.

Order- Chlorococcales

Family-Hydrodictyaceae

Hydrodictyon reticulatum **(Linnaeus) Lagerheim Fig. 33.**

Colonies reticulate, meshes pentagonal or hexagonal, cells elongated cylindrical, cells 15-16 μm in diameter.

***Pediastrum duplex* Meyen Fig. 34.**

Colonies free floating, disc shaped to stellate, generally 16 celled, polygonal cells arranged in a single layer, outer face of marginal cell with two projections.

***Pediastrum boryanum* (Turpin) Menegh Fig. 35.**

Outer face of marginal cells with two projections, cell wall smooth or punctate, long cell 12-16.5 µm.

***Pediastrum simplex* Meyen var. duodenarium Fig. 36.**

Colonies of 4-8-16-32-64 cells, interface of marginal cells concave, outer face prolonged into a single delicately tapering process. Cells 10-45 µm long.

***Pediastrum tetras* Ralfs var. *excisum* (Rabenh) Hansgrig Fig. 37.**

Colonies 4 celled, lobed adjacent to incision more or less deeply concave, cell wall smooth, diameter of colonies 25-28 µm, cells 14-18 µm in diameter.

Family- Selenastraceae

***Ankistrodesmus falcatus* (Corda) Ralfs Fig. 38.**

Cells acicular to narrowly fusiform with the end tapering to actual apices usually in fasciculate bundle of 2-4-8 or more, rarely solitary, chloroplast single, cells, 6-8 µm broad, 170 µm long.

***Kirchneriella lunaris* (Kirchner) Moebius Fig. 39.**

Cells lunate to sickle-shaped with pointed ends, cells flattened, strongly curved with pointed ends.

***Kirchneriella obesa* (W. West) Schmidle Fig. 40.**

Cells strongly lunate with their bluntly rounded ends placed very near to each other, long cell 7-13.5 µm.

Family- Coelastraceae

***Coelastrum cambricum* Archer var. *intermedium* (Bohlin) Fig. 41.**

Cells 10-12 sided with short connecting processes. Subspherical at periphery with slightly thick, blunt and rounded projections.

***Cloelastrum microsporum* Naegeli in A. Braun Fig. 42.**

Coenobium spherical, composed of 8-64 sheathed globose cells, closely adjoined. Cells 8-20 µm in diameter.

***Coelastrum sphaericum* Naegeli Fig. 43.**

Coenobium ovoid, composed of conical cells, the narrow end directed outward.

Family- Scenedesmaceae

***Scenedesmus acutiformis* Schroeder Fig. 44.**

Colonies 2,4 celled, cells cylindrical fusiform and arranged in a single linear series, cell wall smooth, median cell with a lateral longitudinal ridge extending from pole to pole on each side. Cells 6-8 µm broad, 17 µm long.

***Scenedesmus armatus* (Chodat) G. M. Smith Fig. 45.**

Colonies usually four celled, rarely two-eight celled, cells oblong-ellipsoid with acute apices and arranged in a series, terminal cells with a single long spine from each pole. Cells 3-8 μm broad.

***Scenedesmus bijugatus* (Turpin) Kuetz. Fig. 46.**

Colonies flat, consisting of 4 cells arranged in a linear series, cell oblong ellipsoid to ovoid with broadly rounded ends.

***Scenedesmus bijugatus* (Turp) Kuetz. f. *irregularis* Fig. 47.**

Cells arranged in a sub-alternating or more or less in double series, inner cells joined to each other basal and lateral portions.

***Scendesmus quadricauda* (Turp) Fig. 48.**

Colonies 4 celled, ovoid –cylindrical, spines on poles of terminal cells slightly curved.

***Scenedemus arcuatus* Lemm. Fig. 49.**

Colonies usually eight celled, curved and with small intercellular spaces, oblong ovoid, cell wall smooth, cells 3.5 μm broad.

Order-Ulotrichales

Family- Schizomeridaceae

***Schizomeris leibleinii* Kuetzing Fig. 50.**

Filaments stout, macroscopic, the cross wall of the lower cells are thick plates.

Order- Cladphorales

Family- Cladophoraceae

***Cladophora* sp. Fig. 51.**

Branched filamentous thallus with basal distal differentiation.

***Pithophora oedogonia* (Mont) Wittrock Fig. 52.**

Filaments slender, branching mostly solitary, akinetes cylindrical or slightly swollen to cask shaped.

Order-Chaetophorales

Family-Chaetophoraceae

***Stigeoclonium subsecundum* kuetzing Fig. 53.**

Filaments elongate and sparingly branched, the branches gradually and gracefully tapering to fine points alternate in origin, cells-12-18 μm in diameter.

Order- Oedogoniales

Family- Oedogoniaceae

***Oedogonium alternans* Wittrock and Lund. Fig. 54.**

Androsporangia hypogynous, epigynous spore wall reticulate costate, oogonia single.

***Bulbochaete setigera* (Roth) C.A. Agardh Fig. 55.**

Freshwater algae of *Chlorophyceae*

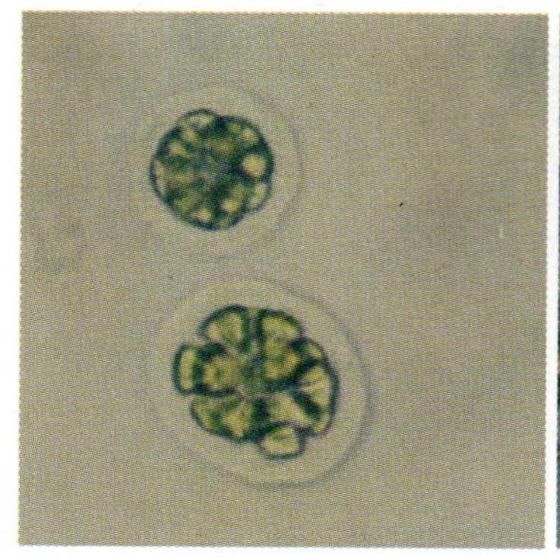

29. *Pandorina morum*

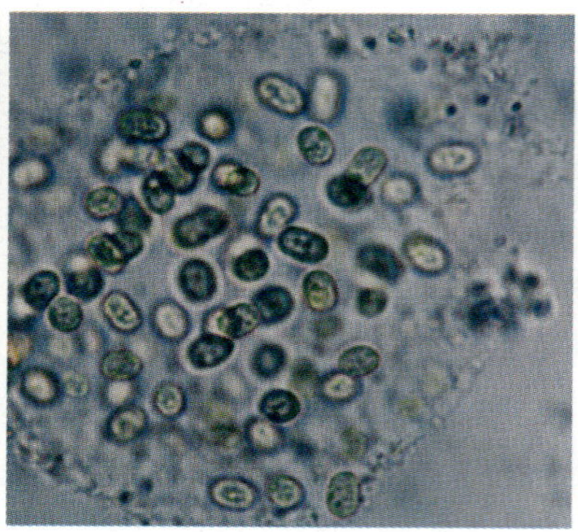

30. *Eudorina elegans*

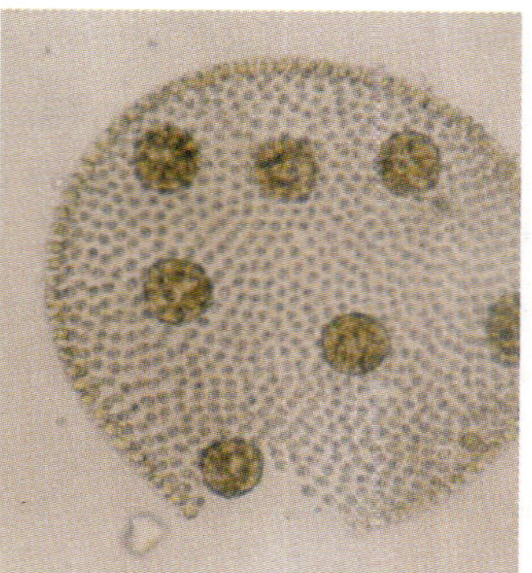

31. *Volvox aureus*

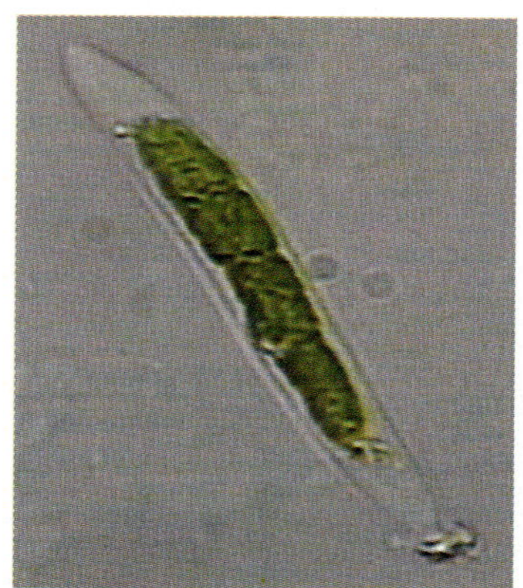

32. *Characium rabenhorstii*

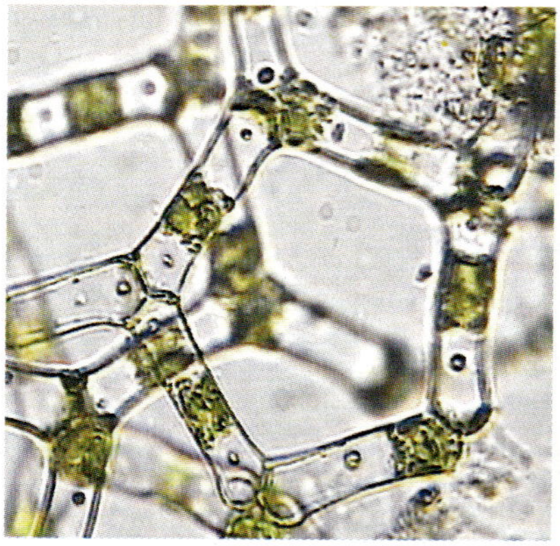

33. *Hydrodictyon reticulatum*

34. *Pediastrum duplex*

35. *Pediastrum boryanum*

36. *Pediastrum simplex*

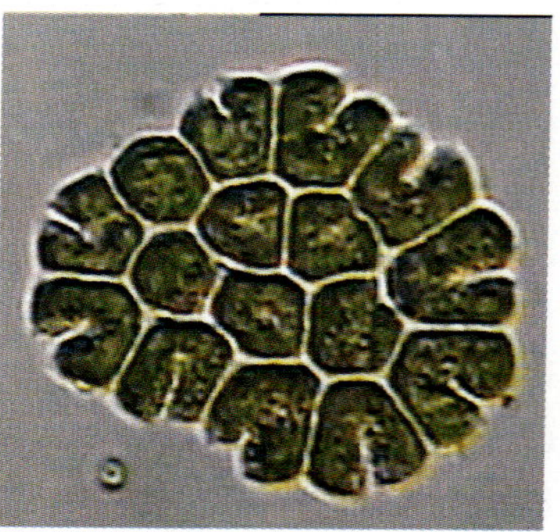

37. *Pediastrum tetras*

Freshwater algae of *Chlorophyceae*

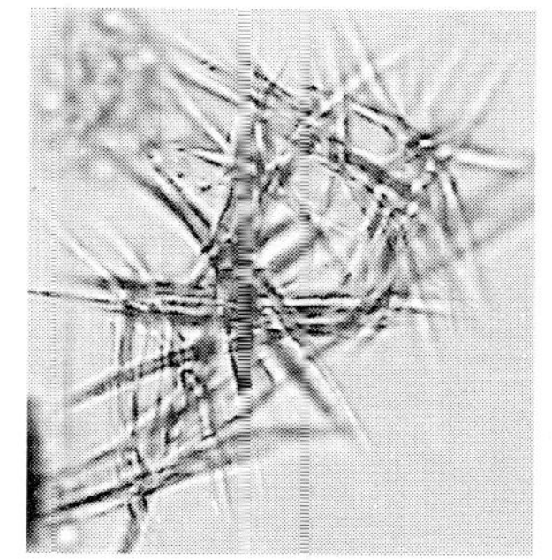
38. *Ankistrcdesmus falcatus*

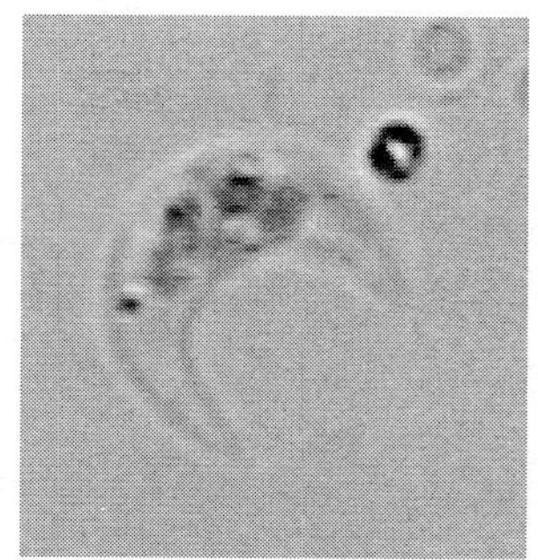
39. *Kirchneriella lunaris*

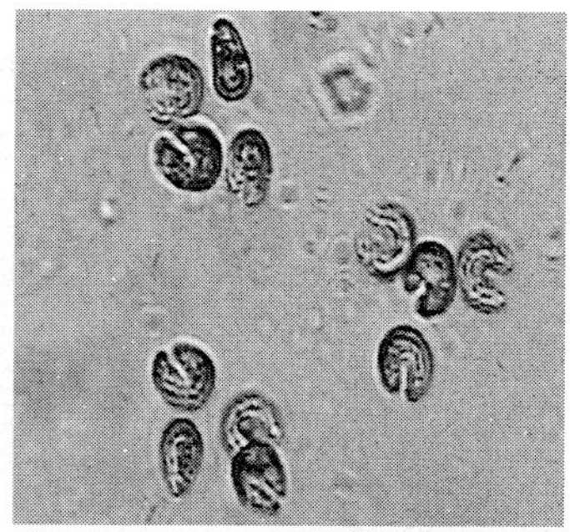
40. *Kirchneriella obesa*

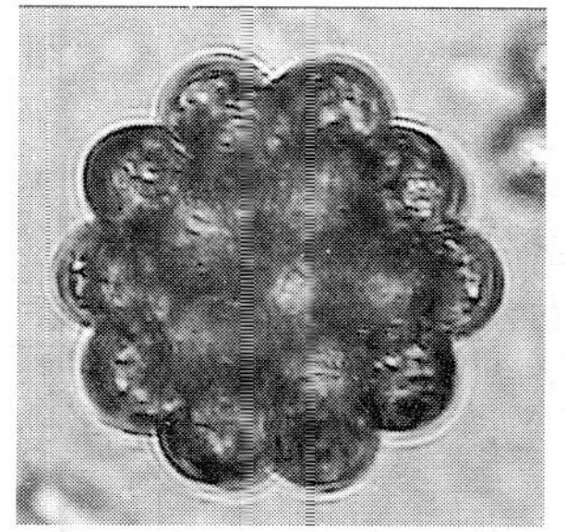
41. *Coelastrum cambricum*

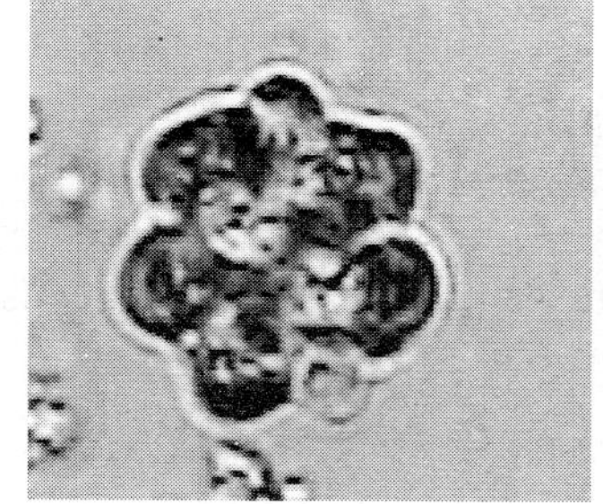
42. *Cloelastrum microsporum*

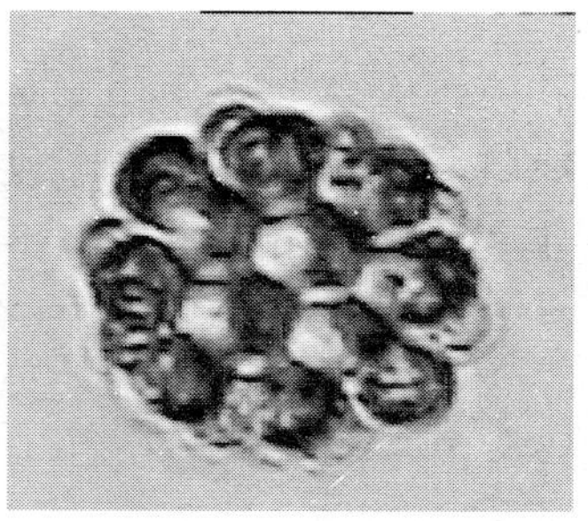
43. *Coelastrum sphaericum*

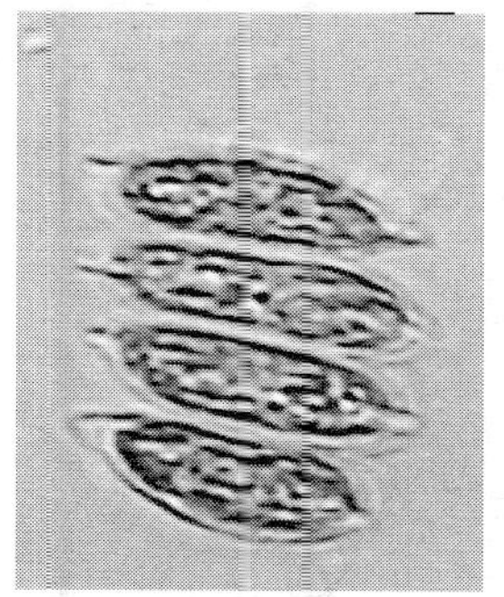
44. *Scenedesmus acutiformis*

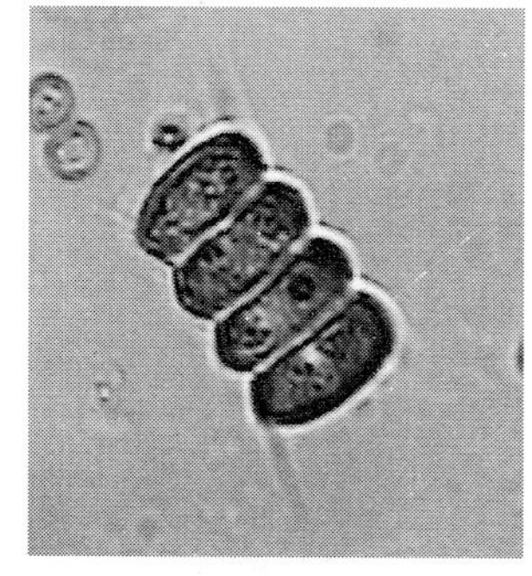
45. *Scenedesmus armatus*

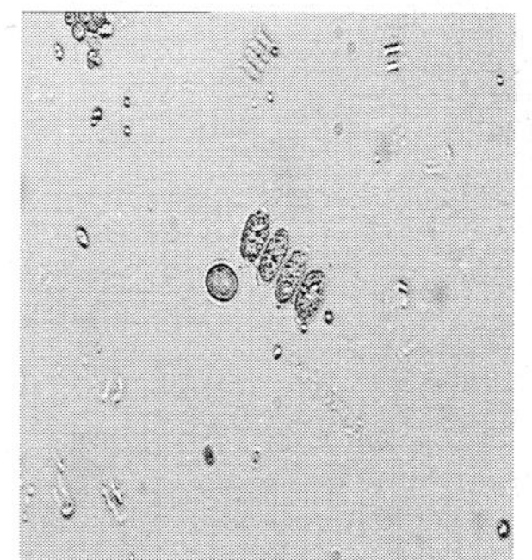
46. *Scenedesmus bijugatus*

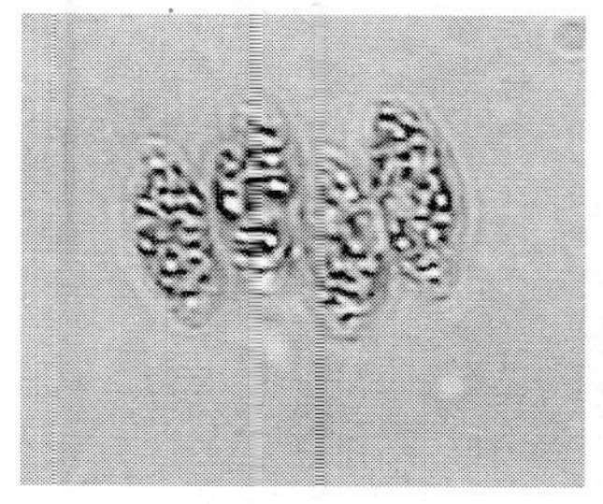
47. *Scenedesmus bijugatus*

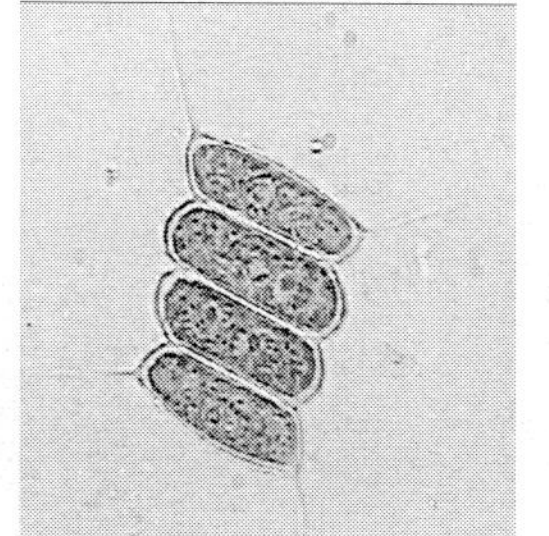
48. *Scendesmus quadricauda*

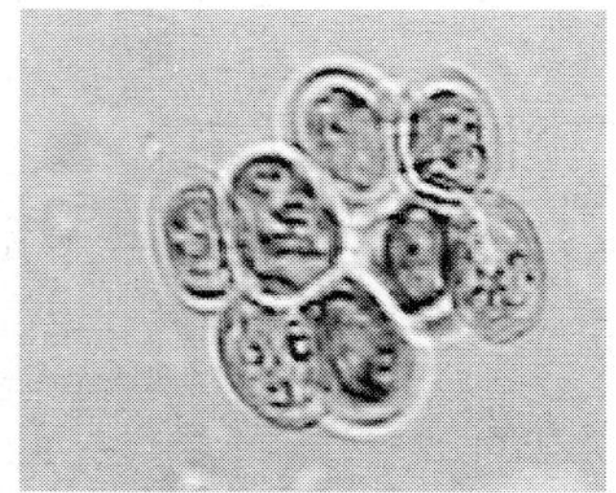
49. *Scenedemus arcuatus*

Freshwater algae of *Chlorophyceae*

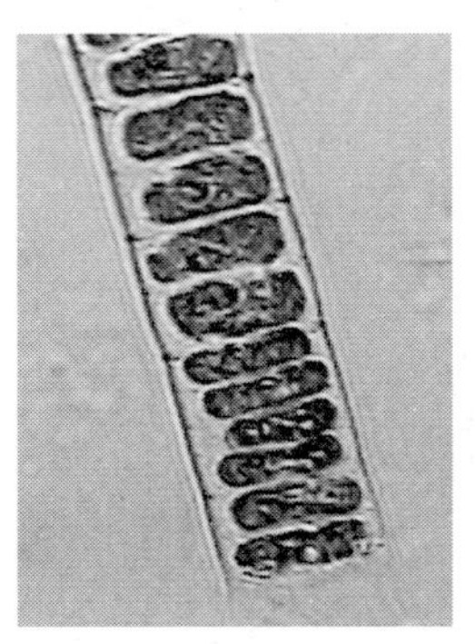

50. *Schizomeris leibleinii*

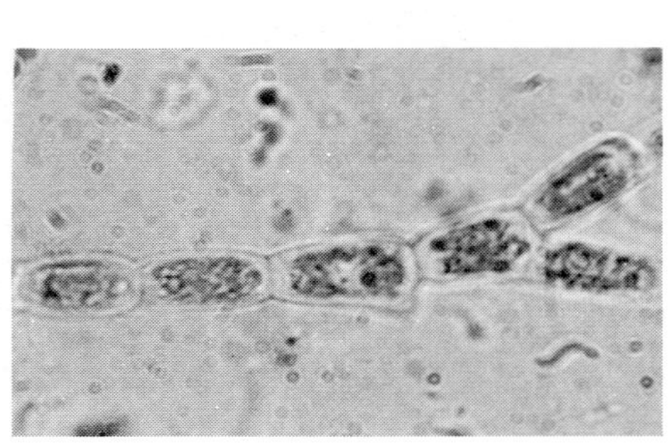

51. *Cladophora* sp.

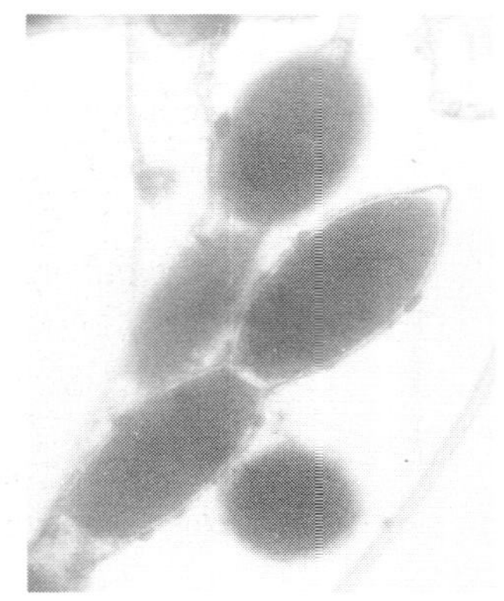

52. *Pithophora oedogonia*

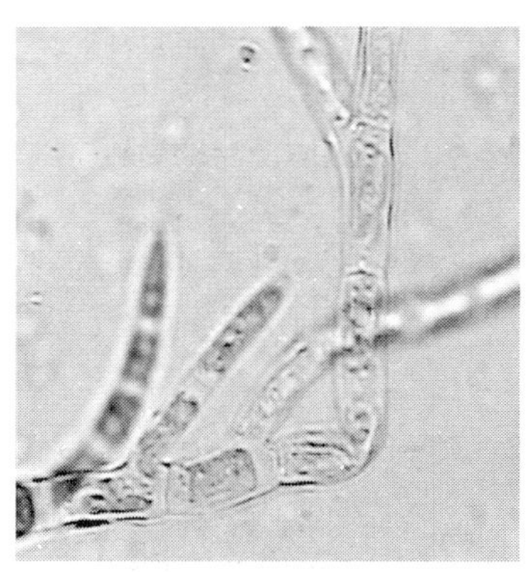

53. *Stigeoclonium subsecundum*

54. *Oedogonium alternans*

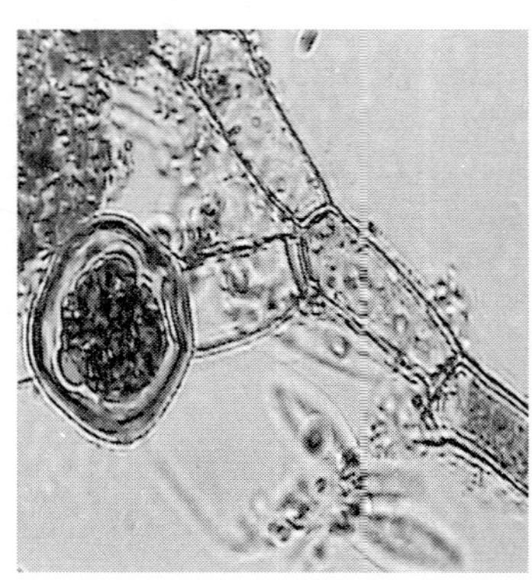

55. *Bulbochaete setigera*

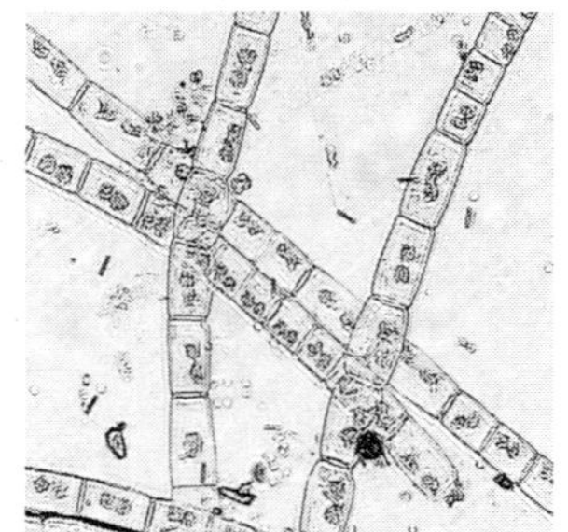

56. *Zygnema vaucherii*

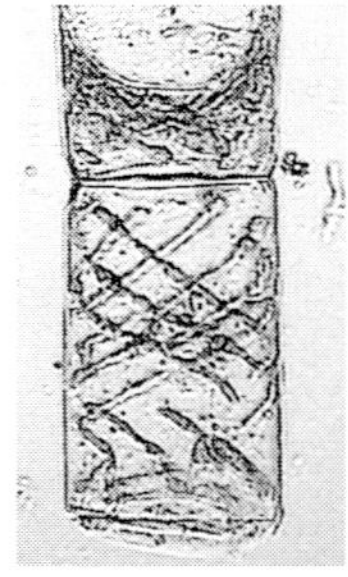

57. *Spirogyra crassa*

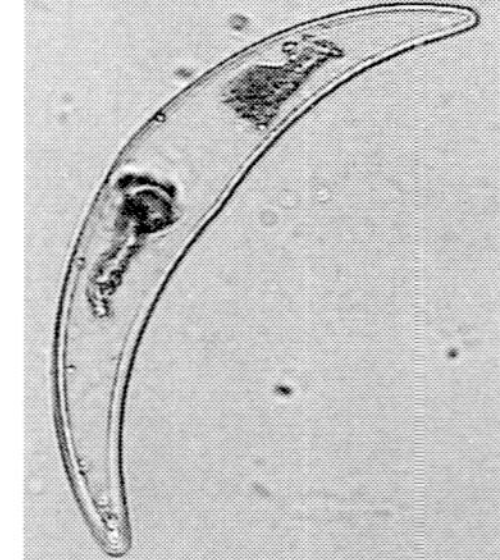

58. *Closterium acerosum*

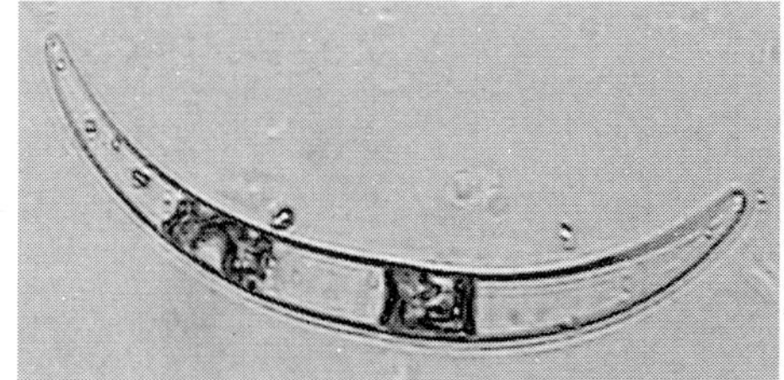

59. *Closterium calosporum*

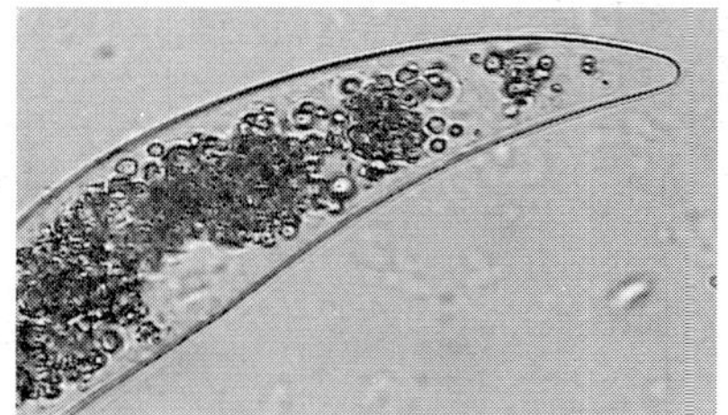

60. *Closterium ehrenbergii*

Freshwater algae of *Chlorophyceae*

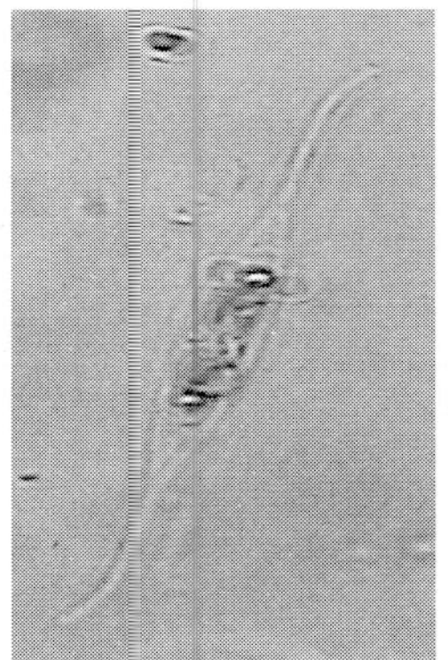
61. *Closterium kuetzingii*

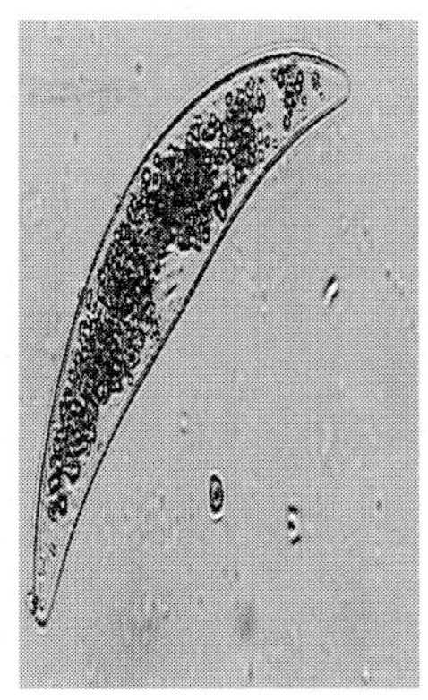
62. *Closterium leibleinii*

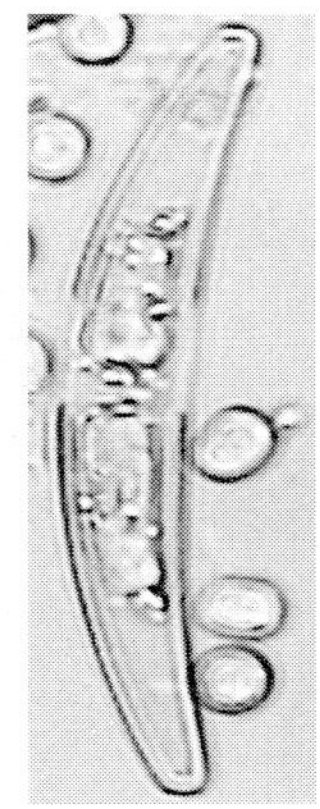
63. *Closterium pritchardianum*

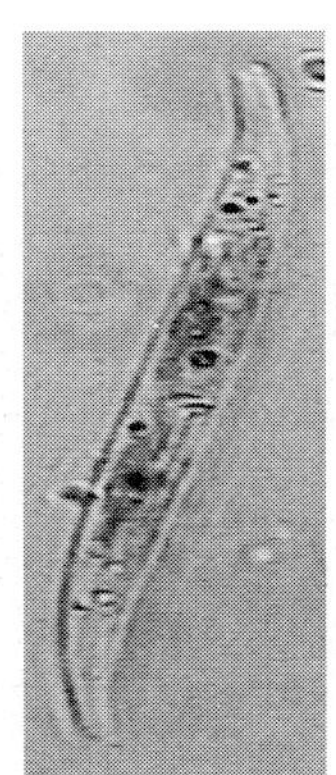
64. *Closterium sigmoideum*

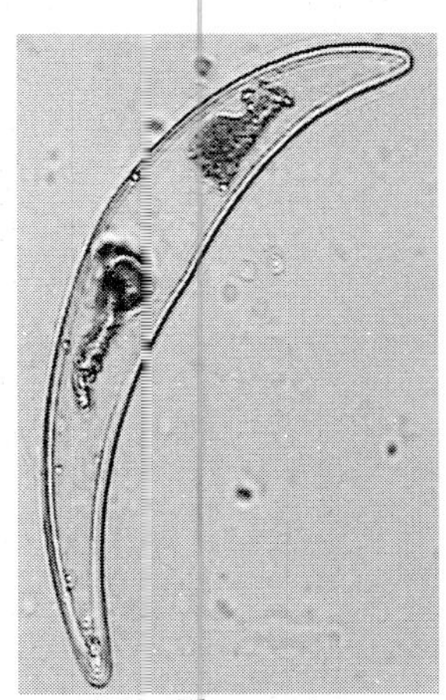
65. *Closterium tumidum*

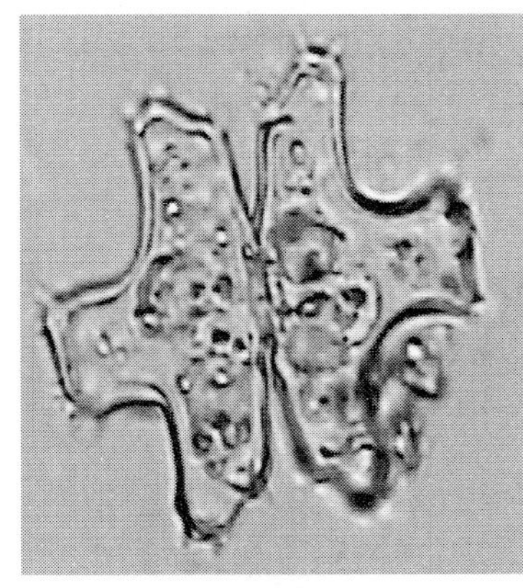
66. *Euastrum ceylanicum*

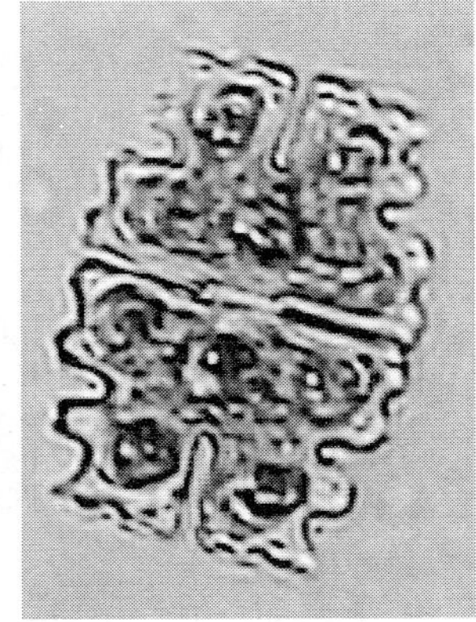
67. *Euastrum coralloides*

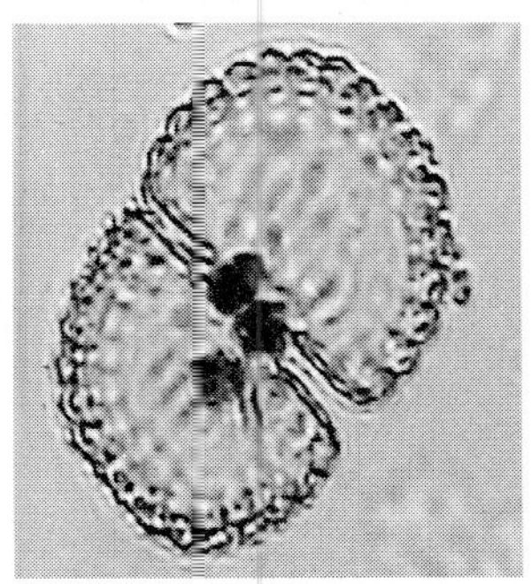
68. *Cosmarium decoratum*

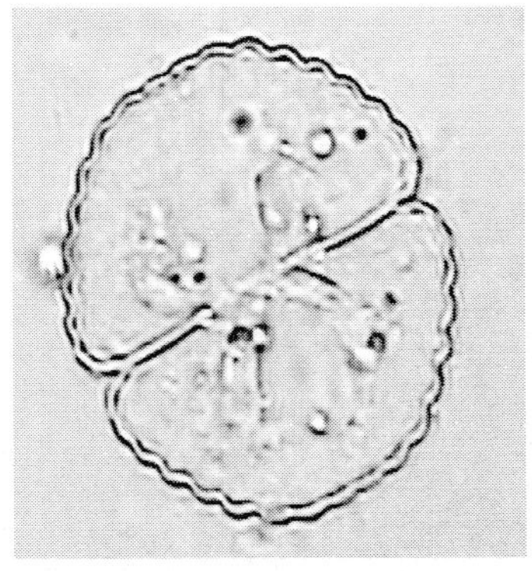
69. *Cosmarium awadhense*

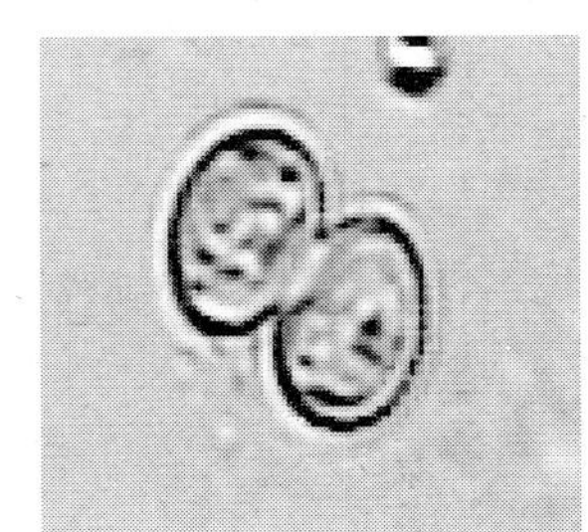
70. *Cosmarium hammeri*

Freshwater algae of *Chlorophyceae*

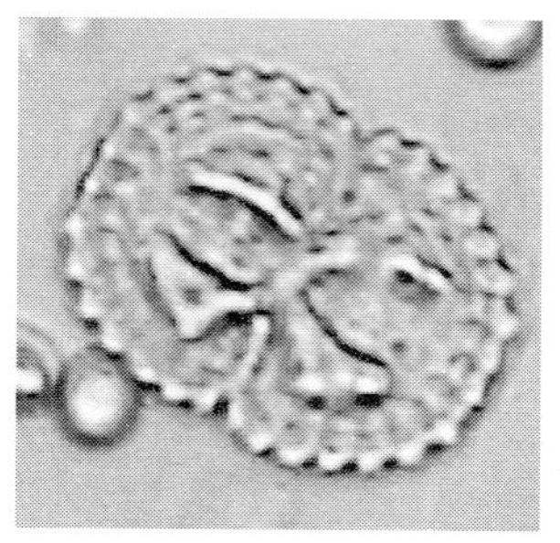

71. *Cosmarium impressulum*

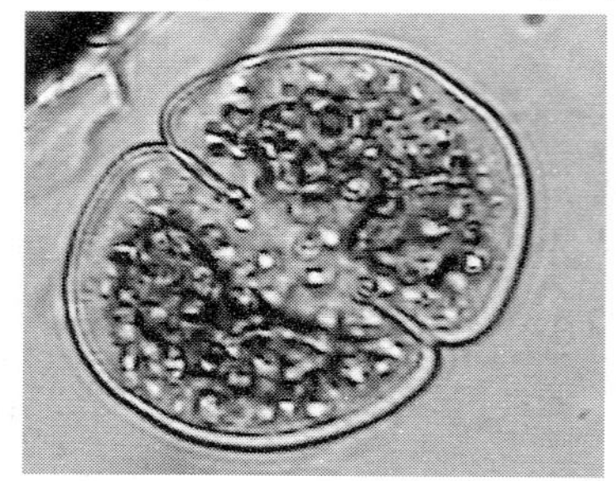

72. *Cosmarium lundellii*

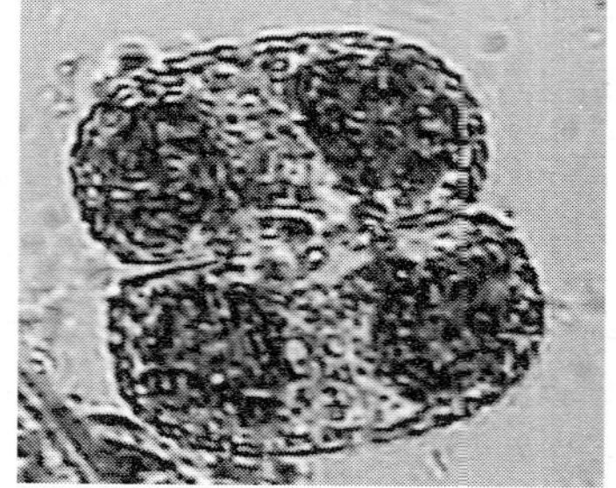

73. *Cosmarium miscellum*

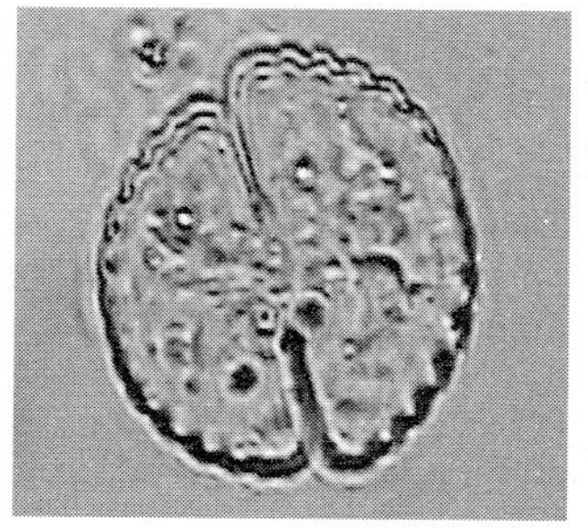

74. *Cosmarium nitidulum*

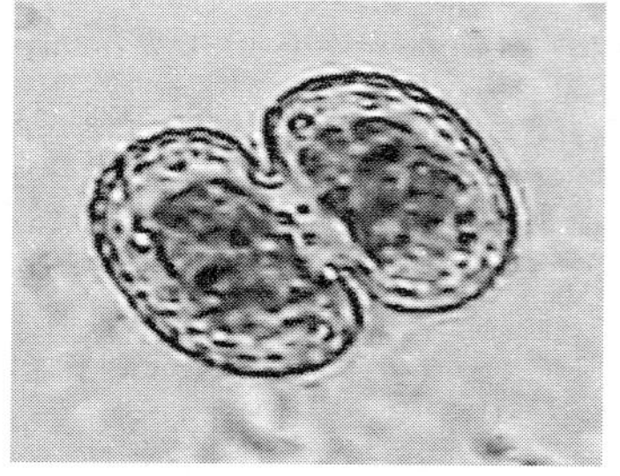

75. *Cosmarium portianum*

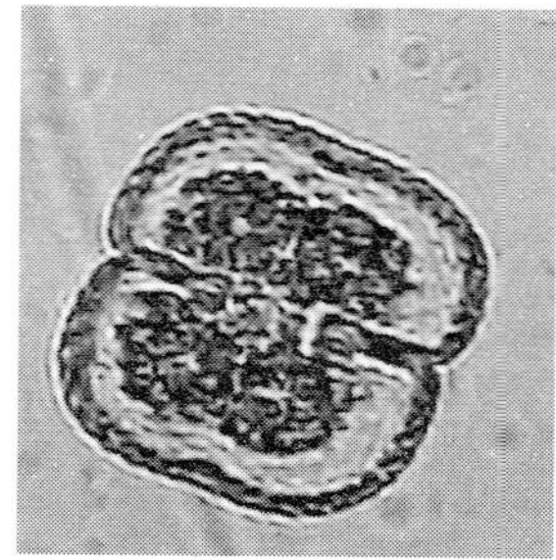

76. *Cosmarium pseudobroomei*

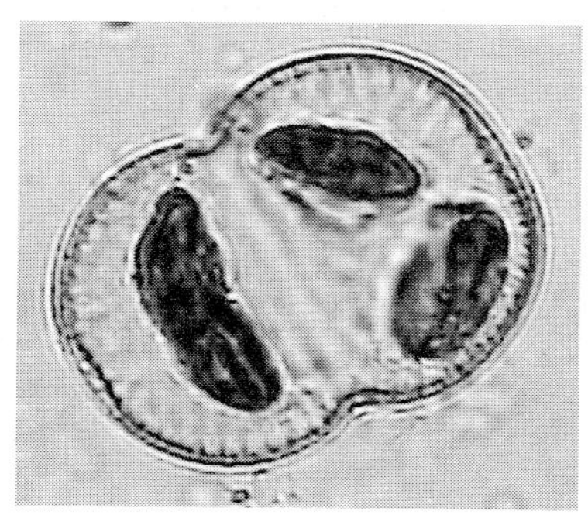

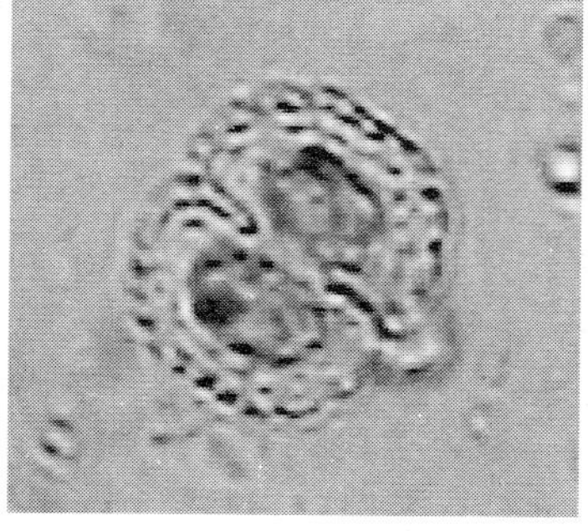

77. *Cosmarium pseudoconnatum* 78. *Cosmarium punctulatum*

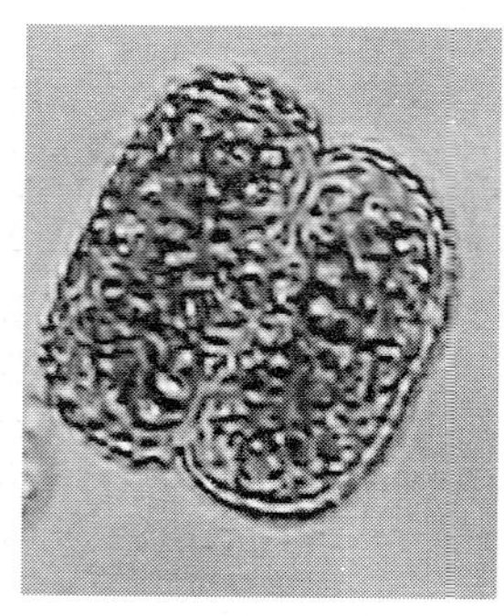

79. *Cosmarium quadrum*

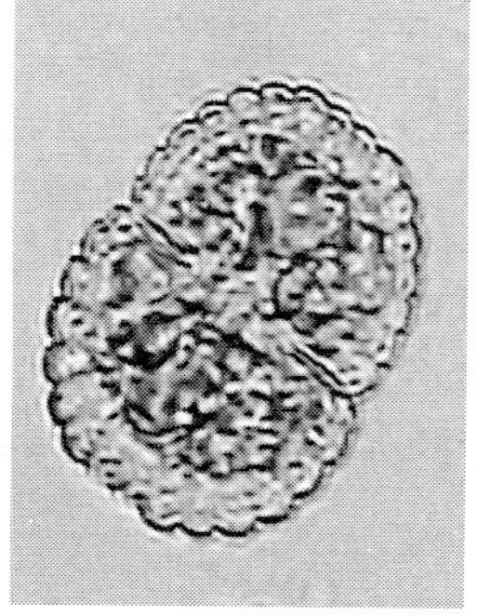

80. *Cosmarium speciosum*

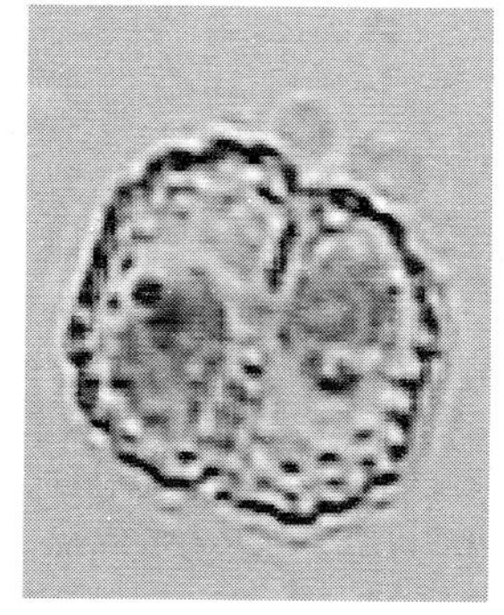

81. *Cosmarium subalatum*

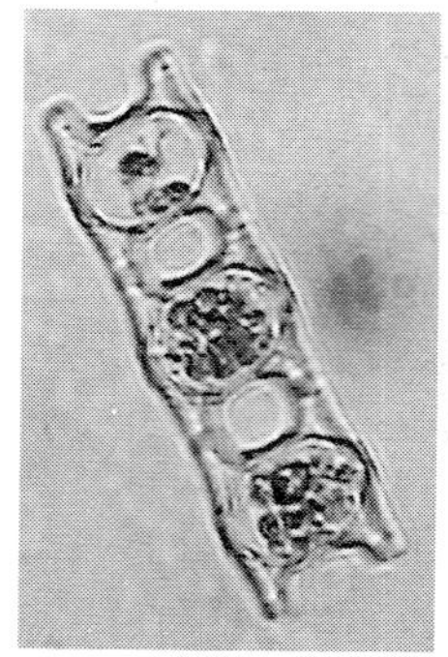

82. *Desmidium baileyi*

Vegetative cells elongate, 20-28 μm in diameter, oospore depressed-globose, 56-65 μm long.

Order- Zygnematales
Family- Zygnemataceae

***Zygnema vaucherii* C. A. Agardh Fig. 56.**

Vegetative cell 22-25 μm broad, conjugation scalariform, zygote ovoid, in one of the gametangia.

***Spirogyra crassa* Kuetzing Fig. 57.**

Filaments coarse, stout, feeling glassy to the touch, chloroplast 6-12 slender, and conjugation by tubes from both gametangia.

Family- Desmidiaceae

***Closterium acerosum* (Sch) Breb. var. tumidum Borge Fig. 58.**

Cells with slightly tumid middle portion and deeply curved outer margin, gradually tapering from mid-region towards truncately rounded apices, cells long 225 μm-236 μm.

***Closterium calosporum* Witt. Fig. 59.**

Cells 73-75.5 μm long, with subacute apices, outer margin strongly curved, cell wall smooth.

***Closterium ehrenbergii* Menegh. Fig. 60.**

Cells large, stout, 6-8 times longer than broad, moderately curved, cells gradually attenuate towards obtusely rounded apices.

***Closterium kuetzingii* Breb. Fig. 61.**

Cells of medium size, almost straight, cells tapering towards each extremity and ending in long setaceous processes with parallel sides and round apices.

***Closterium leibleinii* Kuetz. Fig. 62.**

Cells of medium size 7-8 times longer than broad, strongly curved, outer margin with 154-180 degree of arc, inner margin concave and slightly tumid in the middle.

***Closterium pritchardianum* Arch. Fig. 63.**

Cell wall striated poles slightly recurved, cells 185-760 μm, long, outer margin 24-40 degrees of arc, inner margin straight or slightly concave.

***Closterium sigmoideum* Lagerh. *et* Nordst. Fig. 64.**

Cells of medium size, about 7 times longer than broad, faintly sigmoid with round apices.

***Closterium tumidum* Johns. Fig. 65.**

Cells 144-150 μm long, 16.5-18 μm broad, outer margin 30-38 degree of arc, inner margin broadly tumid in the middle.

***Euastrum ceylanicum* (West and West) Krieg. Fig. 66.**

Cells longer than broad constricted sinus, semicells three-lobed, isthmus linear, lateral wall with short spines, long cell 35 μm, isthmus 8-9 μm.

***Euastrum coralloides* Josh. var. trigibberum Lagerh. Fig. 67.**

Cells small with deep constriction, semicells 3-lobed, polar lobe short and broad with deep median incision, lateral lobes bilobulate, each lobule truncate emarginated with a small spine in the furrow, cells 33.5-39 μm long.

***Cosmarium decoratum* W. and G.S. West Fig. 68.**

Cells of medium size, slightly longer than broad, deeply constricted, sinus narrowly linear and bilipped, semicells, semi-elliptic, apex flattend truncate with round angles, margin deeply crenate. Length of cell 71 μm, isthmus 21 μm.

***Cosmarium awadhense* Prasad and Mehrotra Fig. 69.**

Cells small, slightly longer than broad, deeply constricted, cell apices truncate and flat with straight margin. Long. cell.16.5 μm.lat.isthmus 5 μm.

***Cosmarium hammeri* Reinsch var. *schmidlei* Gronblad and Scott. Fig. 70.**

Cell small, longer than broad, sub-hexagonal, deeply constricted, sinus narrowly linear with dilated apex, Long cell.38.5 μm, lat.isthmus 7 μm.

***Cosmarium impressulum* Elfv. var. crenulatum (Naeg.) Fig. 71.**

Cells wall margin undulate or crenate, semicells vertically sub-quadrate with tri-undulate sides, cell 29.5 μm.

***Cosmarium lundellii* Delp. Fig. 72.**

Cells of medium size, a deeply constricted, sinus linear open, semicells sub-semicircular to sub-pyramidate, cell wall coarsely punctate. Long cell 62.5-71 μm.

***Cosmarium miscellum* Skuja Fig. 73.**

Cell large, margin undulate or crenate, semicells compressed semicircular with 23-26 marginal crenulations, cell wall with granules on periphery, cell 78 μm long, 61 μm broad.

***Cosmarium nitidulum* De Not. Fig. 74.**

Cells small, deeply constricted, sinus narrowly linear with slightly dilated apex, semicells truncate-sub-semicircular, tapering evenly from rounded basal angles to flattened apex with straight margin, long cell. 33-37.5 μm.

***Cosmarium portianum* Arc. var. nephroideum Witter. Fig. 75.**

Cells small, deeply constricted, sinus gradually opening from a rounded extremity, isthmus slightly elongated, 24-29 visible margin of each semicell, long cell 21-24.5 μm.

***Cosmarium pseudobroomei* Wolle Fig. 76.**

Cells small, very deeply constricted, semicells oblong-rectangular cells 34-41 μm long, 31-38.5 μm broad.

***Cosmarium pseudoconnatum* Nordst. Fig. 77.**

Cell of medium size, slightly constricted, semicells semielliptic with circular outline, cell 51.56 μm broad.

***Cosmarium punctulatum* Breb. Fig. 78.**

Cells small, deeply constricted, sinus narrowly, semicells oblong-trapeziform, cells 28-34 μm long, 27.5-31 μm broad.

***Cosmarium quadrum* Lund Fig. 79.**

Cells rather small, quadrate, deeply constricted, semicells sub-rectangular, cells 34.5-37 μm.

***Cosmarium speciosum* Lund Fig. 80.**

Semicells sub-rectangular or sub-pyramidate, margin with 4 apical 7 lateral crenations, cells 43-51 μm long, 31-36 μm broad.

***Cosmarium subalatum* W. *et* G.S. West. Fig. 81.**

Cells small, slightly longer than broad, deeply constricted, sides tricrinate, angles rounded, apex with two small crenations. Long cell 24.5 μm.

***Desmidium baileyi* (Ralfs) Nordst. Fig. 82.**

Cells small, united to form straight filaments lacking gelatinous sheath, little broader than long, median constriction reduced to a faint undulation, long cell 14.5 μm, intercellular spaces prominent.

***Micrasterias mahabuleshwarensis* Hobson Fig. 83.**

Cells 80-85 μm in diameter, constriction deep, sinus open with acuminate extremity, semicells three-lobed, with symmetry in three planes, margin with small and acute spines.

***Micrasterias pinnatifida* (Kuetz) Ralfs Fig. 84.**

Semicells with two lateral and sub-rectangular polar lobes with bifid extremities.

***Staurastrum bieneanum* Rabenh. Var.ellipticum Wille Fa.Skuja Fig. 85.**

Cells slightly broader than long, very deeply constricted, sinus widely open with acute and acuminate apex. Long cell 25 μm.

***Staurastrum recurvatum* Turner Fig. 86.**

Cell small, deeply constricted with open sinus, semicells fusiform with dorsal margin convex undulate.

Class: Bacillariophyceae

Vegetative cells of the diatoms are uninucleate, without flagella, unicellular or forming simple colonies and have a silicified cell wall (Frustuls) of two distinguishable 'halves' one of which overlap the other, the cell wall is composed of pectic compound impregnated with silica. The longer 'helf' of the wall, the epivalve (epitheca) and the shorter 'half' the hypovalve (hypotheca). Each half has valve and Cingulum. Valves are the main surface of theca formed flattened plate like structure. Cingulum (connecting band), it is an incurved margin of valve. Cingulum of epitheca and hypotheca jointed together form girdle. When frustules lie in such a way that valve side is uppermost it is known as valve view or topview. When connecting band are uppermost is

Freshwater algae of *Chlorophyceae*

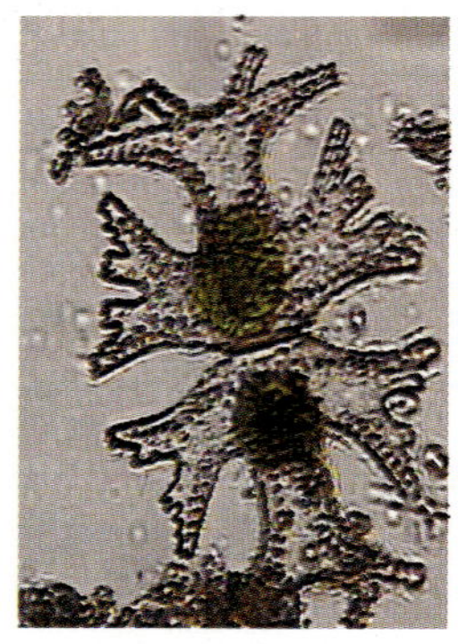

83. *Micrasterias mahabuleshwarensis*

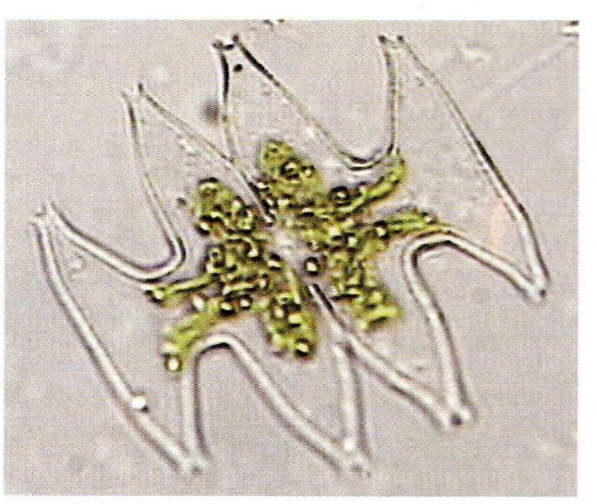

84. *Micrasterias pinnatifida*

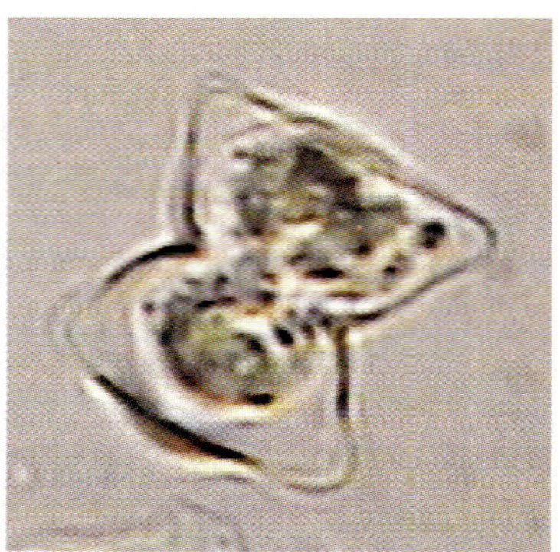

85 *Staur astrum bieneanum*

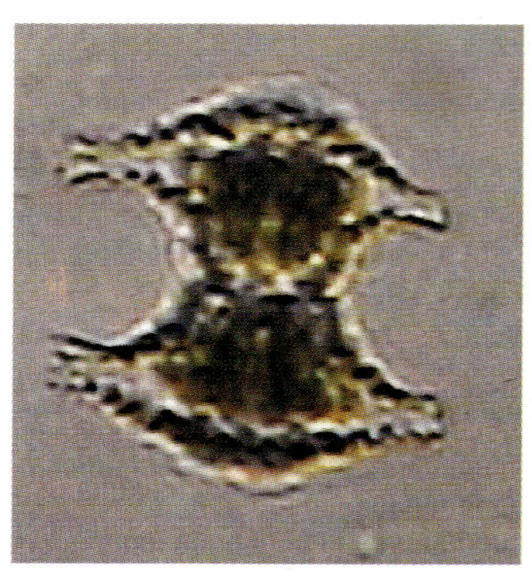

86. *Staurastrum recurvatum*

Freshwater algae of *Bacillariophyaceae*

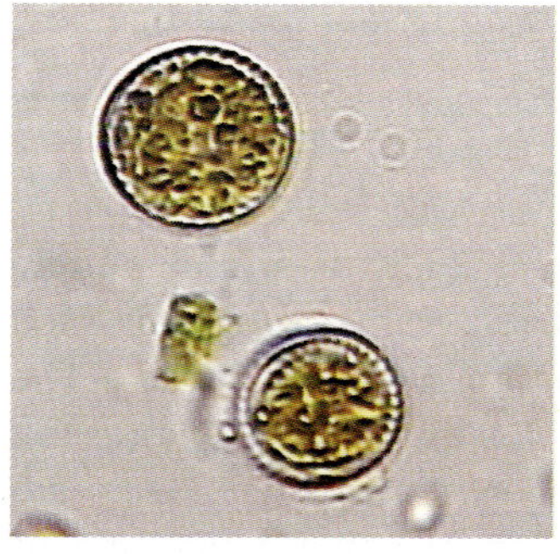

87. *Cyclotella glomerata*

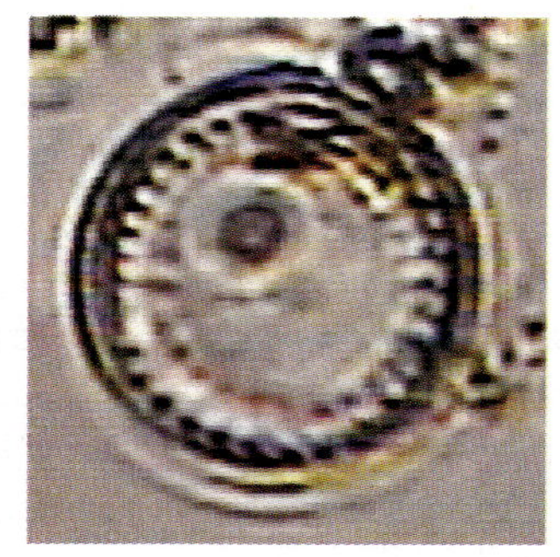

88. *Cyclotella meneghiniana*

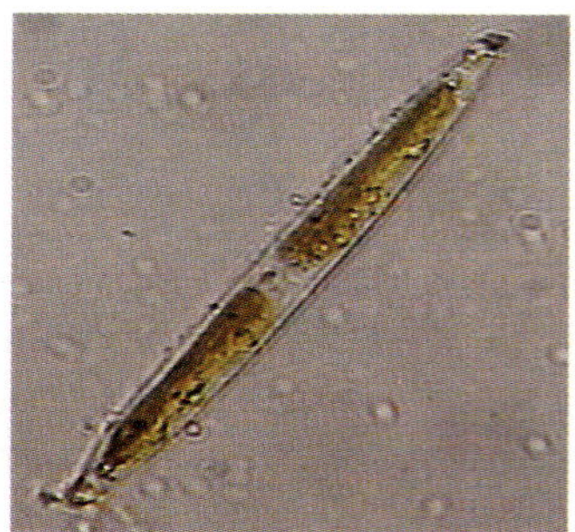

89. *Synedra ulna*

Freshwater algae of *Bacillariophyaceae*

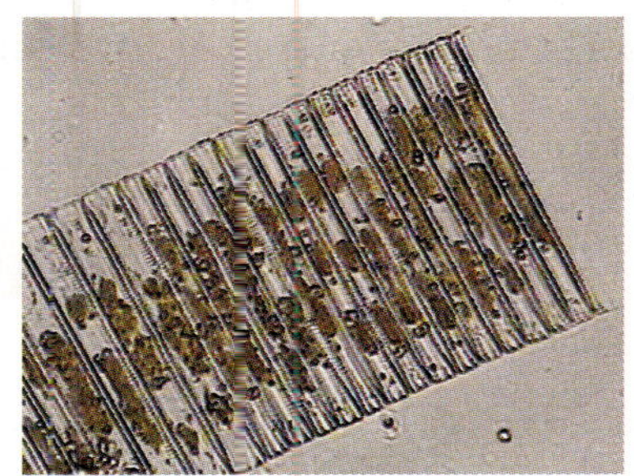

90. *Fragilaria fonticola*

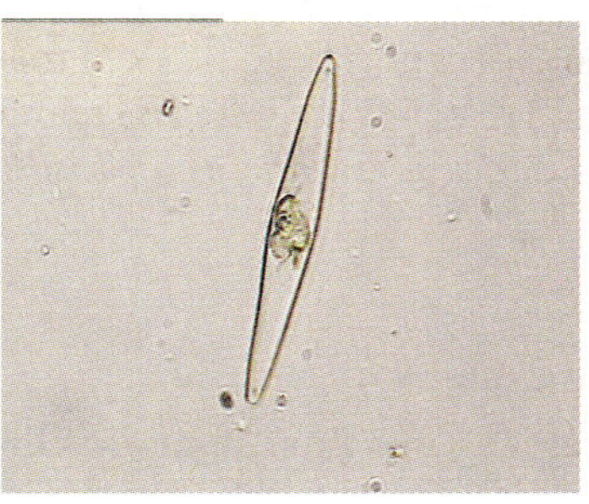

91. *Nitzschia dissipata*

92. *Hantzschia amphioxis*

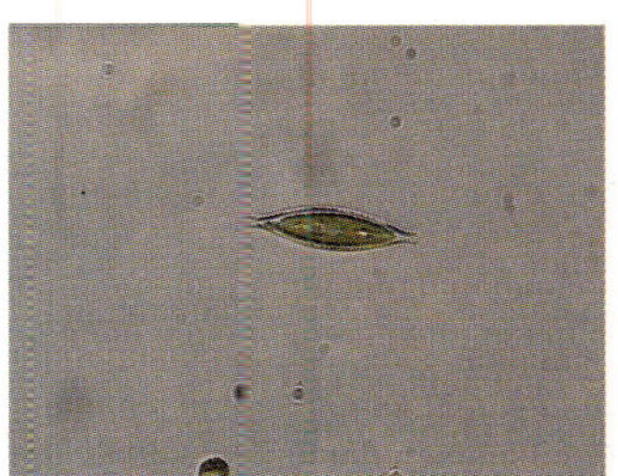

93. *Navicula cuspidate*

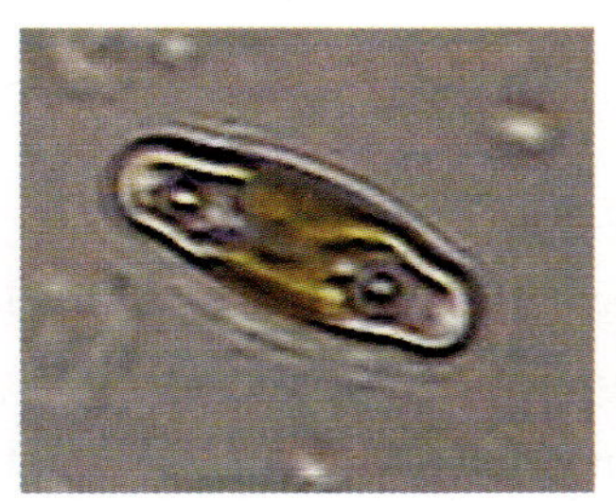

94. *Pinnularia interrupta*

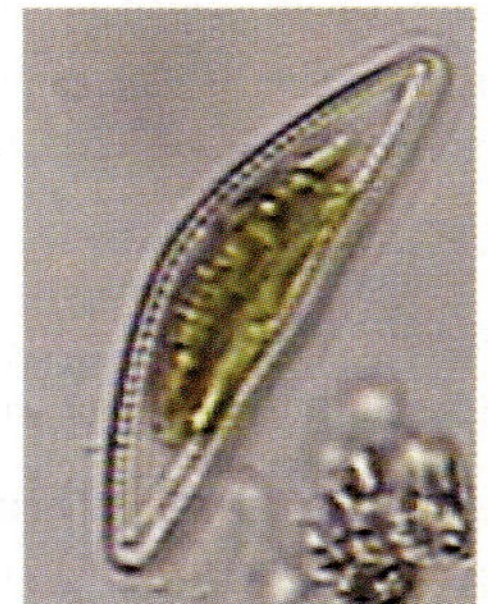

95. *Cymbella ventricosa*

Freshwater algae of *Euglenophyceae*

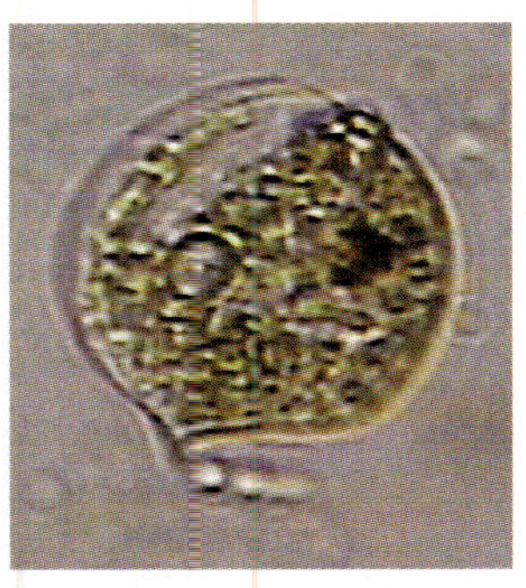

96. *Phacus acuminatus*

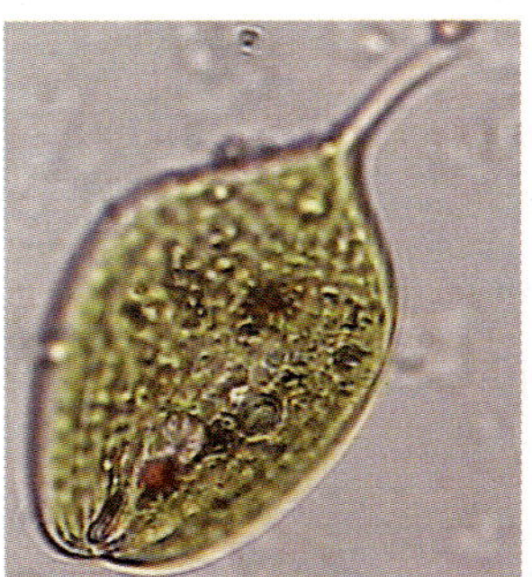

97. *Phacus longicauda*

called girdle view or side view, the valves are variously ornamented with coarse or very fine striae; punctae or areolae that one radially disposed about central point in centric genera or bilaterally with respect to an axial field in pinnate genera. The axile field may be symmetric or asymmetric with respect to the longitudinal axis and may be perforated by complicated longitudinal slot or raphe. In the absence of raphe, the axile field is referred to as a pseudoraphe.

Order- Bacillariales

Family- Coscinodiscaceae

***Cyclotella glomerata* Bachmann. Fig. 87.**

Frustules small, radially symmetrical, marginal striae fine lineate, central field with seven radial striae in a ring.

***Cyclotella meneghiniana* Kuetz. Fig. 88.**

Frustules small, in girdle view more less rectangular, valve view discoid, radially symmetrical marginal striae, wedge-shaped, central zone smooth.

Family- Fragilariaceae

***Synedra ulna* Ehr. Fig. 89.**

Valves solitary, linear, much elongated, striae lineate, parallel.

***Fragilaria fonticola* Hust. Fig. 90.**

Valves linear with parallel margins united by their margins to form long bands, ends gradually attenuated, slightly constricted, raphe absent.

Family- Bacillariaceae

***Nitzschia dissipata* (Kuetz.) Grun. Fig. 91.**

Valves small, linear lanceolate, gradually attenuated towards bluntly rounded ends, keel punctae excentric, striae not visible.

***Hantzschia amphioxis* Ehr. Fig. 92.**

Keel margin on one half, lies opposite to that of other margin.

Family- Naviculaceae

***Navicula cuspidate* Kuetz. Fig. 93.**

Valves surface striated, striae transverse and longitudinal, sometimes craticular plates present.

***Pinnularia interrupta* W. Smith Fig. 94.**

Ends constricted, slightly produced rounded.

Family- Cymbellaceae

***Cymbella ventricosa* Kuetz. Fig. 95.**

Valve asymmetrical, ends acutely rounded, raphe thick, slightly undulate, valve lunate, 36 μm long, and striae coarse lineate, radiate.

Class: Euglenophyceae

***Phacus acuminatus* Heuber Fig. 96.**

Cells more or less flattened with striations and terminating in a spine.

***Phacus longicauda* (Ehr) Fig. 97.**

Single cells, motile, cells more or less flattened with striations and long terminating in a spine.

RESULTS AND DISCUSSION

Random sampling method has been applied to get the maximum information regarding number of desmidian algae and their distribution in Kanchipuram district. 13 genera found (+) in different districts have been briefly given (Table 2).

Morphological features of taxa, viz. *Euastrum* sp-1, *Euastrum* sp-2 and *Cosmarium* sp., etc., have been described under open nomenclature system. In these taxa variations from known species may be due to environmental conditions, hence, new taxon cannot be created without doing culture of these specimens.

Table 2

Genera	Tiruvallur	Kanchipuram	Tiruvnnamalai
Chroococcus	+	+	+
Gloeothece	+	+	+
Lyngbya	+	+	+
Spirulina	+	+	+
Nostoc	+	+	+
Scenedesmus	+	+	+
Kirchneriella	+	+	+
Zygnema	+	+	+
Euastrum	+	+	+
Micrasterias	+	+	+
Cosmarium	+	+	+
Spirogyra	+	+	+
Navicula	+	+	+

Genus richness in three regions

Overall Tiruvallur district has less number of Cyanobacteria, Kanchipuram district has richness in desmids, Tiruvnnamalai district has Bacillariophyceae and during collection time *Phacus* sp. is found only in Triuvallur district.

ACKNOWLEDGEMENTS

Authors are thankful to the Ministry of Environment & Forests for financial support.

REFERENCES

Anand, N. 1998. Indian freshwater micro algae: B. Singh and M.P. Singh Dehradun, India, pp. 94.

Anand, N., Shantha, R.S. and Kumar, H. 1995. Distribution of blue-green algae in rice field of Kerala state, India. *Phykos.* 34 (1&2) 55-64.

Anand, V. K. 1979. Blue-green algae of Chandigarh stream (Miran Sahib, Jammu)-I *Phykos* 18 (1&2): 21-24.

Biswas, K. P. 1949. Common fresh and brackish water algae flora of India and Burma. *Rec. Bot. Surv. India.* Pt. 15(1): 1-105.

Chaturvedi, U. K. and Habib I. 1996. A systematic account of Chlorococcales from Nepal. *Phykos.* 35 (1&2) 129-137.

Desikachary, T. V. 1959. Cyanophyta: ICAR. Monograph on algae, New Delhi.

Desikachary, T. V. 1962. The diatom. *Current Science*, 31: 43-45.

Foged, N. 1976. Freshwater diatoms in Sri Lanka (Ceylon) *Biblitheca Phycologia, J. Cramer Band* 23.

Fritsch, F. E. 1935. The structure and reproduction of algae, Vol. I: Cambridge Univ. Press, London.

Gandhi, H. P. 1999. Freshwater diatom of Cental Gujarat: B. Singh and M.P. Singh, Dehradun, India.

Iyengar, M. O. P. and Desikachary, T. V. 1981. Volvocales I.C.A.R. monograph on algae: New Delhi.

Philipose, M. T. 1967. Chlorococcales I.C.A.R. monograph on algae: New Delhi.

Prasad, B. N. and Singh.Y. 1996. Algal indicator of water pollution: B. Singh and M.P. Singh, Dehardun, India.

Prescott, G. W. 1961. Algae of Western great lakes area: Wm.C. Brown Co. Publishers Dubuque Iowa.

Prasad, B. M. and Srivastava, M. M. 1992. *Fresh water algal flora of Andaman and Nicobar Islands*, Vol. I. B. Singh and M. P. Singh, Dehradun, India.

Williamson, D.B. 1999. A proposed new desmids genus *Cruciangulum* and descriptions of three new desmids species from rock pools in the Western Cape Town, South Africa, Algological Studies 93: 55-62.

29

UV-B Radiation and High Light Induced Oxidative Damage in *Phormidium corium*

Rupali R. Bhandari and Prabhat K. Sharma
Department of Botany, Goa University, Goa-403 206

ABSTRACT

There has been report of extensive coral bleaching and death of corals of the Lakshadweep Islands, India, which could probably be due to the global climatic changes. In this work we have studied the effect of UV-B and PAR flux at 30°C in *Phormidium corium* (Agardh) Gomont, a cyanobacterium isolated from coral, *Porites lutea* (Milne-Edwards and Haime) on the Kavaratti reef of the Lakshadweep Islands, (11° N; 71°E), India, with an objective to see whether these factors might be responsible for the bleaching? Exposure of *P. corium* to UV-B and PAR decreased photosynthetic efficiency and increased oxidative damage measured as F_v/F_m ratio and lipid peroxidation of cell membrane. Data indicate little activity of superoxide dismutase (SOD) and ascorbate peroxidase (APX) in response to the UV-B and PAR treatment. Activity to quench reactive oxygen species (ROS) was less in *P. corium* as compared to other cyanobacteria and higher plants. Though mycosporine like amino acids (MAAs) increased significantly as a result of UV-B treatment, PAR caused decrease in the MAAs content, thus neutralising the beneficial effect of MAAs. Results suggest that one of the factor of bleaching of coral reefs off the Lakshadweep Islands, may be due to oxidative damage caused to it by production of ROS by associated cyanobacterium, *P. corium* which lacks efficient antioxidant system.

Keywords: Antioxidant enzymes, coral bleaching, cyanobacteria, lipid peroxidation, mycosporine like amino acids, reactive oxygen species, UV-B.

INTRODUCTION

Corals are small animals those live in immense colonies, harvesting nourishment and energy from micro- and macro-algae, which inhabit their cells (Goreau and Goreau, 1960; Muscatine, 1973). The pigments of associated micro- and macro-algae in combination with pigments of various hosts such as zooxanthallae lend their coral hosts a spectacular variable coloured appearance (Kleepel et al., 1989). Apart from protecting the shoreline from the damaging effects of the ocean, coral reefs provide habitats and shelter for many organisms and are the sources of nitrogen and other nutrients essential for the food chain supporting thousands of marine species in reefs. Potential pharamaceutical products are also being studied from coral reef ecosystems (Hay and Fenical, 1996).

Corals are one of the most sensitive ecosystems to climate change on the earth (Grigg et al., 1984; Reaka-Kudla, 1996). Lately, this ecosystem is suffering from coral bleaching which is a consequence of stress conditions in reef corals that involves a breakdown of the symbiotic relationship and a reduction in photosynthetic pigment concentrations in zooxanthellae residing within the corals. When stressed, corals may lose much of their symbiotic algae. This loss exposes the white calcium carbonated skeletons of the coral colony referred to as bleached. Coral bleaching events have been increasing in both frequency and extent worldwide in the past 20 years resulting in a decline in size or live coral cover. Preliminary assessments indicate that the Indian Ocean is severely impacted region, greater than seventy per cent mortality observed off the coasts of the Maldives, the Andamans, and the Lakshadweep and Seychelles Marine Park System (Wilkinson et al., 1999). The outer-atoll seaward slopes of Kadmat in the Lakshadweep Islands, India, had heavy mortality with only 3% live coral cover and 87% dead branching forms.

Global climate change may play a role in the increase in coral bleaching events. Coral reef bleaching is caused by various anthropogenic and natural variations in the reef environment including sea temperature, solar irradiance, increased levels of ultraviolet radiation, sedimentation, pollution, disease, excess shade, salinity changes, etc. (Hoegh-Guldberg et al., 1997; Wilkinson and Buddemeier, 1994). Solar radiation and temperature stresses are reported to be principal causes of coral bleaching (Lesser, 1997). Solar radiation, in the form of both photosynthetic active radiation (PAR) and ultraviolet (UV), can elicit a stress response in symbiotic associations. UV-B penetrates oceanic waters sufficiently to have both direct and indirect effects (Menon et al., 2005). UV-B is probably more harmful than high levels of PAR (Gleason and Wellington, 1995), and when high light is combined with high temperature synergistic effects may be observed.

In this study, we present data to show that both UV-B and high PAR results in increased level of oxidative damage due to inability of *Phormidium corium* (Agardh) Gomont, one of the cyanobacteria found in association with coral *Porites lutea* on the Kavaratti reef of the Lakshadweep island (11°N; 71°E), India, to metabolize reactive oxygen species (ROS) generated due to the stress may be one of the reasons for the bleaching.

MATERIALS AND METHODS

Culture Conditions

Phormidium sp. was isolated in association with normal and bleached coral, *Porites lutea* (Milne-Edwards and Haime) on the Kavaratti reef of the Lakshadweep group of island (11°N; 71°E), India by Ravindran and Raghukumar (2002). Histological examination revealed that *Phormidium* filaments were found in corals suffering from pink line syndrome (PLS) in the calicoblastic epithelium cells in polyp of the coral and suggested that cause of death of corals due to acidification of extracellular space by the presence of the cyanobacterium, *Phormidium* (Ravindran and Raghukumar, 2006). The culture was routinely grown in autoclaved liquid ASN III culture medium according to Rippka et al. (1979) on a environmentally control shaker set to a temperature of 30°C under cool white fluorescent light tubes providing 120 µmol m^{-2} s^{-1} PAR at the culture level with a 14th of photoperiod. *P. corium* was allowed to grow for 30 days to obtain its logarithmic phase of growth (balance growth). All experiments were set with 30 days old culture. The isolated cyanobacterium was identified as *Phormidium corium* through National Facility for Marine Cyanobacteria (NFMC), Tiruchirapalli, Tamil Nadu, India.

Exposure to UV-B Radiation

The algal culture was transferred to a Petri plate. The UV-B treatment of 0.8 ± 0.1 mW cm^2 (Vilbour-Lourmat, France T-6M source with a λ-max at 312 nm) was given in a BOD chamber at 30°C up to 6 h while keeping the algal culture continuously stirred using a magnetic flea (0.1 × 1 cm).

Exposure to High Light

The algal suspension was exposed to a relatively high irradiance of 500 µmol m^{-2} s^{-1} PAR (Li-cor, Model Li-189) at 30°C up to 6 h in a double walled cuvette circulated with temperature controlled water. The light treatment was given using two slide projectors from opposite sides. Culture was kept constantly stirred during the treatment at a slow speed in order to avoid mechanical damage.

Photosynthesis Measurements

Photosynthesis measurements was taken using chlorophyll fluorometer (PAM 101-102, Walz, Germany) according to Sharma et al. (1998a). Culture was dark adapted for 10 minutes prior to measurements at room temperature. The dark adapted cultures were exposed to a modulated light with an intensity of 4 µmol m^{-2} s^{-1} to measure initial fluorescence (F_o). This was followed by an exposure to a saturating pulse of white light of 4000 µmol m^{-2} s^{-1} to obtain the maximum fluorescence (F_m). Variable fluorescence (F_v) was determined by deducting the F_o from F_m ($F_v = F_m - F_o$) and the F_v/F_m ratio was calculated.

Peroxidation of Cell Membrane Lipids

Lipid peroxidation was determined by the production of TBA-MDA adduct formation according to method described by Sharma et al. (1998b). Algal culture was harvested by centrifuging at 8000 g for 15 min, pellet was homogenized, and resuspended in fresh culture medium with a ratio of 1:5 (w/v). Resuspended algal culture (5 ml) was recentrifuged and pellet was homogenized in 0.5% TCA. The homogenate was made up to 5 ml and centrifuged at 8000 g for 15 minutes. The supernatant was collected and used for measuring the peroxidation of membrane lipids. 1 ml of the supernatant was added to the test tube containing 2.5 ml of freshly prepared (0.5%) TBA in (20%) TCA and allowed to incubate for 30 min at 90°C in a water bath. After incubation, it was allowed to cool at room temperature and centrifuged for 2 min at 1000 g to settle the debris and non-specific precipitation. The optical density was taken at 532 nm (Schimadzu, UV-250). Peroxidation of lipids was measured using an extinction coefficient of 155 mM^{-1} cm^{-1}.

Production of Reactive Oxygen Species

Production of ROS was determined using epinephrine method (Boveris, 1984). Epinephrine (4.5 mM) was added to the buffer containing 6 mM EDTA in 10 mM sodium carbonate (pH 10.2) with or without the algal tissue. This resulted in production of ROS, which was determined at 480 nm using double beam spectrophotometer (Shimadzu, UV-2450). The extent of quenching of ROS was determined by adding homogenate/protein.

Assay of Superoxide Dismutase (EC 1.15.1.1)

Activity of superoxide dismutase (SOD) was assayed according to Sankhalkar and Sharma (2002). The assay medium contained 50 mM potassium phosphate (pH 7.8), EDTA (disodium 0.1 mM), cytochrome c (10 mM), xanthine (0.1 mM) and the algal homogenate equivalent to 25 µg of protein. The reaction was followed spectrophotometrically as reduction of cytochrome c at 550 nm. One unit of SOD activity was defined as the amount of enzymes required to inhibit the rate of cytochrome c reduction by 50%.

Assay of Ascorbate Peroxidase (1.11.1.7)

Ascorbate peroxidase (APX) activity was assayed according to Sankhalkar and Sharma (2002). Pellet of alga (0.5 g) was extracted in 5 ml of 100 mM sodium dihydrogen phosphate buffer (pH 7.5), 1 mM EDTA and 5 mM ascorbate using tissue homogenizer. The extract was centrifuged for 15 minutes at 6000 × g at 4°C. Supernatant was used for peroxidase assay. Activity was assayed in a reaction mixture containing potassium hydrogen phosphate (100 mM), EDTA (1 mM), sodium ascorbate (1.5 mM), hydrogen peroxide (10 mM) and tissue extract. The decrease in absorbance was recorded for 3 min at 290 nm.

Extraction and Purification of Mycosporine Like Amino Acids (MAAs)

Extraction and purification of UV-B absorbing compounds such as mycosporine like amino acids (MAAs) was carried out according to the method by Sinha and Häder (200). Lyophilized algal cells (0.1 g) was homogenized in 5 ml of 20% (v/v) aqueous methanol (HPLC grade) and incubated at 45°C for 2.5 h. The homogenate was centrifuged at 4000 × g for 15 minutes. The pellet was re-extracted and supernatant was pooled together and evaporated to dryness using lyophilizer. The dried supernatant was dissolved in 1.5 ml of 0.2% acetic acid. The samples were analyzed using an HPLC system with reverse phase column (Waters Spherisorb ODS C18, 25 µm × 4.6 mm × 250 mm specifications) using isocratic mobile phase of 0.2% acetic acid with a flow rate of 1.0 ml/min and detected using a PDA detector (Waters 2996). The HPLC profile was extracted at 330 nm from spectral range of 200-400 nm.

RESULTS

A linear decrease in the photosynthesis (PS II efficiency) measured as F_v/F_m ratio in response to UV-B as well as high light treatment was observed. Six hours of UV-B treatment to *P. corium* resulted in complete inhibition of F_v/F_m ratio while high light treatment to the same duration resulted in a decrease of 97% in the efficiency of photosynthesis (Fig. 1).

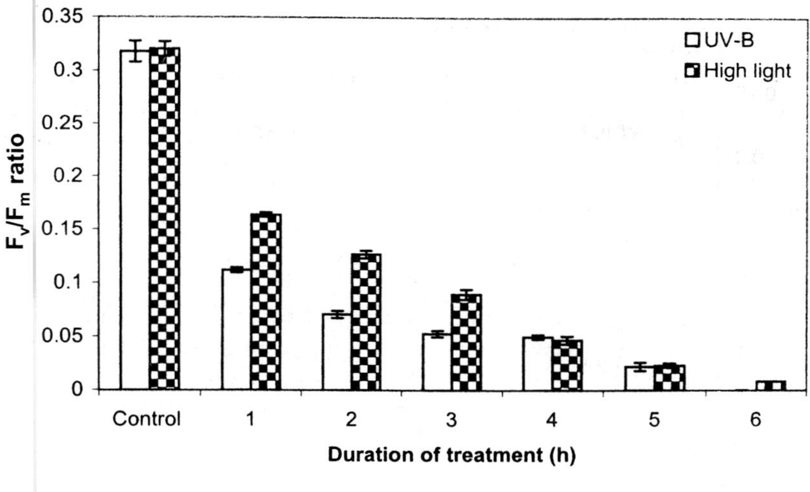

Fig. 1

Peroxidation of membrane lipids was studied by measurement of malondialdehyde (MDA) formed as a product of membrane peroxidation. A 6 hour of UV-B and high light treatment resulted in an increase of peroxidation of cell membrane lipids by 73% and 44% respectively (Fig. 2).

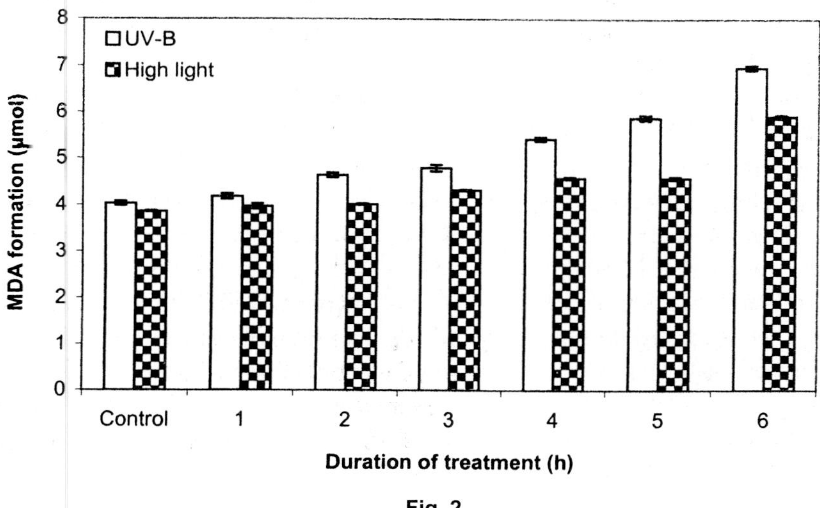

Fig. 2

It was seen that *P. corium* grown under control growth condition did not show significant ability to prevent generation of ROS (Fig. 3). We compared this ability of ROS quenching of *P. corium* with *Nostoc spongiaeforme*, another cyanobacterium (freshwater) and *Cassia tora*, a terrestrial weed. It was seen that tissue homogenate of *P. corium* could quench the ROS only by

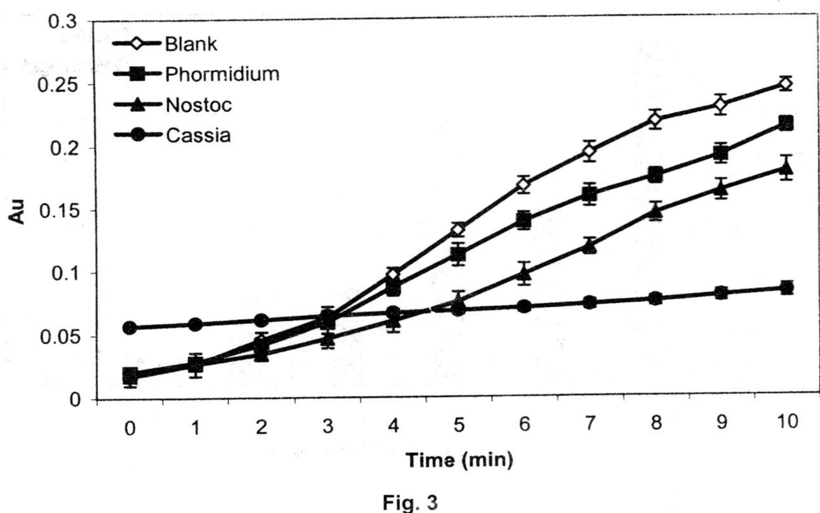

Fig. 3

13% as compared to 33% seen in *N. spongiaeforme* and 70% observed in *C. tora* (Fig. 3). Since the quenching of ROS is mainly due to enzymatic (SOD and APX) and non-enzymatic (ascorbic acid) antioxidant system we also assayed the activities of SOD and APX; and determined the ascorbate content in *P. corium* under control as well as under stress conditions.

A slight increase in the activity of SOD (Fig. 4a) as well as APX (Fig. 4b) was observed in *P. corium* as a result of the UV-B and high light treatment. Six hours of UV-B or high light treatment resulted in approximately 20% increase in the activity of SOD as compared to their control (Fig. 4a). APX activity was slightly higher as compared to SOD. UV-B or high light treatment up to 3 h resulted in an increase in APX activity by 32% and 46% respectively as compared to control. However, further increase in the duration of the UV-B treatment (up to 6 h) reduced the increase in the APX activity to only 28% (Fig. 4b). When compared the increase observed in SOD and APX in *P. corium* with wheat it was seen that the UV-B treatment resulted in 55% increase in SOD and approximately 400% increase in the APX activity (Sharma et al., 1998b). Ascorbate content using titration method according to Reiss (1994) using DCIP remained constant in both control as well as treated *P. corium* at 1.2 mg g^{-1} F. W. (data not shown).

After exposure to UV-B radiation for 6 h, considerable increase (80%) in the amount of UV-B absorbing compound such as MAAs was observed as compared to control. These MAAs could absorb in the range of 200-323 nm (Fig. 5). However, high light treatment (500 PAR) for the same duration resulted in 40% decrease in the amount of MAAs as compared to control (Fig. 5).

DISCUSSION

Our results indicate that UV-B and high light both affected photosynthesis (Fig. 1) and thereby growth and productivity of the cyanobacterium. Decrease in the F_0 is an indicator of decrease in the excitation energy reaching the photosynthesis reaction centre II largely due to loss of pigments in the light harvesting complex II, while decrease in the F_m is an indicator of damage to

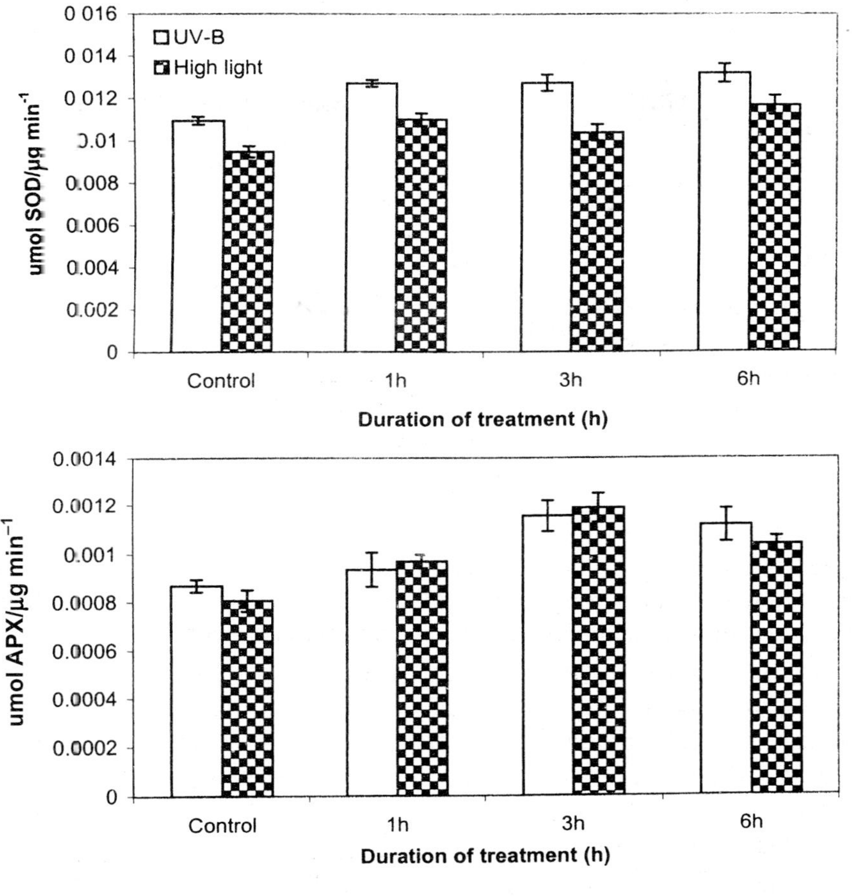

Fig. 4

the PS II reaction centre itself (Krause, 1998). The results obtained may represent direct damage to key components within the photosystem such as D1 and D2 protein of PS II (Jansen et al., 1996) as well as loss photosynthetic pigments due to generation of ROS (Bischof et al., 2002). It is known that both oxidizing as well reducing conditions in the photosynthetic electron transport are source of generation of ROS, which may lead to bleaching of surrounding pigments and peroxidation of lipid membranes (Sharma, 2002; Bischof et al., 2002). This study indicates significant oxidative damage to the cyanobacteria under our experimental conditions as seen with level of lipid peroxidation, which is an indicative of oxidative damage, was significantly higher in UV-B and PAR treated *P. corium* (Fig. 2). Lipids are some of the oxidative targets attacked by the elevated ROS and lipid peroxidation occurs especially at sites where polyunsaturated fatty acids occur in high concentrations. Ultraviolet radiation has been shown in induce-free oxygen radicals (Palmer et al., 2002) which are very effective to inducing lipid oxidation of biological membranes (Kochevar, 1990; Malanga et al., 1997), polyunsaturated fatty acids (Yamashoji et al., 1979) and

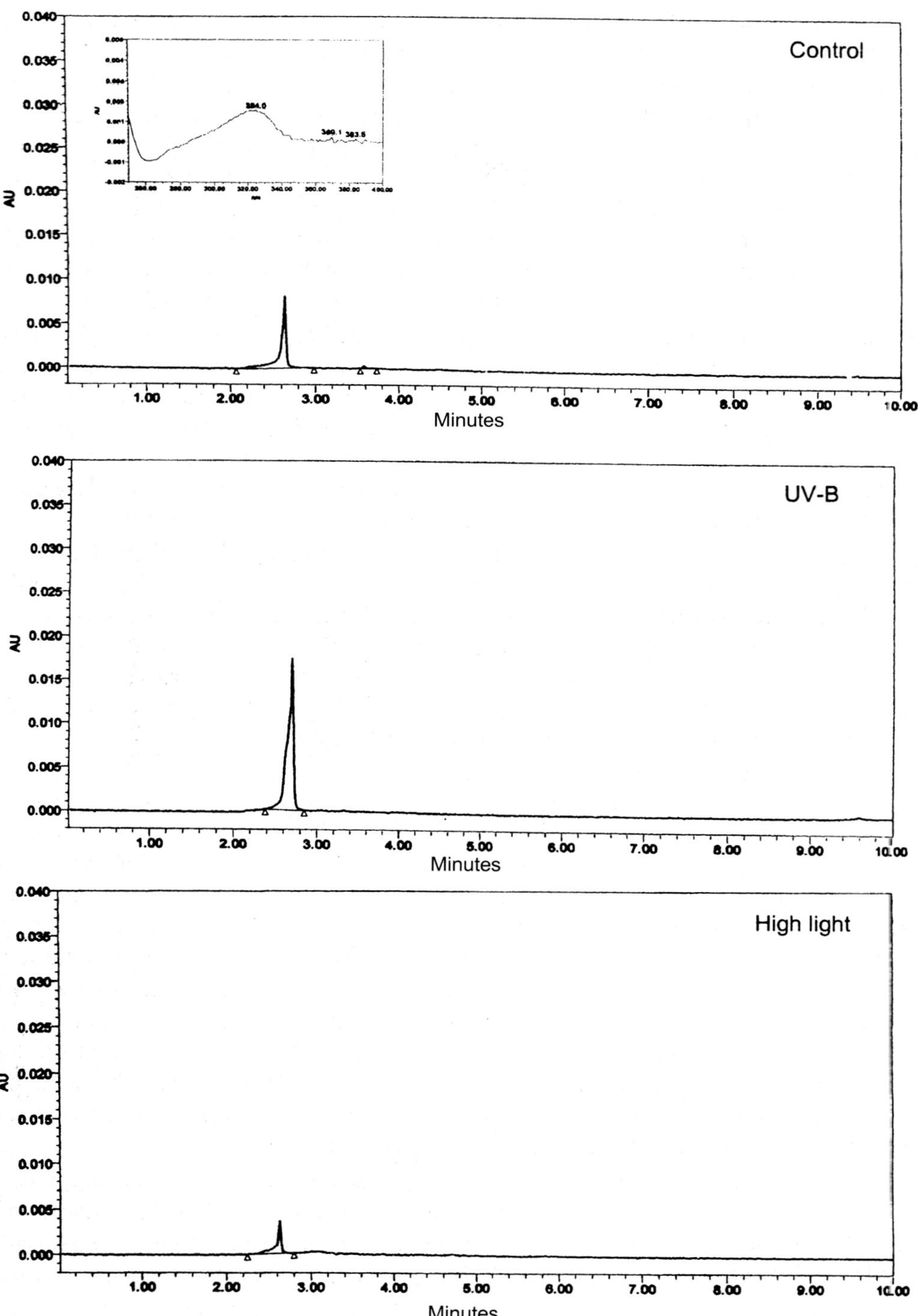
Control
UV-B
High light
AU
Minutes
0.040
0.035
0.030
0.025
0.020
0.015
0.010
0.005
0.000
1.00
2.00
3.00
4.00
5.00
6.00
7.00
8.00
9.00
10.00
324.0

Fig. 5

phospholipid liposomes (Pelle et al., 1990) and damage to the photosynthetic antennae and photobleaching of the cells of cyanobacteria (Xenopoulos et al., 2002).

He and Häder (2002) observed induction of ROS under *in vivo* condition in cyanobacterium, *Anabanena* sp. and reported that impaired photochemical reactions of PS II can enhance the production of ROS in cyanobacterium. UV-B was found to reduce the amounts of photosynthetic pigments and cause photoinhibiton (a reduction in the rate of photosynthesis) in the red alga *Porphyra leucosticta* (Figueroa et al., 1997). Shick et al. (1991) reported that the octocoral *Clavularia* exhibited a 50% decrease in photosynthesis when exposed to high levels of UV-B. Gleason and Wellington, (1993) used an underwater spectroradiometer to determine that increased dosages of UV could induce bleaching in the stony coral *Montastraea annularis*.

In this study, we observed little ability of quenching of ROS existed in *P. corium* (Fig. 3) which leads to increased in the formation of ROS, a natural product of aerobic system, exacerbated under UV-B and high light conditions due to impaired photochemical reaction. This is substantiated by increased oxidative nature (observed as lipid peroxidation) of damage to *P. corium*, which may be one of the reasons why corals, largely associated with *P. corium*, are getting bleached. It will also be interesting to study the other micro- and macro-algae found in association with corals for their antioxidant activity and extent of photobleaching.

Lesser (1997) had reported generation of reduced oxygen intermediates within both the algal symbionts and host, resulting in oxidative stress which causes decrease in photosynthesis and subsequent bleaching. Lesser and Shick (1989) showed generation of ROS under conditions which lead to bleaching of symbiotic *Anthopleura elegantissima*. Brown et al. (1994) cited solar irradiation as a possible cause of bleaching in corals either acting alone or in conjunction with other environmental factors. Downs et al. (2002) also observed strong positive correlations between accumulation of oxidative damage products and bleaching in corals.

Our study showed that UV-B treatment and high light though resulted in slight increase in the activity of SOD and APX over control, but declined further on exposure to longer duration (Fig. 4), but was insufficient to provide protection against oxidative damage (Fig. 3). Level of increase in the SOD and APX was considerably less and was found to be for short duration. Increase to the extent of 400% has been observed in APX activity in wheat in response to UV-B treatment in order to protect against UV-B damage (Sharma et al., 1998b). This may explain greater damage to growth and oxidative damage in this study under longer duration and accumulative effect of the UV-B. Ascorbic acid content, a non-enzymatic antioxidant, was also found to be less in *P. corium* (1.2 mg g^{-1} F. W.) and UV-B and light treatment slightly decreased the ascorbate content (data not shown). It seems that inability of *P. corium* to quench ROS species on account of low inherent level of antioxidant enzymes and ascorbic acid and limited induction of antioxidant enzymes on imposing of the stress resulted in greater formation of ROS that led to oxidative damage and subsequent bleaching and death of the coral reefs.

Reef–building corals also contain UV absorbing compounds capable of blocking potentially damaging UV radiation as a first line of defence. Our study also showed that *P. corium* was able to synthesize MAAs having absorbance in UV region, in response to UV-B radiation but was inhibited by high light (Fig. 5). Kawaguti (1994) speculated that some coral pigments offered protection against strong sunlight. Many reef animals can produce natural sunscreens (MAAs) to protect themselves against UV but MAAs concentrations are possibly a response to the amount of UV to which they are subjected. Dunlap et al. (1998) reported that shallow-water corals produce large quantities of MAAs. Vernet and Whitehead (1996) and Karsten et al. (1998) observed synthesis of mycosporine stimulated by UV radiation in a dinoflagellate *Lingulodinium* and red alga, *Chondrus crispus*. The amount of MAAs in a corals tissues helps to determine how much UV it can withstand without bleaching (Gleason, 1993). Lesser and Farrell (2004) observed that both photosynthetic pigments and MAAs are depressed after experimental exposure to high solar radiation and thermal stress to common Caribbean coral, *Montastraea faveolata*. Our results showed that when cyanobacterium was exposed to UV-B radiation there was considerable increase in the MAAs content, however, high light resulted in considerable decline in the MAAs content (Fig. 6). This antagonistic effect of UV-B and high light on MAAs may probably explain greater to *P. corium* under *in vivo* conditions even when having the ability to produce higher MAAs under UV-B conditions.

In conclusion, the present study suggests that bleaching of coral reefs off the Lakshadweep Islands, India, which are largely associated with *P. corium,* may be due to two reasons, one because of oxidative damage caused as a result of generation of ROS under high UV-B and PAR conditions and anther due to inefficient enzymatic (SOD and APX) and non-enzymatic (ascorbic acid) antioxidant system present in the associated cyanobacterium. The two processes together lead to formation of excess ROS and result in repelling of algae from symbiont corals as a protection to itself against ROS loading resulting in starving and death of the corals, as autotrophic algae are the source of providing food to corals.

ACKNOWLEDGEMENTS

Dr. C. Raghukumar, NIO, Goa for the *Phormidium* sp. culture and Council of Scientific and Industrial Research Grant No. 38(1030)/02/EMR-II, are greatly acknowledged.

REFERENCES

Bischof, K., Krabs, G., Wiencke, C. and Hanelt, D. 2002. Solar ultraviolet radiation affects the activity of ribulose-1,5-bisphosphate carboxylase-oxygenase and the composition of photosynthetic and xanthophyll cycle pigments in the intertidal green algae *Ulva lactuca* L. *Planta* 215:502-509.

Boveris, A. 1984. Determination of the production of superoxide radicals and hydrogen peroxide in mitochondria. *Method. Enzymol* 105:429-435.

Brown, B. E., Dunne, R. P., Scoffin, T. P. and LeTissier, MDA. 1994. Solar damage in intertidal corals. Mar Ecol.*Prog. Ser.* 105:219-230.

Downs, C. A., Fauth, J. E., Halas, J. C., Dustan, P., Bemiss, J. and Woodley, C. M. 2002. Oxidative stress and seasonal coral bleaching. *Free. Radical. Biol. Med.* 33:533-543.

Dunlap, W. C., Chalker, B. E., Bandaranayake, W. M. and Wu Won, J. J. 1998. Nature's sunscreen from the Great Barrier Reef, Australia. *Int. J. Cosmetic. Sci.* 20:41-45.

Figueroa, F. L., Salles, S., Aguilera, J., Jimenez, C., Mercado, J., Vinegla, B., Flores-Moya, A and Altamirano, M. 1997. Effects of solar radiation on photoinhibition and pigmentation in the red alga *Porphyra leucosticta*. *Mar. Ecol.Prog. Ser.* 151:81-90.

Gleason, G. M. 1993. Effects of disturbance on coral communities: Bleaching in Moorea, French Polynesia. *Coral. Reefs.* 12:193-201.

Gleason, D., Wellington, G. M. 1993. Ultraviolet radiation and coral bleaching. *Nature.* 65:836-838.

Gleason, D. F. and Wellington, G. M. 1995. Variation in UV-B sensitivity of Planula larvae of the coral *Agaricia agaricites* along a depth gradient. *Mar Biol.* 123:693-703.

Glynn, P.W. 1996. Coral reef bleaching: Facts, hypotheses and implications. *Global. Change. Biol.* 2:495-509.

Goreau, T. F. and Goreau, N. I. 1960. Distribution of labelled carbon in reef building corals with and without zooxanthellae. *Science* 181:668-669.

Grigg, R. W., Polovina, J. J. and Atkinson, M. J. 1984. Model of a coral reef ecosystem. III. Resource limitations community regulation, fisheries yield and resource management. *Coral. Reefs.* 3: 23-27.

Hay, M. E., Fenical, W. 1996. Chemical ecology and marine biodiversity: Insights and products. *Oceanography.* 9:10-20.

He, Y. Y. and Häder D. P. 2002. Involvement of reactive oxygen species in the UV-B damage to the cyanobacterium *Anabaena* sp. *Photochem. Photobiol. B:Biol.* 66:73-80.

Hoegh-Guldberg, O., Berkelmans, R. and Oliver, J. 1997. Coral bleaching: Implications for the Great Barrier Reef Marine Park, CRC January 1997 conference in research and reef management proceedings (pp. 21-43).

Jansen, M. A. K., Aba, V., Greenberg, B. M., Matto, A. K. and Edelman, M. 1996. Low threshhold levels of UV-B in a background of photosynthetically active radiation trigger rapid degradation of D_2 protein of PS II. *Plant.* J 9:693:699

Karsten, U., Franklin, L. A., Luning, K. and Wiencke, C. 1998. Natural ultraviolet radiation and photosynthetically active radiation induce formation of mycosporine like amino acids in the marine macroalga. *Chondrus cripus* (Rhodophyta). *Planta* 205:257-262.

Kawaguti, S. 1994. On the physiology of reef corals VI. Study of the pigments. Palao Tropical Biological Station Studies 2:617-674.

Kleppel, G. S., Dodge, R. E. and Reese, C. J. 1989. Changes in pigmentation associated with the bleaching of stony corals. *Limnol. Oceanogr.* 34:1331-1335.

Kochevar, I. 1990. UV induced protein alterations and lipid oxidation in erythrocytes membranes. *Photochem. Photobiol.* 52:795-800

Krause, G. H. 1988. Photoinhibition of photosynthesis. An evaluation of damaging and protective mechanisms. *Physiol. Plant.* 74:566-574.

Lesser, M. P. 1997. Oxidative stress causes coral bleaching during exposure to elevated temperatures. *Coral. Reefs.* 16: 37-192.

Lesser, M. P. and Farell, J. H. 2004. Exposure to solar radiation increases damage to both host tissues and algal symbionts of corals during thermal stress. *Coral. Reefs.* 23: 367-377.

Lesser, M. P. and Shick, J. M. 1989. Effects of irradiance and ultraviolet radiation on photoadaptation in zooxanthellae of *Aiptasia pallida:* Primary production, photoinhibition, and enzymatic defenses against oxygen toxicity. *Mar. Biol.* 102:243-255.

Malanga, G., Calmanovici, G. and Puntarulo, S. 1997. Oxidative damage to chloroplasts from *Chlorella vulgaris* exposed to UV-B radiation. *Physiol Plantarum.* 101:455-462.

Menon, H. B., Lotliker, A. and Nayak, S. R. 2005. Pre-monsoon bio-optical properties in estuarine, coastal and Lakshadweep waters. *Estuarine Coastal and Shelf Science* 63:211-223.

Muscatine, L. 1973. Nutrition of corals. In: Jones OA, Endean R (Eds.) Biology and Geology of Coral Reefs. Academic Press, New York, Vol. 2, pp. 77-115.

Palmer, H., Ohta, M., Wantanbe, M., Suzuki, T. 2002. Oxidative stress induced cellular damage caused by UV and methyl viologen in *Euglena gracilis* and its suppression with rutin. *Photochem Photobiol B:Biol*. 67:116-129.

Pelle, E., Maes, G. A., Paulo, E. K. and Smith, W. P. 1990. An *in vitro* model to test relative antioxidant potential: Ultraviolet induced lipid peroxidation in liposomes. *Arch. Biochem. Biophys*. 283:234-240

Ravindran, J. and Raghukumar, C. 2006. Histological observation on the scleractinian coral *Porites lutea* affected by pink-like syndrome. *Curr. Sci*. 90:720-724.

Ravindran, J. and Raghukumar, C. 2002. Pink-like syndrome (PLS) in the scleractinian coral *Porites lutea*. *Coral Reefs* 21:252.

Reaka-Kudla, M. L. 1996. The global biodiversity of coral reefs: A comparison with rainforests. In: Reaka-Kudla ML, Wilson DE, Wilson EO (Eds.) Biodiversity II: Understanding and Protecting Our Natural Resources. Joseph Henry/National Academy Press, Washington, D.C. pp. 83-108.

Reiss, C. 1994. Determination of ascorbic acid content. In: Reiss C (Ed.) Experiments in Plant Physiology. Prentice-Hall. Inc, pp. 1-8.

Rippka R., Deruelles J., Waterpery J. B. Herdman M. and Stanier R. Y. 1979. Gen assignments, strain histories and properties of pure cultures of cyanobacteria. *J Gen Microbiol* 111: 1-61.

Sankhalker, S. and Sharma, P. K. 2002. Protection against photooxidative damage provided by enzymatic and non-enzymatic antioxidant system in sorghum seedlings. *Indian J Exp Biol* 40:1260-1268.

Sharma, P. K. 2003. Response of plants to ultraviolet-B radiation: Impact on photosynthesis and productivity. *J. Plant. Biol*. 30:271-283.

Sharma, P. K. 2002. Photoinhibition of photosynthesis and mechanism of protection against photodamage in crop plants. *Everyman's Science* 36: 237-252.

Sharma, P. K., Anand, P., Sankhalkar, S., Shetye, R. 1998a. Photochemical and biochemical changes in wheat seedlings exposed to supplementary UV-B radiation. *Plant Sci*. 132: 21-30.

Sharma, P. K., Anand, P. and Sankhalkar, S. 1998b. Oxidative damage and changes in activities of antioxidant enzymes in wheat seedlings exposed to ultraviolet radiation. *Plant Sci*. 75: 359-366.

Shick, J. M., Lesser, M. P., Stockaj, W. L. 1991. Ultraviolet radiation and photooxidative stress in zooxanthellate anthozoans: The sea anemone semoni and the octocoral *Clavularia* sp. *Symbiosis* .10: 145-173.

Sinha, R. P. and Häder, D. P. 2000. Effects of UV-B radiation on cyanobacteria. *Rec Res Devel Photochem. Photobiol*. 4:239-246.

Stapleton, A. 1992. Ultraviolet radiation and plants: burning questions. *The Plant*. Cell. 4:1353-1358.

Vernet, M., Whitehead, K. 1996. Release of ultraviolet-absorbing compounds by the red tide dinoflagellate *Lingulodinium polydra*. *Mar. Biol*. 127:35-44.

Wilkinson, C., Linden, O., Cesar, H., Hodgson, G., Rubens, J. and Stong, A. E. 1999. Ecological and socioeconomic impacts of 1998 coral bleaching in the Indian Ocean: An ENSO impact and a warning of future change? *Ambio*. 28:188-196.

Wilkinson, C. R. and Buddemeier, R.W. 1994. Global Climate Change and Coral Reefs: Implications for People and Reefs. Report of the UNEP-IOC-ASPEI-IUCN Global Task Team on the Implications of Climate Change on Coral Reefs. IUCN, Gland, Switzerland, 124

Xenopoulos, M. A., Frost, P. C. and Elser, J. J. 2002. Joint effect of UV radiation and phosphorus supply on algal growth rate and elemental composition. *Ecology*. 83(2): 23-435

Yamashoji, S., Yoshida, H. and Kajimoto, G. 1979. Photooxidation of linoleic acid by ultraviolet light and effect of superoxide anion quencher. *Agric. Biol. Chem. Tokyo*. 43:1249-1254

30

Intracellular Localization and Gene of Rubisco in *Enteromorpha clathrata* (Ulvales, Chlorophyta)

Peimin He[1*], Shunjie Yin[1], Charles Yarish[2], Huan Zhang[3] and Senjie Lin[3]

[1]Aqua-life Science and Technology College, Fisheries University of Shanghai, Shanghai, P.R. China

[2]Ecology and Evolutionary Biology, University of Connecticut, Stamford, CT 06901, USA

[3]Department of Marine Sciences, University of Connecticut, Groton, CT 06340, USA

ABSTRACT

Enteromorpha clathrata was a marine macroalga characterized with rapid growth. In this paper, the pyrenoid ultrastructure, Rubisco intracellular localization and its coding gene were studied and analyzed. Ultrastructure showed there were two similar pyrenoids in chloroplasts. There was a longitudinal channel through pyrenoids, which was composed of two thylakoid membranes. With a specific antibody raised against the large subunit of Rubisco, the immunogold electron microscopic analysis showed that Rubiscos were highly concentrated in pyrenoids and their starch sheath. The proportion of the total gold particles in the pyrenoid, the starch sheath and the thylakoid region were 71.86%, 27.94%, 0.2% respectively. It indicated that pyrenoid and starch sheath was the site for Rubisco activation and CO_2 fixation, which supported the suggestion that pyrenoids perform photosynthesis function. Western blot prepared with the native-PAGE and SDS-PAGE showed that the molecular masses of holoenzyme and large subunit of Rubisco were below 550 kDa and 56 kDa respectively. The partial gene and its cDNA (1053 bp) coding for the large subunit of Rubisco, *rbcL*, were amplified with PCR. Further sequence analysis showed that the identities in nt and aa sequences between *E. clathrata* and *E. intestinalis* were 98.4% and 99.7% respectively, and it indicated that *E. clathrata rbcL* was closely related to the counterpart in *E. intestinalis*.

Keywords: Chlorophyte, cDNA, *Enteromorpha clathrata*, gene, immunogold localization, pyrenoid, Rubisco.

INTRODUCTION

Enteromorphya was a cosmopolitan chlorophyte seaweed (Lüning et al., 1990). As previously reported, *Enteromorpha* was characterized with high growth rate (Li et al., 1982; He et al., 2000). This seaweed can form green blooms (or green tides) along the coast that may cause problems to the marine environment and have major ecological and economic impacts (Lüning et al., 1990; Lotze and Schramm 2000; Blomster et al,. 2002). These observations suggest that photosynthetic rate may be very high in this genus. Since Ribulose-1, 5-bisphosphate carboxylase (Rubisco) is a key enzyme for photosynthesis, it is of interest to characterize this enzyme and its encoding gene.

In pyrenoid-containing algae, Rubisco has been found to be mainly located in the pyrenoids (Holdsworth 1971; Salisbury and Floyd 1978; Kerby and Evans 1981; Satoch et al., 1984; Caers et al., 1987). The pyrenoid-localization of Rubisco has been widely confirmed in microalgae with gold—and fluorescent—immunolocalization (McKay and Gibbs 1991; McKay et al., 1992; Nozaki et al., 1994; Ascaso et al., 1995; Süss, et al., 1995; Miyamura et al., 1996; Lin and Carpenter 1997a; Morita et al., 1997, 1999; He et al., 2001, 2002a). In *Chlamydomonas reinhardtii*, a Rubisco gene mutation resulted in disappearance of pyrenoid (Rawat et al., 1996). It has also been shown that the abundance of Rubisco in the pyrenoid may be associated with activities of carbon concentrating mechanism (CCM) or CO_2 fixation (Kuchitsu 1988; Tsuzuki and Myyachi 1991; Ramazanov et al., 1994; Lin and Carpenter 1997b; He, et al., 2001, 2002a, b). Although only one pyrenoid is found in many macroscopic green algae, *E. clathrata* has multiple pyrenoids (usually more than two) in each cell (Graham and Wilcox 2000; He et al., 2002a). To date, the majority of work on pyrenoid-localization of Rubisco in chlorophyte has been done for microscopic species, and relatively little information was available for macroscopic counterparts. It also remains a question whether there are multiple copies of the gene coding for Rubisco in this multi-pyrenoic containing lineage. In this paper, we will report results of the measurements of Rubisco intracellular localization, and *rbc*L gene sequence in *E. clathrata*.

MATERIALS AND METHODS

Collection and Culture

Unialgal cultures of *Enteromorpha clathrata* were established from blades collected from the sea field of Hai Feng, Qidong County, Jiangsu Province. They were initially maintained in a 1-L flask at 191°C under a photon flux of 30-45 µmol photons $m^{-2}s^{-1}$, in a 12:12, L:D photoperiod. From these, four clonal cultures were achieved. One of these cultures (H3) was selected for this study and maintained in 1-L flasks with VSE medium (von Stosch's enriched sea water as cited by Ott 1966) at 15°C and a photon flux of 15-20 µmol $m^{-2}os^{-2}$ in a 12:12, L:D photoperiod. Blades were collected for DNA and protein extraction.

Rubsico Gel Electrophoresis and Western Blotting

Total proteins were separated on SDS-PAGE and western blot analysis was carried out as described in the Sambrook *et al.* (1989). The antibody against the large unit of *Chlorella protothecoides* Rubisco (Zhang *et al.* 1994) was employed at a dilution of 1:500. A sheep antibody raised against rabbit IgG conjugated with phosphoesterase (Huamei Co., Shanghai,

China) was used as the secondary antibody. Results were visualized with colorimetric reaction of the enzyme with 5-bromo-4-chloro-3-indolyl phosphate and nitro blue tetrazolium chloride as the substrates.

Immunogold Electron Microscopy

The tissue of *E. clathrata* was fixed in the cold 3% glutaraldehyde prepared with seawater for 1-2 days at 4ºC. The fixed tissue was transferred to 2% osmium tetraoxide solution prepared by PBS (pH 7.2) for 2 h after being rinsed with 0.1 mol/L phosphate buffer solution 3 times. Following three more rinses with PBS, ultrathin sections were prepared for Rubisco gold immunolocalization as described previously (Monita et al., 1997; He, et al., 2001). The primary antibody as mentioned above was used at a dilution of 1:200. The secondary antibody was sheep antibody raised against rabbit IgG and labeled with gold particle (10 µm in diameter, product of Sigma Co). Immunolabeling was assessed by estimating the density of the gold particles precipitated over the areas of interest.

Gene Cloning and Sequencing

Fresh alga blades were homogenized with a mortar and pestle on dry ice. The cell slurry was diluted with 3 mL of DNA extraction buffer (10 mM Tris, 100 mM EDTA, 1% SDS, 20 µg/mL proteinase K) at 55° with vigorous shaking. Phenol and chloroform extraction and DNA precipitation were conducted as reported (Sambrook et al., 1989). Similarly, fresh tissue was homogenized and suspended in Trizol solution. RNA was purified following instructions of Trizol manufacturer. First strand cDNA was synthesized using Preamplification cDNA kit (Invitrogen).

Both genomic and cDNA preparations were used to amplify *rbc*L gene using universal *rbc*L primers designed based on a multi-alignment of form I Rubisco sequences. The primers used were RubI-1F (forward): 5'-ggtytwtcwgstaaraactayggwcg-3' and RubI-2R (reverse): gtgrataccrccwgaagcwacwgg. Each PCR product was subcloned to a TA vector. Five clones from each PCR product were sequenced with a 377 ABI sequencer.

Phylogenetic Analysis

Nucleotide (nt) and deduced amino acid (aa) sequences of *E. clathrata rbc*L were compared with those of other organisms. A phylogenetic tree was constructed using deduced amino acid sequences of *rbc*L by the neighbor-joining method (Saito and Nei, 1987) and corrected by the Kimura method (Kimura 1980), rooted with bacterium *rbc*L (P22849). Sequences were aligned using the CLUSTAL W (v1.8) server at the DNA Data Bank of Japan (DDBJ) (http://www.ddbj.nig.ac.jp/Welcome-e.html) on default values, and the alignment was adjusted using the Sea View program to maintain codon integrity prior to analysis in the Phylo_win package. The reliability of the tree topology was evaluated by bootstrap analysis with 1,000 replicates.

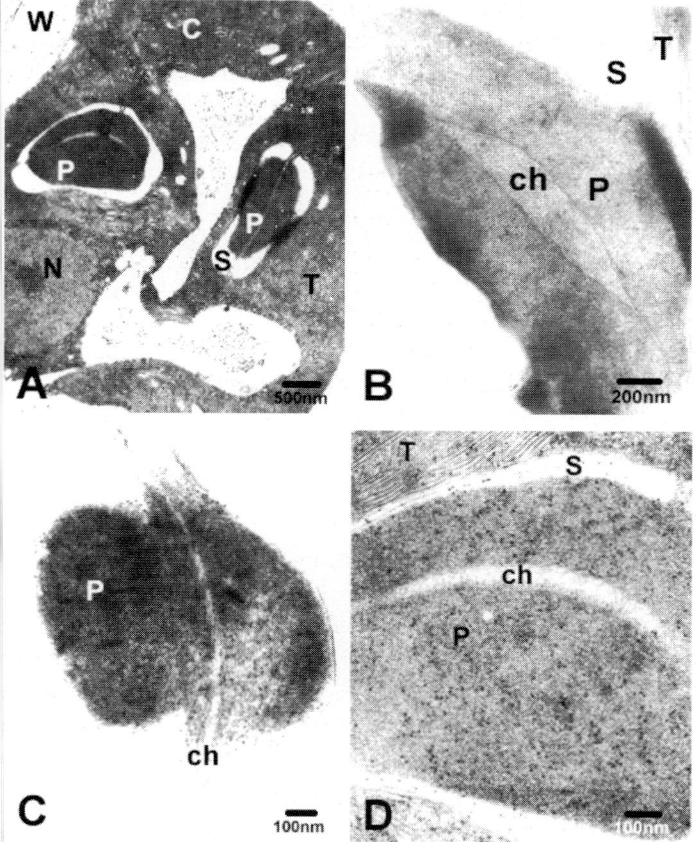

Fig. 1. The micrograph of the ultrastructure of *Enteromorpha clathrata* cells and immunogold labeling. A) Shown are two pyrenoids (P) in the cell of *E. clathrata* thalli; B) A portrait channel (ch) with two thylakoid membranes and the enlargement of the channel in the middle of pyrenoid (P); C) The pyrenoid (P) and immunogold labeling of Rubisco large subunit in the pyrenoids (P); D) Immunogold labeling of Rubisco large subunit in the pyrenoid (P) and its portrait channel (ch), starch sheath (S). Other labels depict chloroplast (C), nucleus (N), thylakoid (T), and cell wall (W).

RESULTS

Pyrenoid Ultrastructure

Ultrastructural analysis showed presence of two similar pyrenoids in the chloroplast of *Enteromorpha clathrata* cells (Fig 1A). The pyrenoids were about 600-1000 x 1200-1500 nm^2 in size, and was cut through by a longitudinal thylakoid channel, which was composed of two thylakoid membranes (Fig. 1B). Sometimes the channel was enlarged in the middle of pyrenoid (see Fig. 1B).

Rubisco Intracellular Localization

Figure 1C & D showed the distribution of gold particles labeled with large subunit of Rubisco in the chloroplast of *E. clathrata*. The specificity of the antibody used was verified on the Western blot (Fig. 2). The Western blot prepared with the native-PAGE showed that there was only one protein band (Rubisco holoenzyme) with molecular mass below 550 kDa (Fig. 2A). The other Western blot prepared from SDS-PAGE also displayed a single protein band, the molecular size was 56 kDa as expected for the large subunit of Rubisco (Fig. 2B). Labeled with this antibody, most of the gold particles were concentrated in the pyrenoid. In the starch sheath that embedded the pyrenoid, there appeared to be some particles, whereas very few gold particles were observed in the thylakoid region (Fig. 1C, D). The density was 660 ± 256.16 gold particles · μm^{-2} in the pyrenoid, 272.22 ± 127.13 gold particles · μm^{-2} in the starch sheath, and only 27.5 ± 11.98 gold particles · μm^{-2} in the thylakoid. After subtracting the background density (25.5 ± 15.5 gold particles · μm^{-2}), proportion of the total gold particles in the pyrenoid, the starch sheath and the thylakoid region were 71.86%, 27.94%, 0.2% respectively.

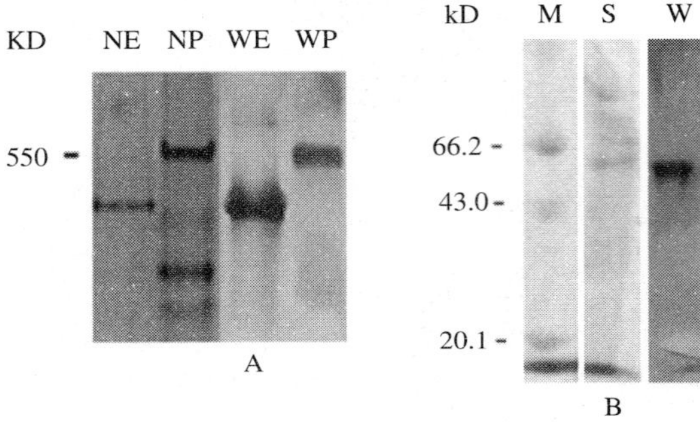

Fig. 2. Native-PAGE, SDS-PAGE, and Western blotting of Rubisco. A. Native-PAGE (Lane NE, *Enteromorpha clathrata* and Lane NP, *Porphyra yezoensis*) performed with 0.8% polyacrylamide gel and Western blotting (Lane WE, *Enteromorpha clathrata* and Lane WP, *Porphyra yezoensis*). The size of the whole Rubisco from *E. clathrata* appeared to be smaller than the size of the whole Rubsico from *Porphyra yezoensis* (550 kDa). B. SDS-PAGE (lane S) performed with 1.2% polyacrylamide gel and Western blotting (lane W). The size of large subunit of Rubisco of *E. clathrata* was about 56 kDa.

Cloning of the Rubisco Gene in *E. clathrata*

With the primers specially designed for form I Rubisco gene, the partial Rubisco gene (1053 bp) was amplified using PCR from both genomic DNA and cDNA of *E. clathrata.* The size of a gene fragment yielded from both genomic DNA and cDNA was the same (Fig. 3), indicating that there was no intron in the gene fragment.

To further analyze the sequence of the gene fragment, each PCR product was subcloned to a TA vector and five of the resulting clones were selected for sequencing from each amplicon. The results showed that all 10 sequences from both DNA and cDNA PCR products were identical.

Phylogenetic Analysis of Rubisco gene

Sequence analysis showed that *E. clathrata rbc*L was most closely related to the counterpart in another *Enteromorpha* species, *E. intestinalis*. The identities in nt and aa sequences between these two macroalgae were 98.4% and 99.7%, respectively. In contrast, the nt and aa identities between *E. clathrata rbc*L and homologs in the other 'green line' organisms were 80-98% and 80-99%, respectively, whereas a lower identity (<60%) was found between *E. clathrata* and the 'red-

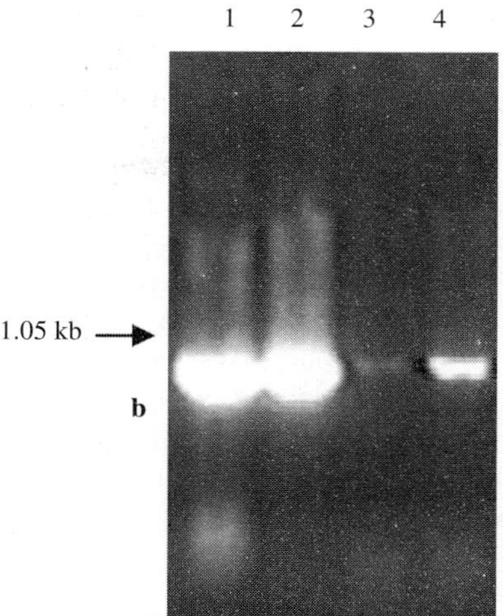

Fig. 3. PCR amplification of *rbc*L gene fragment. Lanes 1-2: PCR amplification from genomic DNA, 3-4, PCR amplification from cDNA. Arrow indicates the specific product of *rbc*L.

like' organisms. Based on the aa sequence data, a rooted neighbor-joining was constructed (Fig. 4). Representative species from major "red-like" and "green-like" lineages were included in the analysis[3]. The tree showed the clustering of chlorophyte taxa in the "green-like" group and stremopile, cryptophytes, and haptophytes, in the "red-like" group. In the "green-like" cluster, *E. clathrata* was most closely related with other *Enteromorpha* species, followed by *Ulva*, and then other chlorophytes.

DISCUSSION

Pyrenoid-localization of Rubisco in *E. clathrata*

In previous studies, it was found that the initial activity of Rubiscos in *Enteromorpha clathrata* was very high and its active ratio of Rubisco (initial activity/total activity) was as high as 77.62% (He et al., 2000). Since pyrenoid was thought to be the site of Rubisco assembly and activation (McKay & Gibbs 1991; Pronina & Semenenko 1992; Portis 1995), the higher initial Rubisco activity observed was possibly related to the structure and biochemical characteristics of the pyrenoid. As early as the 1970's, Holdsworth (1971) first discovered that the pyrenoid from green alga *Eremosphaera* was composed of large and small subunits of Rubisco. Then the same was found in other green alga, such as *Bryopsis maxima, Chlamydomonas reinhardtii, Micromonas sguamata* (Kuchitsu et al., 1988). With the development of immunochemistry technique, the immunogold labeling of Rubisco has demonstrated that the Rubisco is concentrated in the pyrenoid. It was thus suggested that the pyrenoid was associated with photosynthesis and CCM. Therefore, information on the fraction of total cellular Rubisco that is distributed in the pyrenoid is insightful. Kuchitsu et al., (1988) first proposed that the pyrenoid function in photosynthesis, and showed with immunolocalization technique that about 50% Rubiscos was present in the pyrenoid. Although lower fraction has been reported (Süss et al., 1995; Morita et al., (1997); and He et al., (2001, 2003)) reinforced the earlier suspicion on the role of pyrenoid in photosynthesis

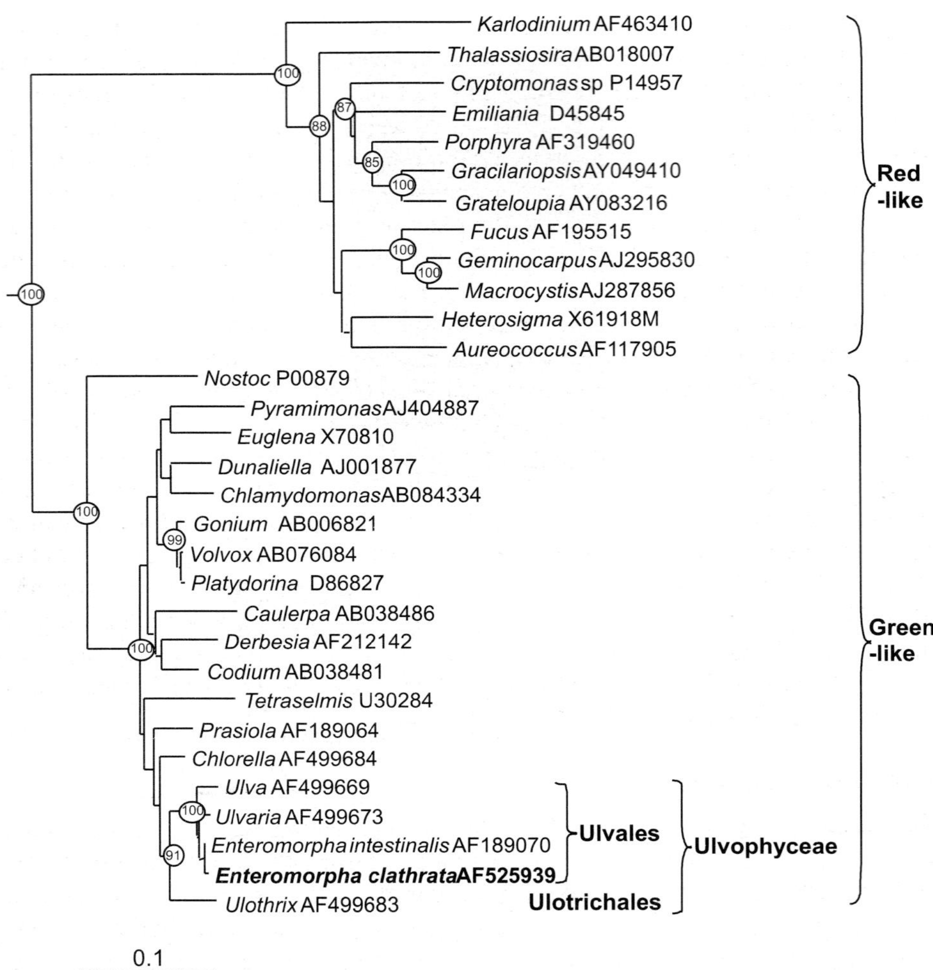

Fig. 4. Neighbor tree of Rubisco based on amino acid sequences. Bacteria *Allochromatium* Rubisco (P22849) was used as an outgroup to root the tree. Scale bar is calibrated in aa replacement per site. Robustness of the tree topology was evaluated by 1000 replicates of bootstraping; only the values >80% are shown. *rbc* L sequences clearly form two groups: 'green-like' and 'red-like', and *E. clathrata* sequence most closely clusters with its cogeneric species.

by showing that up to about 99% of Rubisco was concentrated in the pyrenoid respectivley. The association between pyrenoid-localization of Rubisco and photosynthesis has also been demonstrated experimentally for microscopic chlorophyte species. For instance, Lin and Carpenter (1997) showed that the fraction of *Dunliella tertiolecta* cell population in which Rubisco was highly concentrated in the pyrenoid (PR-index) increased in the light period and decreased dramatically in the dark period. The PR-index also decreased when photosynthesis was inhibited or when the culture entered the stationary growth phase.

Enteromorpha is one of the algal lineages that contain multiple pyrenoids, whereas usually only one pyrenoid is observed in most algal species examined (e.g., Pickett-Heaps 1975; Brawley & Wetherbee 1981; Lin & Carpenter 1997a; Jenks & Gibbs 2000; Calado & Moestrup 2002; Chepurnov et al., 2002; Van Lenning et al., 2003). In *Enteromorpha clathrata*, the two pyrenoids observed appear to be similar in ultrastructure. It is not possible at present to determine whether presence of multiple pyrenoids in this species is responsible for higher Rubisco activity and hence higher growth rate. However, it is interesting to note that both pyrenoids were heavily labeled with Rubsico antibody, indicating that both pyrenoids were rich of Rubisco. The immunogold labeling result showed that in the chloroplast of *Enteromorpha clathrata*, about 71.86%, 27.94% and 0.2% of Rubisco was distributed in the pyrenoid, starch sheath, and thalykoid stroma, respectively. In other words, 99.8% of Rubisco was concentrated in the pyrenoid and its starch sheath. Possible artifact of immunolabeling as the reason to observe such high fraction of Rubisco in the pyrenoid was ruled out by prior tests for reactivity and specificity of the Rubisco antibody used. We chose to use the antibody raised against Rubisco from *Chlorella protothecoides*, the chlorophyceae that is closely related to *E. clathrata*, to maximize cross-reaction and the probability to detect various forms of Rubisco that may occur. The Western blot using total proteins of *E. clathrata* showed that the antibody only reacted with one single protein in both the native- and the SDS-PAGE gels, confirming the specificity of the antibody and indicating the presence of a single form of Rubisco. Furthermore, the possible association of the pyrenoid localization of a large fraction of Rubisco with photosynthesis was also supported by an earlier observation that Rubisco activase, the enzyme catalyzing the activation of Rubisco, was also highly concentrated in the pyrenoid (He et al., 2002a).

The Sequence of *rbcL* in *E. clathrata*

Genetic association of pyrenoid morphology has been documented for chlorophyceae. *In Chloromonas*, nucleotide substitution in *rbc*L facilitates aggregation of the encoded Rubisco in the pyrenoid (Nozaki et al., 2002). Our gene sequence analysis did not indicate unique substitutions that may be associated with the presence of two pyrenoids. Rather, the *E. clathrata rbc*L showed functional domains highly conserved among green algae. As one of the few attempts to reconstruct phylogeny with *rbc*L gene for *Enteromorpha*, the result reveals the closest relationship of *E. clathrata* with its congeneric species *E. intestinalis* (Sherwood et al., 2000), followed by *Ulva laculata* and other chlorophytes. This result is congruent with the prevailing notion that chloroplasts can be divided into groups, the "green-like" and the "red-like" (Tabita 1999). The nt and aa identities between *rbc*L of *E. clathrata* and that of the other 'green-like' organisms were 80-98% and 80-99%. The identity is significantly low between *E. clathrata* and the 'red-like' algae (<60%).

*E. clathrata rbc*L is different from counterparts in many other chlorophytes in its absence of introns. In other chlorophytes, numerous introns have been observed in *rbc*L. For instance, in *Euglena gracilis*, there were 9 introns in the *rbc*L sequence (Hallick et al., 1993). Although the gene cloned in this study is not full-length, it is not likely that there are introns with the short stretches of sequence near the ends.

From the multiple clones sequenced for both genomic DNA and cDNA, only one identical sequence was found. This did not likely result from bias of PCR, since the primer set used in this study was designed from highly conserved regions of this gene and it has facilitated cloning of *rbc*L from chlorophyte, diatoms, gymnoidinoid dinoflagellates, and cryptophytes (Lin and Zhang, unpublished data). Isolation of only one identical gene sequence is also in accordance to the result of Western blotting as mentioned above. This result is similar to that in other green chlorophyceae algae such as *Chlamydomonas* and *Chromonas*. In contrast, the red alga *Porphyra purpurae*, possesses two copies of *rbc*L in chloroplast DNA genome (Reith and Munholland 1993; Reith 1995). In the chlorophyte *Euglena*, which has only one pyrenoid in the chloroplast, the nuclear-encoded *rbc*S occurs in multiple copies (Tessier et al., 1995). Taken together, these observations suggest that the copy number of Rubisco genes does not seem to correlate with the number of pyrenoids present in the chloroplast. Further, the dense immunogold labeling in both pyrenoids suggest that they contain high amounts of this single Rubisco large subunit.

ACKNOWLEDGEMENTS

Support for this project was provided by funding in part by E-Institute of Shanghai Municipal Education Commission, and in part by the Key Principle of Aquaculture, Shanghai Education Committee, and the Twilight Plan of Shanghai Education Committee (No:SG-98-032) as well as the National Nature Foundation (No:30371101).

REFERENCES

Aizawa, K. and Miyachi, S. 1986. Carbonic anhydrase and CO_2 concentrating mechanisms in microalgae and cyanobacteria. *FEMS Microbiol. Rev.* 39: 215-223.

Blomster, J., Back, S., Fewer, D. P., Kiirikki, M., Lehvo, A.J., Maggs C. A. and Stanope, M. J. 2002. Novel morphology in *Enteromorpha* (Ulvophyceae) forming green tides. *Amer. J. Bot.* 89: 1756-1763.

Brawley, S. H. and Wetherbee, R. 1981. Cytology and ultrastructure, in: *The Biology of Seaweeds,* edited by C. S. Lobban & M. J. Wynne, pp. 248-299.

Calado, A.J. and Moestrup, O. 2002. Ultrastructural study of the type species of *Peridiniopsis, Peridiniopsis borgei* (Dinophyceae), with special reference to the peduncle and flagellar apparatus. *Phycologia.* 41: 567-584.

Chepurnov, V.A., Mann, D. G., Vyverman, W., Sabbe, K. and Danielidis, D. B. 2002. Sexual reproduction, mating system, and protoplast dynamics of Seminavis (Bacillariophyceae). *J. Phycol.* 38: 1004-1019.

Hallick, R. B., Hong, L., Drager, R., Favreau, M. R., Montfort, A., Orsat, B., Spielmann, A. and Stutz, E. 1993. Complete sequence of *Euglena gracilis* chloroplast DNA. *Nucl. Acids Res.* 21: 3537-3544.

He, P., 2000. Molecular localization of photosynthesis key enzymes in algae and stress physiology in *codA* transgenic plant, Ph.D. thesis, Nanjing Agriculture University, China.

He, P., Zhang, R., Zhang, D., Liang, W., Zhao, J. and Wu, W. 2001. Gold immunolocalization of Rubisco in Chloroplast in *Chlorella. Acta Biolog. Experim. Sinca.* 34: 17-23.

He, P., Wu, W., Zhao, J., Chen, G. and Zhang, D. 2002a. Studies on ultrastructure of pyrenoid from several algae. *Acta Hydrobiol. Sinca.* 26: 327-334.

He, P., Xu, C., Yu, J., Yu, Y., Wu, W. and Zhang, R. 2002b. Molecular localization and activity inducement of perplasmic carbonic anhydrase in *Chlamydomonus reinhardtii. Chin. Bull. Bot.* 19: 317-321.

He, P., Zhang, D., Chen, G., Liu, Q. and Wu, W. 2003. Immunogold localization of Rubisco and Rubisco activase in pyrenoid of *Chlamydomonas reinhardtii. Algae.* 18: 121-127.

Holdsworth, R. H. 1971. The isolation and partial characterization of the pyrenoid protein of *Eremosphaera viridis*. *J. Cell Biol.* 51: 499-513.

Jenks, A. and Gibbs, S. P. 2000. Immunolocalization and distribution of form II Rubisco in the pyrenoid and chloroplast stroma of *Amphidinium carterae* and form I Rubisco in the symbiont-derived plastids of *Peridinium foliaceum* (Dinophyceae). *J. Phycol.* 36: 127-138.

Kajikawa, H., Okada, M.I., Shikawa, F., Okada, Y. and Nakayama, K. 1988. Immunochemical studies on ribulose-1, 5-bisphosphate carboxylase/oxygenase in the chloroplasts of the marine alga *Bryopsis maxima*. *Plant Cell Physiol.* 29: 549-556.

Kimura, M. 1980. A simple method for estimating evolutionary rates of base substitutions through comparative studies of nucleotide sequences. *J. Mol. Evol.* 16:111-120.

Kuchitsu, K., Tsuzuki, M. and Miyachi, S. 1988. Characterization of the pyrenoid isolated from unicellular green alga *Chlamydomonus reinhardtii*: Particulate form of Rubisco protein. *Protoplasma*. 144: 17-24.

Lacoste-Royal, G. and Gibbs, S. P. 1987. Immunocytochemical localization of ribulose-1,5-bisphosphate carboxylasei in the pyrenoid and thylakoid region of the chloroplast of *Chlamydomonas reinhardtii, Plant Physiol.* 83: 602-606.

Lin, S. and Carpenter, E. J. 1997a. Pyrenoid-Rubisco in relation to the cell cycle and growth phase in *Dunaliella tertiolecta* (Chlorophyceae). *Phycologia*. 36: 24-31.

Lin, S. and Carpenter, E. J. 1997b. Rubisco of *Dunaliella tertiolecta* is redistributed between the pyrenoid and the stroma as a light-shade response. *Mar. Biol.* 127: 521-529.

Lotze, H. K. and Schramm, W. 2000. Ecophysiological traits explain species dominance patterns in macroalgal bloom. *J. Phycol.* 36: 287-295.

Lüning, K., Yarish, C. and Kirkman, H. 1990. Seaweeds, (John Wiley & Sons, Inc.).

Maribel, L. D., Mikio, T. and Shigetoh, M. 1989. Light requirement for carbonic anhydrase induction in *Chlaydomonas reinhardtii*. *Plant Cell Physiol.* 30: 207-213.

Mckay, R. M L. and Gibbs, S.P. 1991a. Composition and function of pyrenoids: cytochemical and immunocytochemical approaches. *Can. J. Bot.* 69: 1040-1052.

Mckay, R. M. L. and Gibbs, S. P. 1991b. Rubisco activase is present in the pyrenoid of green algae. *Protoplasma*. 162: 38-45.

Mckay, R. M. L., Lichitle, C. and Gibbs, S. P. 1992. Immunocytochemical characterization of the intrapyrenoid thylakoids of cryptomonads. *J. Phycol.* 35: 156-160.

Miyamura, S., Hori, T., Ohya, T., Tohma, T. and Ikawa, T. 1996. Co-localization of chloroplt DNA and ribulose-1,5-bisphosphate carboxylase/oxygenase in the so-called pyrenoid of the sihonous green alga *Caulerpa lentillifera* (Caulerpales, Chlorophyta). *Phycologia*. 35: 156-160.

Morita, E., Kuroiwa, H., Huroiwa, T. and Nozaki, H. 1997. High localization of ribulose-1,5-bisphosphate carboxylase/oxygenase in the pyrenoids of *Chlamydomanas reihardtii*(Chorophyta), as revealed by cryofixation and immunogold electron microscopy. *J. Phycol.* 33: 68-72.

Morita, E., Abe, T., Tsusuki, M., Fujiwara, S., Sato, N., Hirata, A., Sonoike, K. and Nozaki, H. 1999. Role of pyrenoids in the CO_2-concentrating mechanium: Comparative morphology, physiology and molecular phylogenetic analysis of closely related strains of *Chlamydomonas* and *Chloromonas* (Vovocales). *Planta*. 208(1999) 365-372.

Nozaki, H., Kuroiwa, H. and Kuroiwa, T. 1994. Light and electron microscopic characterization of two types of pyrenoids in *Genium* (Goniaceace, Chlorophyta) *J. Phycol*. 30: 279-290.

Nozaki, H., Onishi, K. and Morita, E. 2002. Differences in pyrenoid morphology are correlated with differences in the *rbc*L genes of members of the *Chloromonas lineage* (Volvocales, Chlorophyceae). *J. Mol. Evol.* 55: 414-430.

Osafune, T., Yokota, A., Ehara, T. and Hase, E. 1990. Immunogold localization of ribulose-1,5-bisphosphate carboxylase with reference to pyrenoid morphology in chloroplasts of synchronized *Euglena gracilis* cells. *Plant Physiol*. 92: 803-808.

Ott, F. D. 1966. A selected listing of xenic algal cultures, Contrib., No. 72. Systematic-ecology Program, Marine Biology Laboratory, (Woods Hole, Mass.)

Pickett-Heaps, J. D. 1975. Green algae, (Sinauer Associate, Inc, Sunderland, Massachusetts) pp. 176, 368, 388, 414, 460, 464.

Portis, A. R. 1992. Regulation of ribulose-1,5-bisphosphate carboxylase/oxygenase activity. *Ann. Rev. Plant Physiol. Plant Mol. Biol.*. 43: 415-437.

Ramazanov, Z., Henk, C., Mason, C., Mathews, S. and Moroney, J. 1994. The induction of the CO_2-concentrating mechanism is correlated with the formation of the starch sheath around the pyrenoid of *Chlamydomonas reinhardt*. *Planta*. 195: 210-216.

Rawat, M., Henk, M., Lavigne, L. L. and Moroney, J. V. 1996. *Chlamydomonas reinhardtii* mutants without ribulose-1,5-bisphosphate carboxylase-oxygenase lack a detectable pyrenoid. *Planta*. 198: 263-270.

Reith, M. 1995. Molecular biology of Rhodophyte and Chromophyte plastids. *Ann. Rev. Plant Physiol. Plant Mol. Biol.* 46: 549-575.

Reith, M. and Munholland, J. 1993. A high-resolution gene map of the chloroplast genome of the red alga *Porphyra purpurea*. *Plant Cell*. 5: 465-475.

Saitou, N. and Nei, M. 1987. The neighbor-joining method: A new method for reconstructing phylogenetic trees. *Mol. Biol. Evol*. 4: 406-425.

Sambrook, J., Fritsch, E. F. and Mamiatis, T. 1989. Molecular Cloning: A Laboratory Manual, Cold Spring Harbor Laboratory Press, New York.

Sherwood, A. R., Garbary, D. J. and Sheath, R. G. 2000. Assessing the phylogenetic position of the *Prasiolales* (Chlorophyta) using *rbc*L and 18S rRNA gene sequence data. *Phycologia*. 39: 139-146.

Satoch, H., Okada, M., Nakayyama, K. and Miyaji, K. 1984. Purification and further characterization of pyrenoid protein and ribulose-1,5-bisphosphate carboxylase-oxygenase from the green alga *Bryopsis maxima*. *Plant and Cell Physiol*. 25: 1205-1214.

Süss, K. H., Prokhorenko, I. and Adler, K. 1995. *In situ* association of Calvin cycle enzymes, ribulose-1,5-bisphosphate carboxylase/oxygenase, ferredoxin-NADP+ reductase, and nitrite reductase with thylakoid and pyrenoid membranes of *Chlamydomonas reinhardtii* chloroplasts as revealed by immunoelectron microscopy. *Plant Physiol*. 107: 1387-1397.

Tessier, L. H., Paulus, F., Keller, M., Vial, C. and Imbault, P. 1995. Structure and expression of *Euglena gracilis nuclear rbc*S genes encoding the small subunits of the ribulose-1,5-bisphosphate carboxylase/oxygenase: A novel splicing process for unusual intervening sequences. *J. Mol. Biol.* 245: 22-33.

Tsuzuki M. and Myyachi S. 1991. CO_2 syndrome in *Chlorella, Can. J. Bot.,* 69: 1003-1007.

Van Lenning, K., Latasa, M., Estrada, M., Saez, A. G., Medlin, L., Probert, I., Veron, B. and Young, J. 2003 Pigment signatures and phylogenetic relationships of the Pavlovophyceae (Haptophyta). *J. Phycol.* 39: 379-389.

Vladmirova, M. G., Markelova, A. G. and Semenenko, V. E. 1982. Use of the cytoimmunofluorescent method to clarify localization of ribulose bisphosphate carboxylase in pyrenoids of unicellular algae. *Soviet Plant Physiol.* 30: 725-734.

Zhang, D., Li, L., Miao, Y. and Wu, Q. 1994. Rubisco purifying and its content change during the conversion from heterotrophy to autotrophy. *Acta Plant Physiol. Sinica*. 20: 346-352.

31

Mariculture of Seaweeds Based on the New "Germling Cluster Method" and Utilizing Deep Seawater in Japan

Masanori Hiraoka[1] and Naohiro Oka[2]
[1]Usa Marine Biological Institute, Kochi University, Inoshiri, Usa, Tosa, Kochi 781-1164, Japan.
[2]Umi-no Kenkyusha (Sea Lab Ltd), 5708-1 Murotomisaki, Muroto, Kochi 781-7107, Japan.

ABSTRACT

Seaweeds are an important health food in Japan. However, wild stocks of edible seaweeds have decreased due to water pollution and over-fishing. As such, there is a growing need to establish a stable and safe, mass-culture means of production of edible seaweeds. Recently, we invented a new "germling cluster method" for tank cultivation of seaweeds in a free-floating form. Furthermore, using deep seawater (DSW) with constant salinity and temperature, inorganic nutrient richness and clearness, we designed a new stable mass-culture tank system. Based on this system, a large-scale plant for mariculture of seaweeds was first built in Muroto, Kochi Prefecture, Japan, in June 2004. The edible green seaweed *Enteromorpha* is now produced at the plant the year round. Natural populations of *Enteromorpha* occur during the winter and into spring, attaining reproductive maturity in high water temperatures (> 20°C) before disappearing during the summer and autumn. In our mariculture system, however, *Enteromorpha* grows well even during the summer and autumn because the DSW source is maintained at about 12°C throughout the year. *Enteromorpha* fronds with an initial weight of 100 g attain more than 1,000 g after one week's culture. The total annual biomass of this alga is estimated to be 3 tons dry weight in a 500 ton tank, to which 1,500 tons of DSW is supplied per day. Cultivated material produced without contamination in this way is of high commercial value in the food market in Japan.
Keywords: Deep seawater (DSW), edible seaweeds, *Enteromorpha*, germling cluster method, mariculture.

INTRODUCTION

Japanese people have used seaweeds as a health food for long time. However, wild stocks of seaweeds have been decreasing year by year due to water pollution (Ohno and Miyanoue, 1980).

Cultivated production of seaweeds has fluctuated due to variable environmental factors including salinity, temperature etc. It is important to establish a stable mass-cultivation technique for seaweeds.

In an experimental mariculture system, we utilized deep seawater (DSW) pumped up from a depth of about 300 m using pipelines positioned 2 km off the coast of Muroto in Kochi, Japan, for the production of edible seaweeds (Hiraoka et al., 2004). DSW has several characteristic properties: the water temperature is constantly low (10-14°C) throughout the year; the seawater is clean, and contains less organic matter, few pathogenic micro-organisms, and a high concentration of major inorganic nutrients, as compared to surface seawater. These characteristics of DSW (constant low temperature, inorganic nutrient richness and clearness) make it suitable for culture of seaweeds. However, mariculture of seaweeds using DSW has not yet been established on a commercial scale.

Recently, we proposed a new "germling cluster method" for large-scale tank cultivation of seaweeds in a free-floating form and designed a stable mass-culture tank system using DSW. In the present paper, we introduce the germling cluster method and the tank system.

Test Cultivation of *Enteromorpha*

The edible green seaweed *Enteromorpha prolifera* was used first in a test cultivation. A high degree of synchronous spore formation was induced by cutting *Enteromorpha* thalli into small fragments. Spores released from the fragments were collected by their phototactic response and concentrated. The dense spore suspension was placed in a Petri dish containing PES medium (Provasoli, 1968), adusted to a density of more than 10,000 spores per 1 ml medium. The Petri dish was then placed in an incubator at 20°C with a 12 h:12 h light/dark cycle under cool white fluorescent light. Under these conditions, germlings grew at high density at the bottom of the Petri dish, attaching to each other and forming aggregations. These aggregations were torn into

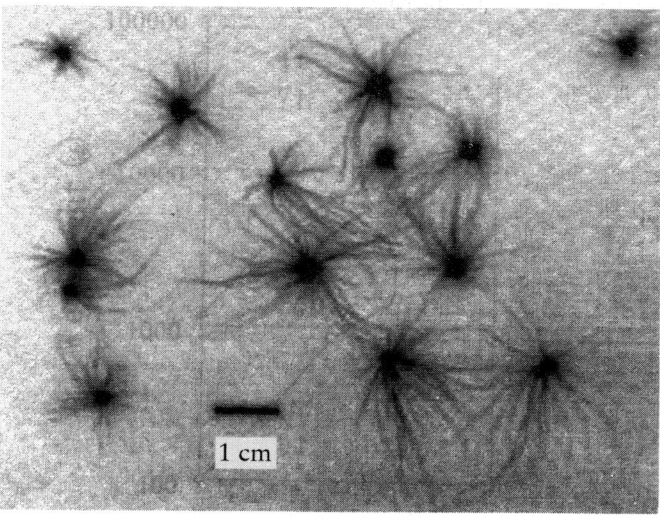

Fig. 1 Germling clusters of *E. prolifera*.

numerous small clusters containing 10-100 germlings. The germling clusters were then cultured with aeration under the same conditions outlined above, freely drifting with the current in the vessel. Once the germling clusters had attained a length of 10 mm or more (Fig. 1), they were transplanted as free-floating forms to an outdoor cylindrical 1 ton tank and cultured with continuous aeration and DSW supply under natural light. The production of *E. prolifera* in 1 ton tank was conducted for one-week period throughout the year using fronds with an initial wet weight of 100 g. Larger scale culture of *Enteromorpha* was carried out using DSW in a rectangular tank (1.5 m × 5.5 m × 0.9 m, 7 ton), after the primary culture of *E. prolifera* was established in a 1 ton tank.

Table 1 shows the monthly final wet weight of *E. prolifera* after one-week culture from an initial wet weight of 100 g in a 1-ton tank using DSW. Natural populations of *Enteromorpha* occur in seawater at temperatures lower than 20°C during the winter and into spring, but typically only attain reproductive maturity in waters above 20°C before disappearing in the summer and throughout the autumn (Hiraoka et al., 1999). However, *Enteromorpha* thalli cultured in the present tank system using DSW grew well throughout the year, even in summer and autumn, because the temperature in the tank was comparatively low. The temperature in the tank was higher in the summer and lower in winter, fluctuating with air temperature, although the DSW was supplied at a constant temperature of about 12°C (10-14°C) throughout the year. Therefore, growth rates fluctuated along with seasonal changes. The highest production cultured in these experiments was 2,620 g for an initial wet weight of 100 g in a one-week period in April. The production of *E. prolifera* at lower temperatures during the winter season was lower, at 540-920 g, than that cultured in the other seasons (Table 1).

The growth curves of *E. prolifera* fronds cultured in the 1 ton and 7-ton tanks are shown in Fig. 2. Fronds with an initial wet weight of 100 g attained more than 1,000 g after one week's culture. The growth curves for material cultured in the 1 ton and 7-ton tanks are similar. The growth curve declines after 7 days, as the frond density in the tank increases. These results

Table 1 Monthly final wet weight in *Enteromorpha prolifera* after 1 week culture from an initial wet weight of 100 g and temperatures ranges in 1 ton tank using DSW (From Hiraoka et al., 2004)

Month	Final wet weight (g)	Temperature (°C)
Jan.	540	8.4-12.2
Feb.	1100	11.1-15.0
Mar.	1260	10.4-14.1
Apr.	2620	13.0-16.5
May	1250	14.6-15.6
Jun.	1330	14.8-16.9
Jul.	1300	15.9-18.8
Aug.	1400	16.0-23.7
Sep.	940	16.1-20.4
Oct.	1710	16.2-21.2
Nov.	920	11.0-14.5
Dec.	860	9.2-15.0
Average	1270	

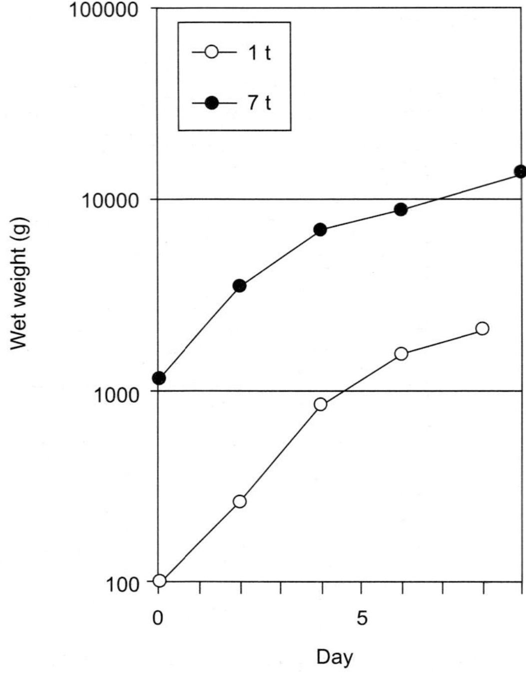

Fig. 2 Growth rate of *E. prolifera* cultured using 1 and 7-ton tanks (From Hiraoka et al., 2004).

demonstrate that the initial wet weight increases by more than 10 times after one week's culture. They also show that when the frond density in the tank is over 1 kg wet weight per 1 ton tank volume, the growth rate decrease.

A Mass-culture Tank System

Based on these results, a mass-culture tank system was designed (Fig. 3). In this system, an initial 10 g cluster of *Enteromorpha* germlings cultured in the laboratory is transplanted to an outdoor 100 L tank and cultivated with continuous DSW supply and aeration. Every week the cultivated *Enteromorpha* clusters are transferred to increasingly larger tanks. The final wet weight of the material cultivated in four different sized tanks for 4-week periods attain 100 kg, at which point the seaweed is harvested. If six sets of this tank system are prepared and set to run simultaneously on consecutive days, it would be possible to harvest 100 kg of *Enteromorpha* everyday for six days. On the basis of this system, a large scale plant for mariculture of seaweeds equipped with 60 tanks of 7-ton, 66 tanks of 1 ton and 6 tanks of 100 L tanks was first built in Muroto, Kochi Prefecture, Japan, in June 2004 (Fig. 4). The total annual biomass of *Enteromorpha* is estimated to be 30-ton in a tons 500 ton tank volume, to which 1,500 tons of DSW is supplied per day. Large-scale cultivation of *Enteromorpha* using this means of production has already begun. Harvested materials are washed in fresh water and dried in cool-wind driers. The *Enteromorpha* seaweed is dried as a "green laver" and lightly toasted for use as a topping for many foods, in soaps and as a coating. An important characteristic of *Enteromorpha* produced using DSW is its

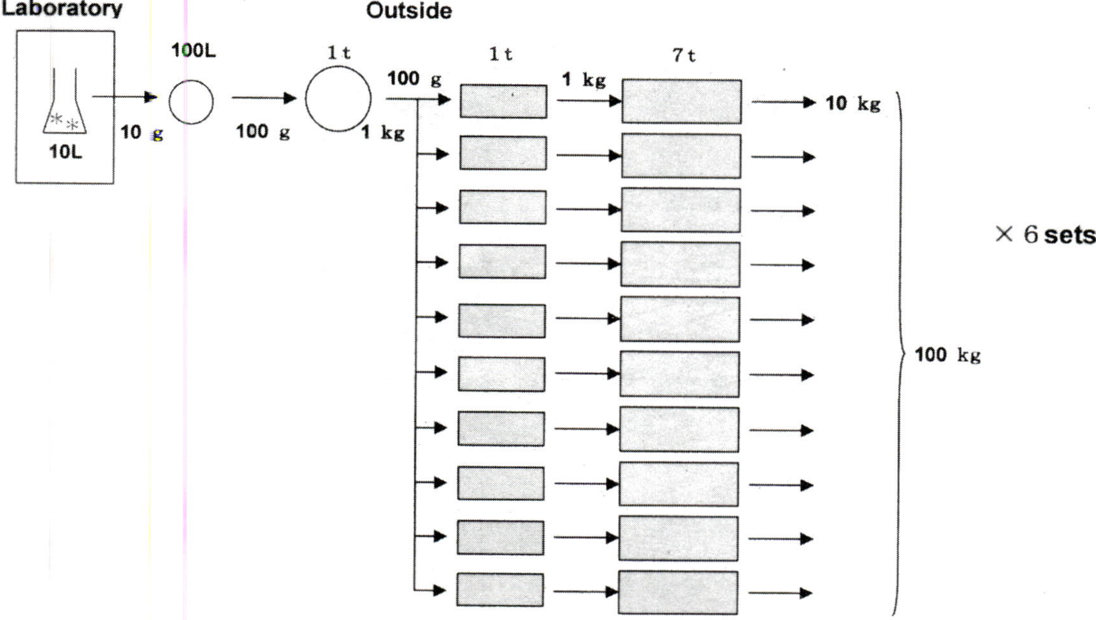

Fig. 3 A mass-culture tank system based on the new "germling cluster method" and utilizing deep seawater. See text for detailed explanation.

Fig. 4 A large-scale plant for mariculture of *Enteromorpha* in Muroto, Kochi Prefecture, Japan.

Fig. 5 Commercial products of dried *Enteromorpha* cultivated using deep seawater.

clearness without contamination. *Enteromorpha* cultivated under natural conditions tends to contain a considerable amount of sand, small organisms and trash, and must be cleaned of such contaminants prior to its processing as a commercial food. In this respect, *Enteromorpha* materials produced using DSW are of higher market value, as compared to natural materials (Fig. 5).

Other Seaweeds

More recently, we tried cultivating other seaweeds with comparable success, suggesting that they too can be grown under the same mass-tank culture system. After the initial trial in *Enteromorpha*, germling clusters of other edible brown and red seaweeds, *Scytosiphon lomentaria* and *Porphyra dentata*, were produced using the same methods as for *Enteromorpha*,

Table 2 Final wet weight in three different species after 1-week culture from an initial wet weight of 100 g and temperature ranges in a one-ton tank using DSW

Species	Final wet weight (g)	Temperature (°C)
Enteromorpha prolifera (green alga)	1250	14.6-15.6
Scytosiphon lomentaria (brown alga)	627	14.0-16.8
Porphyra dentate (red alga)	1222	12.7-17.5

and cultured in an outdoor cylindrical 1 ton tank as described above. Weekly production of the examined seaweeds are shown in Table 2. These productivities are similar to maximum growth rates observed under natural conditions (Hiraoka et al., 2004). In the near future, these culture techniques for stable production of seaweeds without contamination will be developed and new plants established not only in Kochi but also at other sites capable of pumping DSW in Japan.

ACKNOWLEDGEMENTS

We thank the staff of Kochi Prefectural Deep Seawater Laboratory for their cooperation in this study, and Mr. Stephan Gale, Kochi Prefectural Makino Botanical Garden for correcting the English. This study was partly supported by a NEDO fellowship.

REFERENCES

Hiraoka, M., Dan A., Hagihira, M. and Ohno, M. 1999. Growth and maturity of clonal thalli in *Enteromorpha prolifera* under different temperature conditions. *Nippon Suisan Gakkaishi*. 65: 302-303.

Hiraoka, M., Ohno, M., Dan, A. and Oka, N. 2004. Utilization of deep seawater for the mariculture of seaweeds in Japan. *The Japanese Journal of Phycology* 52 (Supplement): 215-219.

Ohno, M. and Miyanoue, K. 1980. The ecology of the food alga *Enteromorpha prolifera. Rep. Usa Mar. Biol. Inst.* 2: 11-17.

Provasoli, L. 1968. Media and prospects for the cultivation of marine algae. In Cultures and Collections of Algae. Proceedings of the U.S.-Japan Conference, Hakone, September. Watanabe, A. and Hattori, A. (Eds.) *Japanese Society of Plant Physiologists, Tokyo*. pp. 63-75.

32

Algal Diversity in Wetlands

Rimpy Madan and A. S. Ahluwalia
Department of Botany, Panjab University, Chandigarh-160014

ABSTRACT

Wetlands are buffer strips between land and water. They are rated among the most productive ecosystems on earth. This high production neither requires irrigation facilities nor application of fertilizers. Freshwater wetlands are nutrient sinks. Wetlands are rich germlplasm reserves as they support a diverse flora and fauna. Algae make up an important component of wetlands and account for a significant portion of their metabolic activities. Algal forms significantly contribute to the biogeochemical cycles, food chain and energy flow in the wetlands. Only 0.31% of Punjab area is occupied by 12 natural and 9 man-made wetlands.

Keywords: Algae, epipelon, metaphyton, plankton, Punjab, wetland.

INTRODUCTION

Wetlands provide major reducing system; hence function as transformers of nutrients and minerals. Wetlands are areas that remain wet for at least a part of season long enough to support vegetation that thrives in saturated soil. Globally, wetlands comprise about 6% of the total land area (Mitsch and Gosselink, 1993). Wetlands are transitions between terrestrial and aquatic ecosystems. Algae are among the fundamental players in the physical, chemical and biological processes that characterize wetland ecosystems. They are rich as germplasm reserves, support high productivity, occupy an important place in the wetland ecosystem, are nesting grounds for a variety of flora and founa and are also known as nutrient sink (Goldsborough and Robinson, 1996; Xiong et al., 2003). In wetlands, epiphytic algae have been reported to contribute 5-60 m T/ha/yr to the primary productivity whereas planktonic algae contribute only to the tune of 2-10 m T/ha/yr (Wetzel, 1983). Planktonic forms improve the quality of the water through provision of oxygen and by contributing to a number of biogeochemical cycles (Janse et al., 2001; Luyigas and Kiwanuka, 2003). The standing crop of algae affects the density of invertebrates, which in turn affect the food supply for the marsh birds in the wetlands (Murkin et al., 1991). Algae having rapid growth rates and tendency to respond quickly to perturbations are reliable indicators of water quality and biological integrity in aquatic ecosystems (Gaiser, 2004). Algal population is determined by many factors comprising hydraulic parameters, nutrition, temperature, illumination, herbivory, etc. (Xiong et al., 2003).

0.31% of Punjab area is occupied by 12 natural and 9 man-made wetlands (PSCST-Report, 2005). Algal diversity of three natural and two man-made wetlands will be discussed in this communication (Table 1). The three natural wetlands include Keshopur Miani, one of the largest natural wetlands of Punjab and another such wetland, Kahnuwan, are located in Distt. Gurdaspur whereas Jastarwal (another natural wetland) lies in Distt. Amritsar (Punjab). Among man-made wetlands, Ropar lake lies in Distt. Ropar and Dholbaha wetland lies in Distt. Hoshiarpur of this state.

MATERIAL AND METHODS

Samples were collected from water column, sediments, submerged sides, wet soil, rocks etc., from the wetlands of Kahnuwan, Dholbaha, Jastarwal, Ropar and Keshopur Miani. A knife was used to scrap epipelic, epilithic and benthic algae. Meta planktons were lifted gently with hand or with help of a stick. Planktonic algae were collected with dropper. Various physical parameters like pH, dominating macrophys and other relevant information were recorded. Samples were preserved in a mixture of 4% formalin and 1% glycerin. Identification of the algal taxa was done following the keys given by Prescott (1951), Tiffany and Britton (1952), Desikachary (1959), Randhawa (1959) and Anagnostidis and Komarek (1985, 1988).

Observations

Total 109 taxa were collected from these wetlands. Among these, 63% were Cyanophytes, 20% Chlorophytes, 15% Bacillariophytes and 1% each belonging to Xanthophytes and Euglenophutes (Fig. 1). On the basis of the studies conducted, pattern indicated that Cyanophytes dominated over other algal groups in all the wetlands except Ropar lake. In the marshy natural wetlands Jastarwal and Kahnuwan, Cyanophytes comprised upto 85% of the total algal taxa. Diatoms contributed to the tune of 13-15% in both the wetlands (Fig. 2). Miani, another natural marsh was found to support up to 69% Cyanophytes, 17% Chlorophytes, 11% Bacillariophytes and 3% Xanthophytes (Fig. 2). In the artificial riverine wetland, Dolbaha, Cyanophytes formed 59% of the total algae whereas Chlorophytes just touched 17%. Ropar lake supported upto 50% of Chlorophytes, 17% Cyanophytes and 11% of Bacillariophytes (Fig. 2). Non-heterocystous cyanophycean forms were found to be much more than the heterocystous forms in Jastarwal, Miani, Ropar and Dholbaha wetlands whereas Kahnuwan showed a reverse trend (Fig. 3).

Cylindrospermum was the most commonly found genus in the wetlands (Plate 1 (Figs. 10-12). It was found growing as epipelon in all the wetlands except Ropar lake. *Cylindrospermum musicola* in particular was found in all the four wetlands. *Chroococcus minutus* species of *Nostoc Pseudanabaena* and *Phormidium*, were found common in marshy wetlands whereas green algae comprising species of *Ulothrix* and *Spirogyara* were common in the form of metaphytic mats, in the riverine wetlands (Plate 1, Figs. 4-8).

Man-made wetlands (Ropar and Dholbaha) were found to support maximum of green algae in the form of metaphytic mats of species of *Spirogyra*, *Cladophora* and *Microspora*. Epilithic algae

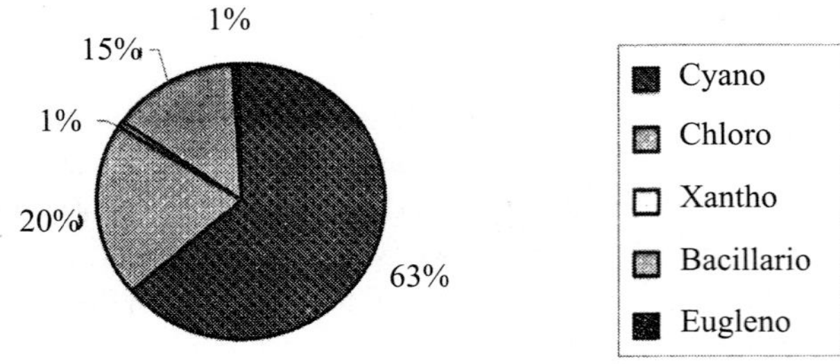

Fig. 1 Frequency (%) of algal groups in wetlands.

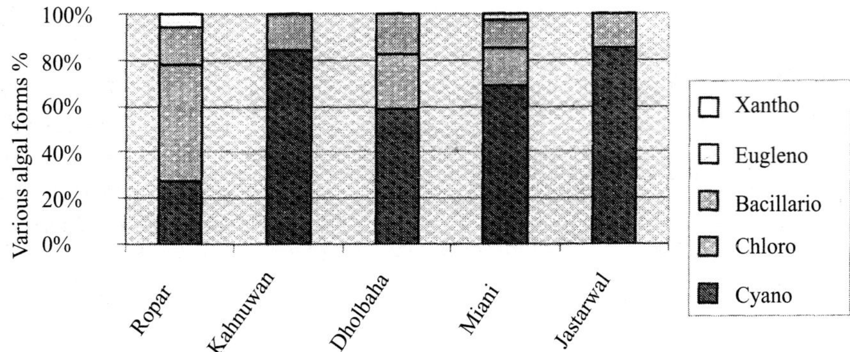

Fig. 2 Species composition in various wetlands.

were found dominating in Dholbaha wetland whereas planktonic algae dominated Miani wetland. Blue-green algae (Cyanobacteria) with well-developed sheath and heterocyst were found growing as epilithic forms (Fig. 3). In addition, filamentous taxa belonging to green algae (Chlorophyta) were also growing as metaphytons. In Miani wetland, diverse algal forms were found present along with angiospermic species of *Nelumbo* and *Typha*. The pollution indicating genera comprising *Scenedesmus bijugatus, Scenedesmus acuminatus, Coelosphearium kuetzianum, Euglena* sp. *Oscillatoria limosa* were found in the Miani wetland indicating degradation of this wetland. In Jastarwal wetland, only cyanoprokaryotes and diatoms were found in abundance. Phytoplankton species were found to dominate the wetland complex of Miani. It also supported metaphyton, epiphytons and epipelons. Kahnuwan, a seasonally flooded wetland was, however, found to support majority of the epipelons. Jastarwal wetland supported phytoplanktons, epiphytons, epipelons and little of metaphytons whereas Dholbaha wetland (Dist Hoshiarpur, Punjab) comprised mainly epilithons, periphutons and metaphytons (Table 1).

DISCUSSION

The interacting variables like nutrient levels, grazing pressure and water level in the wetlands decide the composition of algal forms (metaphytons, planktons, epipelons, periphytons, etc.) and

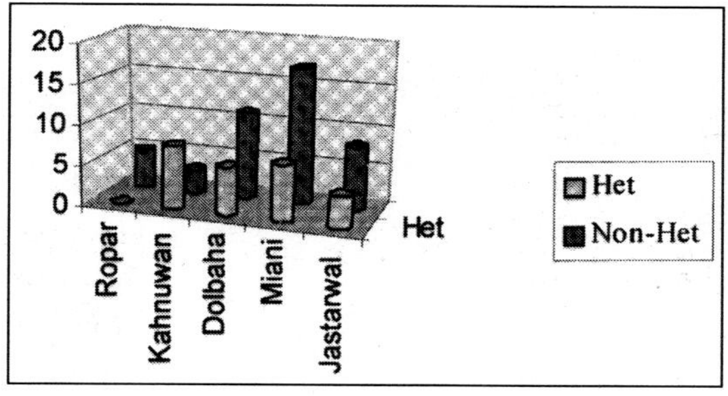

Fig. 3 Heterocystous and non-heterocystous Cyanophytes.

Table 1 General information about the wetlands

Name of wetland	*District*	*Area (sq km)*	*Status*	*Natural/ Depth*	*pH*	*Emergent macrophytes*	*Dominating algal forms*
Kahnuwan Chhamb	Gurdaspur	1.28	P/N	Swamp, Waterlogged	8–8.5	Species of *Arandu*, *Typha*, *Cyanodon*	Epipelons
Keshopur – Mian Jheel	Gurdaspur	4.08	P/N	Shallow (1.5-2.5 m)	7.5–8.5	Species of *Lemna*, *Nelumbo*, *Eichornia*, *Typha*, *Cypress*, *Polygonum*.	Planktonic
Jastarwal Jheel	Amritsar	0.55	P/N	Shallow (upto 3 m)	8–8.5	Species of *Nelumbo*, *Cyanodon* *Lemna*	Planktonic
Ropar Lake	Ropar	13.65	P/M	Riverine (up to 6 m)	7–8	Species of *Typha*, *Cypress*	Mataphytons
Dholbaha Dam	Hoshiarpur	1.32	P/M	Riverine (up to 6 m)	7.5–8.5	Species of *Typha*, *Xanthium*, *Cyanodon*	Epilithic

PERMANENT – P, NATURAL – N, MAN-MADE – M

also the species that will exist (Tilman, 1977; Goldsborough and Robinson, 1996). Algal communities in wetland differ in their ecological requirements and their physical location within the water column (Goldsborough and Robinson, 1996).

Kahnuwan exhibits a dry marsh state during most part of the year. Water level during summer becomes very low to nearly dry causing decrease in the number of macrophytes, metaphytons and planktonic algal forms. Substratum gets irradiance that causes proliferation of epipelons (Goldsborough and Robinson, 1996). Waterlogged soil in Kahnuwan, being deficient in nitrogen supported maximum of heterocystous cyanophytes as suggested by Ahluwalia (2004). Dholbaha reservoir and Ropar wetland, artificially constructed wetlands, are deep enough (up to 6 m) for

Plate 1 Fig. 1-13 (1) View of Miani Wetland – A natural wetland. (2) Ropar Lake – An artificial wetland. (3) Dholbaha dam– An artificial wetland. (4) *Ul t rix* sp. (5) *Mi r sp ra* sp. (6) *P r idiu breve* (Kutzing ex Gomont) Anagnostidis and Komarek, (7) *P r idiu autu nale* (Agardh) Trevisan ex Gomont (8) *C ru s inutus* (Kutz.) Nag (9) *Pinnularia* sp. (10) *Cylindr speru d ryp ru.* Bruhl et Biswas (11) *Cylindr speru* sp. (12) *C. u s i la.* Kutzing ex Born. et Flah. (13) *N st sp ngiaef r e* Agardh ex Born and Flah.

rooting of emergent macrophytes (Plate 1 Fig. 2&3). Water is released occasionally from the reservoirs for irrigation purpose. Hence, the algal mats with trapped oxygen remained intact on the surface of water, in the still or slow moving water for a long time period. Mats of *Spirogyra webrii*, *Ulothrix* sp, *Phormidium autumnale* and other *Spirogyra* sp. were abundant near the boundary parts of the wetland. This was the preferred habitat of the algal members and these got established in such habitats to avoid faster current of water towards the interior of the wetland. Metaphytons had little competition from planktons and submerged aquatic plants (Goldsborough and Robinson, 1996). Sandy rocks that are threatened by weathering surround Dholbaha wetland. As a result of this, more and more silt is entering the reservoir causing turbidity of water, which in turn, affects phytoplanktonic and benthic communities. Wetland was found rich in the epilithic forms with well-developed sheath. It is known that extracellular polysaccharides confer mechanical protection to the cells to form a strong and sticky frame work (Sutherland, 2001). It has been suggested that under highly diazotrophic conditions the excess fixed carbon is channeled to exopolysaccharides (EPS) synthesis (Stal, 1995). In Ropar wetland, water was found to be neutral to slightly alkaline (pH-7-7.8) and was observed to support more of the green algae (about 50%), (Plate-1, Fig. 2). In Miani, a shallow wetland, a part of the wetland that was found to support the cover of emergent macrophytes (species of *Nelumbo, Typha*) almost lacked metaplanktons and phytoplanktons (Plate 1 (Fig. 1). Competition for nutrients and shading effect caused by macrophytes could have resulted in their elimination. Another part of the wetland being used for fish culture was found supporting phytoplanktons in large number. Excreta of fishes and other nutrients may have increased the organic matter in the wetland water resulting in increased population of phytoplanktons (Schindler et al., 1971). Jastarwal wetland is located very close to residential area. People living there have adopted *Nelumbo* (Lotus) as cash crop. It was found that people residing in the wetland area were using insecticides for the last two years to protect their crop from insect infection. It could have resulted in alkalinity of the water and soil of the wetland. Here also non-heterocystous algae dominated heterocystous algae because of the addition of nitrogenous compounds in the form of pesticides and insecticides. Nearly 85% algae were found to be cyanophytes. In the recent years, human interventions to overexploit the precious land have accelerated the loss of wetlands (Sinha and Jha, 1997; Worm et al., 2002). It is our duty to protect and conserve the wetlands by using them wisely in the environmentally safe and sustainable manner and thus protecting the biodiversity in general and algal diversity in particular, these lands are supporting.

ACKNOWLEDGEMENTS

Authors thank the University Grants Commision, New Delhi for financial assistance through SAP programme and the Chairman, Dept. of Botany for research facilities.

REFERENCES

Ahluwalia, A. S. (2004) Cyanobacterial differentiation. *Pb. Univ. Res. J. (Sci).* 54: 7-11.

Anagnostidis, K. and Komarek, J. (1985) Modern approach to the classification system of cyanophytes. 1

Anagnostidis, K. and Komarek, J. (1988) Modern approach to the classification system of cyanophes. 3-

Desikachary, T. V. (1959) Cyanophyta. Indian Council of Agricultural Research, New Delhi.

Fong, P., Donohoe, R. M. and Zedler, J. B. (1993) Competition with macroalgae and benthic cyan bacterial mats limits phytoplankton abundance in experimental microcosms. *Mar. Ecol. Progr. Ser.* 100: 97-102.

Goldsborough L. G. and Robinson, G. C. (1996) Pattern in wetlands. In: Stevenson R. J. Bothwell M. L. and Lowe R. L. (Eds.) Algal ecology in freshwater benthic ecosystems. Academic Press, San Diego, California.

Janse, J. H., Ligtvoet, W., Van, Tol. S. and Bresser, A. H. (2001) A model study on the role of wetland zone in lake eutrophication and restoration. *Scientific World Journal.* (Suppl) 27: 605-614.

Luyigas, S. and Kiwanuka, S. (2003) Plankton composition, distribution and significance in a tropical integrated pilot constructed treatment wetland in Uganda. *Water Sci Technol.* 48: 241-248.

Mitsch, D. F. and Gosselink J. G. (1993) Wetlands. Second edition. Van Nostrand Reinhold, New York.

Murkin, H. R, Stainton, M, P, Boughen, J. A, Pollard, J. B and Titman, R. D. (1991) Nutrient status of wetlands in the Interlake Region of Manitoba, Canada. Wetlands 11: 105-122.

Presscott G.W. (1951) Algae of Western Great Lake Areas. Cranbrook Intitute of Science Bulletin No. 33.

PSCST Report (2005) State of Environment, Punjab – 2005. Sponsored by Ministry of Environment and Forests, Govt. of India: 94

Randhawa, M. S. (1959) Zygnemacese. Indian Council of Agricultural Research, New Delhi.

Schindler, D. W, Armstrong, F. A. J., Holmgren, S. K. and Brunskill, G. J. (1971) Eutrophication of Lake 227.

Sinha, M. and Jha, B. C. (1997) Ecology and fisheries Ox, bow lakes (mans) of North Bihar. CIFRI, Bulletin No. 74, Barrackpore.

Stal, L. J. (1995) Physiological ecology of cyan bacteria in micronial mat and other communities. *New Phytol.* 131: 1-32.

Sutherland I (2001) Biofilm exopolysaccharides: A strong and sticky framework. *Microbiology*. 147: 3-9.

Tiffany, L. H. and Britton, M. E. (1952) The Algae of Illinois. Pub. The University of Chicago Press, Chicago.

Tilman, D. (1977) Resource competition between planktonic algae: An experimental and theoretical approach. *Ecology*. 58: 338-348.

Wetzel, R. G. (1983) Limnology. Second Edition. Saunders College Publishing, New York

Worm, B. Lotze, H. K., Hillebrand, H., Sommer, U. (2002) Consumer versus resource control of species diversity and ecosystem functioning. *Nature*. 417: 848-51.

Xiong Xie, L, Sheng, X., Wu, Z. and Xia, Y. (2003) A review on algal ecology in wetland. Ying Yong Sheng Tai Xue Bao. 14: 1007-1011.

33

Primary and Secondary Metabolites in *Caulerpa sertularioides* and *C. peltata*

R. M. Mulani and D. G. Shah
Botany Department Sheth L.U.J. & Sir M.V. College
Dr. S. Radhakrishnan Marg, Andheri (E), Mumbai-400 069.

ABSTRACT

The present paper deals with the assessment and the impact of seasonal changes in marine environmental conditions on variations in primary and secondary metabolites content in *Caulerpa sertularioides* and *C. peltata* growing along the sea coast of Mumbai, Maharashtra. *C. sertularioides* occurs for six months i.e., from November to April while *C. peltata* occurs for four months i.e., from January to April. Both the species occur seasonally.

There is a variation in primary metabolite content and nutritive values in selected species, while there is slight variation in the various water parameters. The total terpenoid content in *C. sertularioides* were low at the early stage of growing season and increase with increase in age of the plant, while in case of *C. peltata* the values were opposite and the total terpenoid content was highest in growing season and decreased at maturity i.e., at the end of growing season.

GC-MS analysis of *C. sertularioides* showed presence of ten terpenoids such as linalool, menthone, isoborneol, β-pinene, geraniol, nerol, linalyl acetate, geranyl acetate, isophytol and phytol while in *C. peltata* only three terpenoids were identified such as linalool, menthone, α-terpineol. The results obtained were discussed with the available literature.

INTRODUCTION

"Marine algae" as the name suggests are present in sea water and found in two regions, one on the surface of the water and another in the narrow littoral zone. As primary producer they are the major resource of the ocean and provides sufficient quantity of organic matter, oxygen as well as hydrocarbons and their sediments get dissolved in seawater.

At Colaba large number of marine algae were found on rocks and along vertical levels on the seashore. The algal distribution also varies along seashore and is governed by the presence of cracks and gravel bed in sea and the salinity of seawater. The algal distribution along Mumbai coast has been reported by various workers like Boergesen, 1930, 1932; Dixit 1940; Biswas & Mitra 1943; Untawale et al., 1979; Deodhar, 1987. Khadilkar, 1988, etc.

Primary and secondary metabolites in marine algae has been worked out by various workers as, by Capon et al., 1981; Rao et al., 1982; De Napoli et al., 1981, 1983; Vest, et al., 1983; Zahid, 1989. Anjaneyulu, et al., 1991 1991b; Meyer & Paul 1992; Dave and Chauhan 1993; Khotimchenko, 1995; Aliya and Shameel 1998; Amade and Lemee 1998; Ramos et al. 2000; Dumay et al. 2002, etc.

The *C. peltata* have been analysed for terpenoids as new bicyclic diterpene by Capon et. al. in 1983 and *C. sertularioides* for two amino acids like lysine and methionine by Ramos et al., 2000. These two species have been selected for detailed investigation of primary and secondary metabolites.

MATERIAL AND METHODS

Caulerpa sertularioides and *C. peltata* were collected monthly during low tides in the growing season from the Colaba sea coast of Mumbai. The water samples were also collected in plastic bottles and analysed for dissolved oxygen, salinity, pH, nitrate-nitrogen, nitrite nitrogen and inorganic phosphates using standard methods of Strickland and Parsons (1972).

The collected algal species were cleaned, sorted out and immediately weighed for fresh biomass and then dried and weighed for dry biomass. The dry powder is analysed for primary metabolites namely, carbohydrates (Dubois et al., 1956), proteins (Johnson, 1941) and lipids (Raymount et al., 1964), nutritive values as organic carbon and caloric values and for secondary metabolites as terpezoids were analysed by using HPTLC and GS-MS.

OBSERVATIONS

Mumbai is situated on an island at a latitude of 18° 55° N and longitude 72° 54°E and the shoreline mainly consist of creeks, cliffs, bays and beaches.

The average maximum temperature was recorded as 34.6°C in November and May and mean of minimum temperature was 28.5°C in May 2002. The average relative humidity was very high in month of October (83 to 89%) and average low humidity range was recorded in the month of January 2002 (71%) (Fig. 2). The rainfall during the study period i.e. November 2001 to April 2002 has been reported to be negligible i.e., 0 mm to 14.8 mm (Fig. 1). Average wind speed recorded during the months of November to January is 1.9 to 3.8 kmph with slight increase from February to April i.e., 4.8 to 6.5 kmph (Table 1) (Fig. 2)

Caulerpa sertularioides occurs from November 2001 to April 2002 and *C. peltata* occurs from January 2001 to April 2002, are the two species of marine algae belonging to family Caulerpaceae. *C. sertularioides* is long, erect upright, attached to the rocky surface, and about 15-20 cm tall, light green with feather-like blades arising from creeping stolen called runner. Each

Table 1 Monthly variation in temperature, relative humidity, total rainfall and average wind speed at Colaba sea coast, Mumbai

Sr. No.	Months	Temperature °C		Relative humidity			Rainfall	Wind speed (km/hr)		
		Maximum	Minimum	Maximum	Minimum	Average	(mm)	Maximum	Minimum	Average
1.	October'01	33.0	25.0	90	72	85	78.3	10.5	2.3	4.9
2.	November 01	34.6	23.7	88	46	77	0	4.9	1.7	3.8
3.	December 01	32.7	21.6	88	51	78	0	5.5	0.3	3
4.	January'02	30.2	19.7	87	46	71	0	5.6	0.9	3.3
5.	Febuary'02	30.6	20.7	88	41	73	0	9.1	1.2	4.8
6.	March'02	32.4	23.4	85	36	73	14.8	14	2	6.5
7.	April'02	33.4	26.0	83	45	74	0	9.1	2.9	5.4
8.	May'02	34.6	28.5	84	70	77	1.9	10.5	2.4	5.9
9.	June'02	32.2	26.8	93	72	84	436.8	20.5	0.7	7
10.	July'02	31.5	26.8	91	82	87	109.5	11.5	2.8	7.7
11.	August'02	29.3	24.9	98	82	89	130.6	14.7	0.9	8.6
12.	September'02	30.4	25.2	89	78	83	604.2	11.4	0.1	3
13.	October'02	34.4	26.0	81	55	74	7	3.5	0	1.5
14.	November'02	33.7	23.5	89	68	82	0	4.6	0	1.9
15.	December'02	31.9	21.0	84	51	76	0.4	9.3	0	2.6
16.	January'03	30.8	20.9	86	40	75	0	5.1	0	2.1

branchlet is cylindrical and needle shaped with pinnately opposite tapering feather-like structures while *C. peltata* luxuriantly growing marine algae shows dark green upright branches which arises from creeping stolon grows about 20-30 cm in height and shows presence of light green flat peltata disc.

The water sample analysis showed highest value of dissolved O_2 in December 2002 which is 13.76 mg/L. and lowest value is 6.08 mg/L and recorded in month of May 2002. The highest salinity value is recorded in April 2002 and lower salinity value was reported as 31.38 ppt in month of August 2002 (Table 2) pH range was 7.86 to 8.6 and values were increasing from October 2001 (8.5) to January (8.6) (Fig. 3).

The primary metabolite values were low at the initial growth season i.e., in the month of November and reaches to its maximum at peak of their growing season i.e., during March. The total carbohydrate values vary from 9.74% to 26.44%, the lowest value was obtained in November 2001 and highest in March 2002 in *C. sertularioides*. The total protein percentage was higher in *C. sertularioides* as compared to that of *C. peltata*. The values of total protein percentage varies from 9.23% in November 2001 to 12.37% in February 2002 in *C. sertularioides* while in *C. peltata* the values varies from 7.97% in March 2002 and 9.83% in February 2002 (Table 3) (Fig. 5).

There was not much difference in total lipid percentage in both the species, the lipid percentage values for *C. sertularioides* varies from 3.00% in December 2002 to 5.3% in February 2002, while in *C. peltata* the values varies from 2% in January 2002 to 7% February 2002. The per cent values for organic carbon varies from 37.44% to 45.84% in *C. sertularioides* and in *C. peltata* it

Table 2 Monthly variation in dissolved oxygen, salinity, pH, nitrate-nitrogen, inorganic phosphate in sea water collected from Colaba sea coast, Mumbai

Sr. No.	Months	D.O. (mg/lit)	Salinity (ppt)	pH	Nitrate N (m*g/lit)	Nitrate N (m*g/lit)	Inorganic phospate¨m*g/lit)
1.	October'01	6.08	38.32	8.15	227.9	28.86	18.9
2.	November'01	8.96	38.22	8.25	226.68	15.52	25.73
3.	December'01	10.24	37.72	8.4	241.81	15.37	29.57
4.	January'02	9.92	36.81	8.6	265.83	13.05	33.43
5.	February'02	7.68	36.61	8.4	354.74	288.84	49.18
6.	March'02	9.6	37.21	7.9	301.23	17.69	22.23
7.	April'02	10.24	41.14	8.55	208.1	10.59	29.05
8.	May'02	6.08	40.53	8.5	241.12	43.5	23.45
9.	June'02	7.04	38.92	8.5	438.6	12.76	42.18
10.	July'02	8.32	35.84	8.6	473.96	19.14	33.95
11.	August'02	8.32	31.38	8.05	308.53	17.26	32.73
12.	September'02	8.32	36.8	7.85	378.76	15.66	22.4
13.	October'02	7.68	39.02	8.6	441.75	10.88	41.68
14.	November'02	12.16	38.62	8.3	375.4	21.03	28
15.	December'02	13.76	39.03	8.5	411.1	18.56	28
16.	January'03	13.12	38.72	8.35	449.73	14.94	33.78

D.O. = Dissolved oxygen.; N = Nitrogen

Table 3 Monthly variation in primary metabolites and nutritive value in *Caulerpa sertularioides* and *Caulerpa peltata* collected from sea coast of Mumbai

Sr. No.	Months	*Caulerpa sertuularioides*						*Caulerpa peltata*					
		C%	P%	L%	OC%	CC cal/gm	TPM	C%	P%	L%	OC%	CC cal/gm	TPM
1.	November'01	9.74	9.23	4.5	45.84	2630.00	23.47	-	-	-	-	-	-
2.	December'01	11.02	11.35	4.0	42.30	4056.20	26.37	-	-	-	-	-	-
3.	January'02	12.88	10.85	2.5	38.88	4807.35	26.23	10.96	9.36	2.0	37.80	2253.44	22.32
4.	February'02	17.69	12.37	5.3	37.44	5032.70	35.36	7.69	9.83	7.0	34.16	3605.51	24.52
5.	March'02	24.44	10.72	4.0	39.60	5858.96	41.16	8.85	7.97	4.0	37.98	4206.43	20.82
6.	April'02	17.81	10.98	3.0	38.23	4446.80	31.79	5.96	8.34	5.0	34.20	3981.08	19.30
7.	November'02	14.23	10.46	4.0	39.92	3981.09	28.69	-	-	-	-	-	-
8.	December'02	16.35	10.76	3.0	41.16	4957.58	30.41	-	-	-	-	-	-
9.	January'03	14.23	10.31	4.0	40.24	4506.89	28.54	-	-	-	-	-	-

C = carbohydrate, P = proteins, L = lipids, OC = organic carbon, CC = caloric content, TPM = Total Primary metabolites

varies from 34.16% to 37.80% (Fig. 5). The average values for caloric content of *C. sertularioides* was found to be 263.00 cal/gm to 5858.96 cal/gm while in *C. peltata* it varies from 2253.44 cal/gm to 4206.43 cal/gm (Fig. 6). The average values of total primary metabolites ranged from 23.47 to 41.16 in *C. sertularioides* while for *C. peltata* it varies from 19.30 to

24.52 (Table 3, Fig. 5).

Acyclic and monocyclic monoterpenoids and sesquiterpenoids were estimated by using HPTLC method. It was not possible to identify each terpenoids, but group of terpenoids were identified as per Rf values and colour of band obtained by using vaniline sulphuric acid as developing range. In *Caulerpa sertularioides* total quantities of terpenoids varied from 35.18% (February 2002) to 60.17% (November 2001). The total quantities of terpenoids were in decreasing from November 2001 to February 2002. While from February 2002 (39.18%) to April 2002 (46.27%). The value of total quantities of terpenoids were in increasing order (Table 4).

Three distinct bands based on Rf values were identified and estimated as terpenoid-I (0.16-0.19), terpenoid-II (0.26-0.27), terpenoid-III (0.35-0.37). The maximum values for terpenoid-I was 32.4% in December 2001 and minimum value was 7.93% and obtained in month of January 2003. The maximum value of terpenoid-II was 25.86% obtained in month of November 2001 and minimum value was 14.44% obtained in February 2002. The maximum value of terpenoid-III was

Table 4 Monthly variations in secondary metabolite in *Caulerpa sertularioides* and *Caulerpa peltata*

Sr. No.	Months	*Caulerpa sertularioides*					*Caulerpa peltata*				
		Secondary metabolites					Secondary metabolites				
		Abbreviation	T-I	T-II	T-III	TSM	Abbreviation	T-I	T-II	T-III	TSM
1.	November'01	CAU-01	21.69	25.86	12.62	60.17	-	-	-	-	-
2.	December'01	CAU-02	32.40	19.49	6.78	58.67	-	-	-	-	-
3.	January'02	CAU-03	17.13	15.36	14.59	47.08	PEL-01	20.68	5.00	13.17	38.85
4.	February'02	CAU-04	16.16	14.44	4.58	35.18	PEL-02	3.83	22.06	30.62	56.51
5.	March'02	CAU-05	8.61	22.51	12.15	43.27	PEL-03	3.73	24.64	12.58	40.95
6.	Aprill'02	CAU-06	16.67	20.20	9.40	46.27	PEL-04	6.21	18.24	8.44	32.89
7.	November'02	CAU-07	22.61	18.56	7.73	48.90	-	-	-	-	-
8.	December'02	CAU-08	15.50	15.59	14.90	45.99	-	-	-	-	-
9.	January'03	CAU-09	7.92	14.75	17.85	40.52	-	-	-	-	-

T-I = Terpenoid-I, T-II = Terpenoid-II, III = Terpenoid-III

17.85% obtained in January 2003 and a minimum value was 6.78% obtained in December 2001 (Table 4, Fig. 7).

GC-MS analysis of *C. sertularioides* showed the presence of ten terpenoids such as linalool, menthone, isoborneol, β-pinen, geraniol, nerol, linalyl acetate, geranyl acetate, isophytol and phytol (Table 5).

In *C. peltata* the total quantities of terpenoids varied from 32.89% to 56.51% the total quantity of terpenoids were less in January 2002 (38.85%) but suddenly increase in February 2002 (56.51%) and again showed decreasing trend from February 2002 to April 2002. Three distinct bands were identified and estimated as terpenoid-I (0.20-0.23). Terpenoid-II (0.26-0.28), and terpenoid-III (0.36-0.40) (Table 4, Fig. 7).

Table 5 Qualitative analysis by GCMS of terpenoids in *Caulerpa sertularioides* and *Caulerpa peltata*, collected from Colaba sea coast, Mumbai

Species	Terpenoids present (RT)										
	Linalool	*Menthone*	*Isoborneol*	*α-Terpineiol*	*β-pinen*	*Geraniol*	*Nerol*	*Linalyl acetate*	*Geranyl acetate*	*Isophytol*	*Phytol*
Caulerpa sertularioides	10.05	11.50	11.81	–	12.58	13.90	14.02	14.46	15.12	30.49	33.98
Caulerpa peltata	9.66	11.05	-	12.09	-	-	-	-	-	-	-

The maximum value for terpenoid-I were obtained as 20.61% and recorded in January 2002 and then the value were decreased. The maximum values of terpenoid-II were estimated as 24.64% and reported in month of March 2002 and minimum value is 5% and obtained in January 2002. Maximum value for terpenoid-III is 30.62% and obtained in February 2002 and minimum value is 8.44% and obtained in April 2002. The total secondary metabolites as terpenoids were maximum in *C. sertularioides* as compared to *C. peltata* (Table 4 Fig. 7).

GC-MS analysis of terpenoids based on comparative analysis of MS spectrum of each terpenoid with wiely library showed presence of Linalool, Menthone and α-terpineol (Table 5).

DISCUSSION

The present paper deals with the assessment and the impact of seasonal changes in marine and environmental conditions on variation in primary and secondary metabolite content in two species of *Caulerpa* as *C. sertularioides* and *C. peltata.*

Mumbai has grown in terms of human settlement, industrialization and trade activities by sea in last fifty to sixty years, this has changed the marine environment in and around Mumbai sea coast. Mumbai sea coast is also found rich in oil and petroleum products and its trade. During loading and unloading of petroleum and its products, sea water in the ports, especially near Colaba gets contaminated (Ingole et al., 1995).

Along with algal collection water samples were also collected assayed for dissolved oxygen, salinity, pH nitrate-nitrogen, nitrite-nitrogen and inorganic phosphate. The dissolved oxygen was high during winter season. The values for pH was varied between 7-8 and showed hardly any fluctuation. Nitrate-nitrogen and nitrate-nitrate did not show seasonal variation and did not affected by seasonal variation changes (Fig. 4).

The availability of selected species of algae was not much affected by various water parameters and the algae were available in selected seasons as *C. sertularioides* was available for six months, while *C. peltata* was available for months. Thus, seasonal availability of both the species was reported for particular season.

During monsoon season besides low salinity of sea water due to inflow of fresh water from soil as a rain water and due to which nutrient concentration is also low this resulted in disappearance of some of the algal species including *C. sertularioides* and *C. peltata*. While during summer season there is increase in nitrate and phosphates in sea water which helps in algal growth (Levring et al., 1969).

The total carbohydrates and proteins in *C. sertularioides* were found increasing from November to March, the protein content was high in December and low in February, while lipid content was maximum in February and least in December. The organic carbon content was high in March and low in November.

In *C. peltata* the total carbohydrates were found highest in January and lowest in April while protein content was highest in February and lowest in March. The lipid content was highest in February and lowest in January while the organic carbon was highest in March and lowest in

February. The caloric content was highest in March and lowest in January.

The carbohydrate content in *Chlorophyta* along Indian coast ranged between 41.9% and 67.87% as reported by Untawale et al., 1982, whereas long Maharashtra sea coast it was ranged between 31.8% to 62.1% reported by Dhargalkar et al., in 1980 while Khadilkar in 1988 reported the total carbohydrate content as 17.01% to 46.44%, while in present investigation the average per cent total carbohydrates is low and it is 15.60% in *C. sertularioides* and 8.37% in *C. peltata*.

Organic carbon percentage was 45.84% in *C. sertularioides* and it is high compared to that of earlier reports by Untawale et al., 1982 and Dhargalkar et al., 1980. They reported organic carbon in *Chlorophyta* along Maharashtra coast in the range of 24.4% to 42.4% while Khadilkar in 1988 reported along Mumbai sea coast in the range of 17.01% to 46.44%. The caloric values reported by Khadilkar (1988) ranging between 3034.0 cal/gm and 4867.4 cal/gm while the values were low in case of *C. peltata* as 253.44 cal/gm while in *C. sertularioides* the values were high in March 2002 and it was 5858.96 cal/gm.

The terpenoids in *C. sertularioides* varied between 39.18% to 60.17% and total quantities of terpenoids were low at early stage of growing season and increases at peak stage of growing season. Similar observations were made by Dumay et al., 2002 and they observed the decline in secondary metabolites during vegetative growth, while in *C. peltata* the values were opposite and the total secondary metabolites were high in growing season and decreased during end of the growing season.

GC-MS analysis of *C. peltata* showed presence of Linalool, Menthone and α-*Terpineol* while in *C. sertularioides* the terpenoids were identified as linalool, menthone, isoborneol, β-pinen, geraniol, nerol, linalyl acetate, geranyl acetate, iso-phytol and phytol. The concentration of secondary metabolites were high during summer season where water temperature is high and similar observation were made by Amade and Lemeein 1998 in *C. taxifolia*.

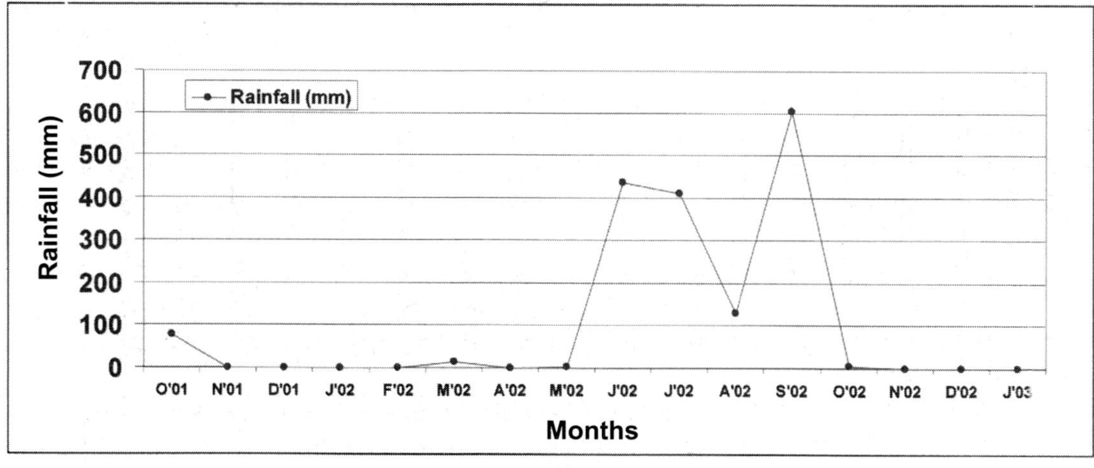

Fig. 1 Monthly variation in total rainfall at Colaba sea coast, Mumbai.

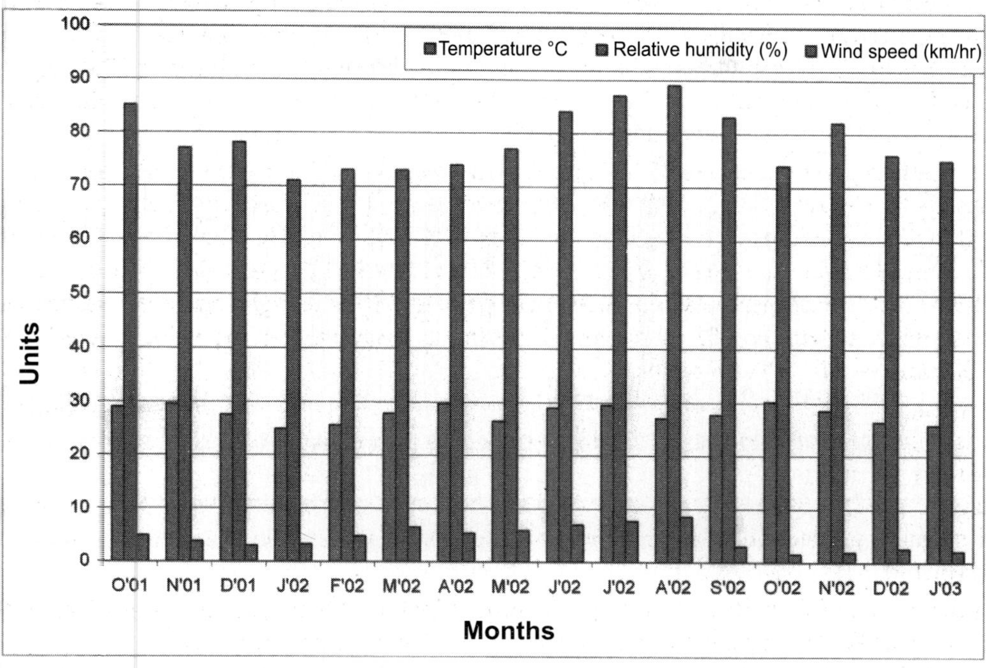

Fig. 2 Monthly variation in temperature, relative humidity, and average wind speed at Colaba sea coast, Mumbai.

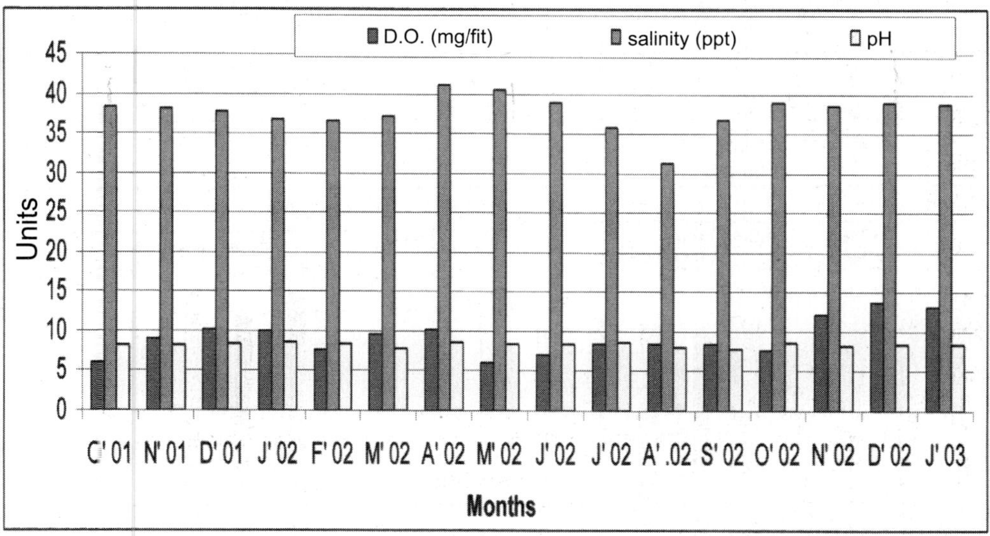

Fig. 3 Monthly variation in dissolved oxygen, salinity and pH in sea water collected at Colaba sea coast, Mumbai.

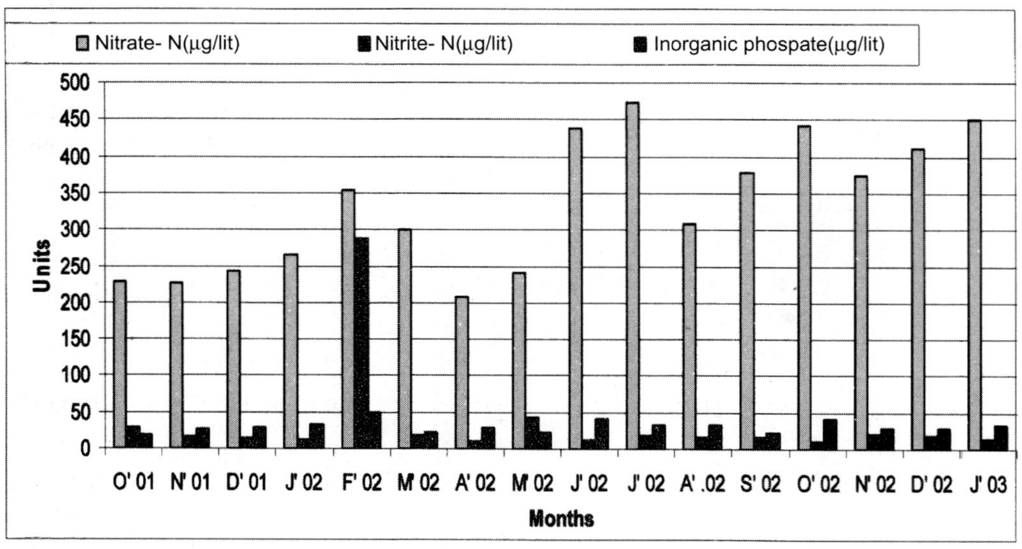

Fig. 4 Monthly variation in nitrate, nitrite and Inorganic phosphate in sea water collected at Colaba sea coast, Mumbai.

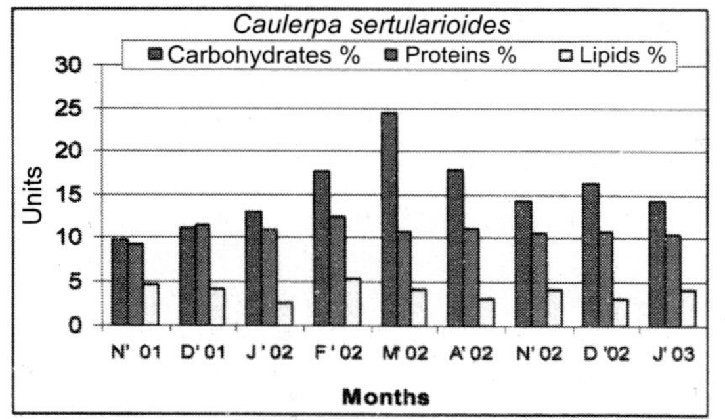

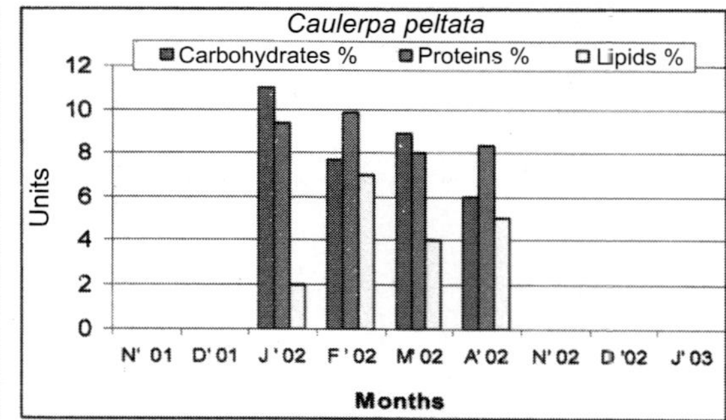

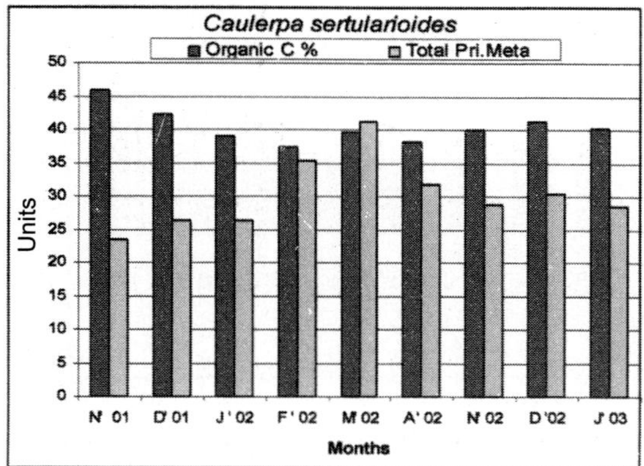

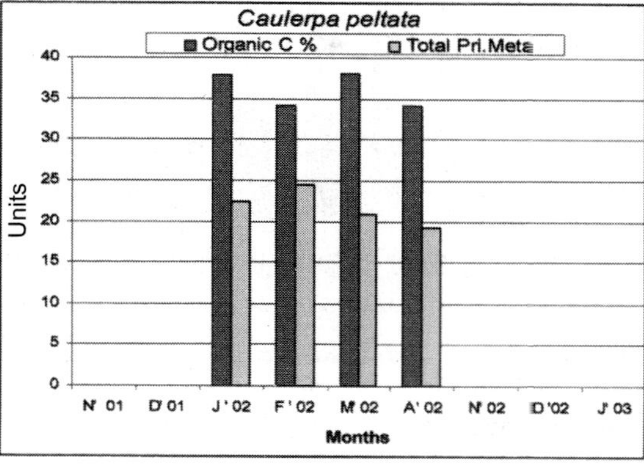

Fig. 5 Monthly variations in primary metabolites in *Caulerpa sertulariodes* and *Caulerpa peltata* collected from sea coast of Mumbai.

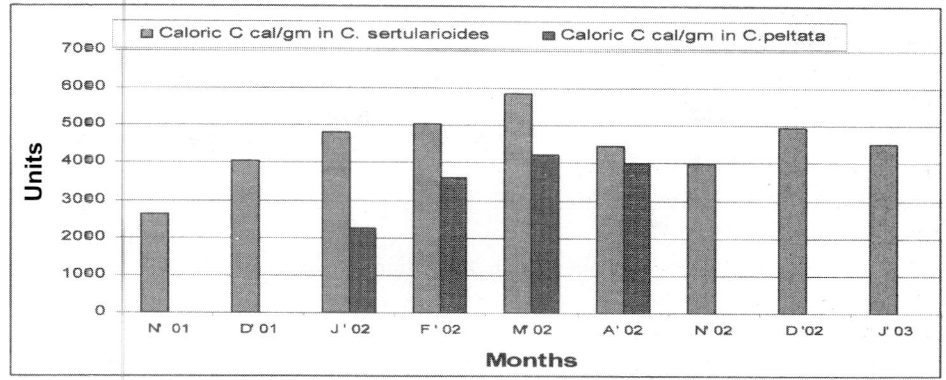

Fig. 6 Monthly variation in Caloric content in *Caulerpa sertulariodes* and *Caulerpa peltata* collected from seacoast of Mumbai.

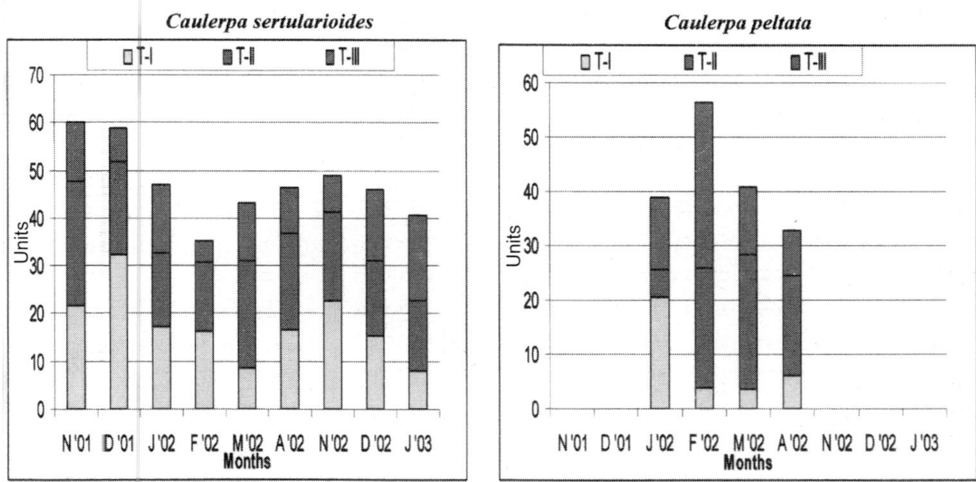

Fig. 7 Monthly variations in secondary metabolites in *Caulerpa*.

REFERENCES

Aliya, R. and Shameel, M. (1998). Phytochemical investigation on air dried material of five species of *Caulerpa* (Bryopsidophyceae). *Botanica Marine*. 41 (2): 125-132.

Amade, R. and Lemee, R (1998). Chemical defence of the Mediterranean alga *Caulerpa taxifolia*: Variation in caulerpenyne production. *Aquatic Toxicology*. 43 (4): 287-300.

Anjaneyulu, A. S. R., Prakash, C. V. S and Mallavadhani, U.V (1991a). Two caulirpin analogues and sesquiterpene from *Caulerpa racemosa*. *Phytochemistry*. 30 (9): 3041-3042.

Anjaneyulu, A. S. R., Prakash, C. V. S and Mallavadhani, U. V (1991b). Sterols and terpenes of the marine green algal species *Caulerpa racemosa* and *Codium decortcatum*. *J. Indian Chem*. Soc. 68 (8): 480.

Biswas, K. and Mitra, G. (1943). Observation on the marine algae from the coast of Bombay. *Sci & Cult.* 9: 251.

Boergesen, F. (1930). Some Indian green and brown algae especially from the shores of the presidency of Bombay II. *J. Indian Bot. Soc.* 9: 151-174.

Boergesen, F. (1932). Some Indian Green and Brown algae especially from the shores of the presidency of Bombay II. *J. Indian Bot. Soc.* 11: 52-70.

Capon, R. J., Ghisalberti, E. L. and Jefferies, P. R. (1981). New sesquiterpenes from *Caulerpa flexilis var. muelleri. Aust. J. Chem.* 34: (8): 1775-1778.

Capon, R. J., Ghisalberti, E. L. and Jefferies, P. R. (1983) Metabolites of green algae *Caulerpa sps. Phytochemistry.* 22 (6): 1465-1467.

Dave, M. J. and Chauhan, V. D. (1993). Amino acid composition of *Caulerpa. Phycos* 32 (1-2): 21-26.

De Napoli, L., Fattorusso, E., Mango, S. and Mayol, (1981). Furocaulerpin, anew acetylinic sesquinterpenoid from green alga *Caulerpa prolifera Experientia.* 37 (1): 1132.

De Napoli, L., Mango, S., Mayol, and Novellino, E. (1983). Further caulerpenyne like eaters from the green algae *Caulerpa proligera. Experientia.* 39 (2): 141-143.

Deodhar, H. D. (1987). *The Biology of Marine Algae of Bombay*, Ph.D. Thesis, University of Poona, Pune.

Dhargalkar, V. K., Jagtap, T. G. and Untawale, A. G. (1980). Biochemical constituents of seaweed along the Maharashtra coast. *Indian J. Mar. Sci.* 9 (4): 279-299.

Dixit, S. C. (1940). Algal investigation in the Bombay Presidency from 1847-1940. *Curr. Sci.* 9 (10): 453-454.

Dubois, M., Giller A., Hamolton. J. K., Rebeus, P.A. and Smith, F. (1956). Colorimetric method for detemination of sugar and related substances. *Anal. Chem.* 28(3): 350-356.

Dumay, O., Pergent, G., Pergent, M. C. and Amade, P. (2002). Variation in Caulerpenyne contents in *Caulerpa taxifolia* and *C. racemosa. J. Chem. Ecol.* 28 (2): 343-352.

Ingole, S. A., Kadam, A. N., Mayadeo, M. S., Dhadke, P. M. (1995) Petroleum hydrocarbons in intertidal ecosystem along Bombay coast. *Indian J. Mar. Sci.* 24: 116-118.

Johnson, M. J. J. (1941). Isolation and properties of pure yeast pesticides *J. Biol. Chem.* 137: 575-586.

Khadilkar, S. G. (1988). *Marine algae of Bombay Coast.* Ph.D. Thesis, University of Mumbai.

Khotimchenko, S. V. (1995). Fatty acid competition of green algae of the genus *Caulerpa. Bot. Mar.* 38 (6): 509-512.

Levring, T., Hoppe, H. A. and Schmid, O. J. (1969) *Marine algae, Bot. Mar. Handbook* 1: 411.

Meyer, K. D and Paul, V. J. (1992). Intraplant variation in secondary metabolite concentration in three species of *Caulerpa*; (Chlorophyta: Caulerpales) and its effect on herbivorous fishes. *Mar. Ecol. Prog. Ser.* 82 (3): 249-257.

Ramos, M. V., Monterio, A. C. O., Moreira, R. A., De Fatima, F. U. and Carvalho, A. (2000) Amino acid composition of some Brazilian seaweed species. *J. Food Biochem.* 24 (1): 33-39.

Rao, C., Bheemshankar and Pullaiah, K. C. (1982). Chemical examination of marine algae of Vishakapatnam coast: Constituents of *Caulerpa.*

Raymount, J. E. F., Austin, A. and Lingford, E. (1964) Biochemical studies on marine zooplankton. The biochemical compositions of *Neomysis interger. J. Con. Per. Explore Mer.* 28: 354-364.

Strickland, J. D. H. and Parsons, T. R. (1972). *Practical Handbook of Seawater Analysis.* Bull. No. 167 (Fish Res. Board, Canada, Ottawa) 203.

Untawale, A. G., Dhargalkar, V. K., Agadi, V. V. and Jagtap, T. G. (1979). Marine algal resources of Maharashtra coast. NIO Publication. 1-48.

Untawale, A.G., Subramaiah, K., Chauhan, V. D. and Dhargalkar, V. K. (1982). *Preliminary report on Integrated R and D Programmers in the area of seaweeds*. CSIR-NIO and CSMCRI report.

Vest, S. E., Dawes, C. J. and Romeo, J. T. (1983). Distribution of caulerpin and caulerlpicin in eight species of green alga caulerpa (Caulerpales) *Bot. Mar.* 26 (7): 313-316.

Zahid, P. B. (1989). On the biochemical composition of ten species of the genus *Caulerpa* (Caulerpales-Chlorophyta), collected from the Karachi coast. *Pak. J. Sci. Ind. Res.* 32 (2): 98-100.

34

Studies on the Algal Flora of Lake Jaisamand in Udaipur, Rajasthan, India

Premlata Vikal and Sandhya Tyagi
Department of Botany, University College of Science, Mohanlal Sukhadia University,
Udaipur-313001 (Rajasthan), India.

ABSTRACT

A study on the algal flora of lake Jaisamand carried out from July, 2004 to December, 2005. During the study period, about 32 genera (including Cyanophyceae, Chlorophyceae, Xanthophyceae and Bacillariophyceae) have been recorded. The dominance of blue-green algae have been reported for two times during the study period. First blooming condition by *Gloeothece palea* was reported in February, 2005 which vanished completely in March, 2005. The reason for its abundant growth might be favourable temperature, high pH and nitrate level. Next blooming condition was reported in Oct., 2005 just after rains. At that time, the bloom forming algae was *Microcystis aeruginosa*. The factors controlled its growth were light, temperature, pH and nutrient status of the water body. Among the attached algae, the major species reported were of *Lyngbya*, *Spirogyra* and *Cladophora* forming dense mats, one side of which is exposed and the other side attached to substrate. *Fragilaria capucina* were abundant in February and March, 2005 found attached to *Cladophora* sp.

Along with the study of algal flora, the physicochemical characteristics have also been studied which include water temperature, pH, conductivity, total dissolved solids, total alkalinity, hardness, chloride, sulphate, nitrate, phosphate, dissolved oxygen and biochemical oxygen demand. Average values of the results showed high value of pH, TDS, conductivity, hardness, sulphate and phosphate.

Keywords: Algal assay, algal bloom, hardness, nitrate, phosphate, water quality.

INTRODUCTION

Lake Jaisamand is Asia's largest sweet water lake and world's second largest man-made lake covering an area of 36 square kilometers. It is nearly 51 kilometers far from Udaipur city. It was constructed by Maharaja Jai Singh in the form of a dam at river Gomati during 1685-1691. Later, it was transformed into lake. The lake is about 15 kilometers long and 2-8 kilometers broad. It is 102 feet deep at the deepest point. The lake is known for fish culture. Its water supplied to the inhabitants of Udaipur city for drinking and other domestic purposes. Water is also supplied for irrigation through canals namely, Shyampura and Bhat.

Water is an essence of life for all the living organisms. We largely depend on surface water for domestic, irrigational and industrial requirements. Therefore, it should be of good quality along with its aesthetic value in the scenic environment. Today surface water pollution is an important problem. Due to direct or indirect interferences of human beings, the water have been found to contain various pollutants that cause harmful effects not only on human beings and animals, but also to aquatic life.

Algae are common and normal inhabitants of surface waters. Today algal assays are commonly used in assessing water quality (Wong, 1989 and Whitton et al., 1991). Micro, as well as, macro algae are known to be the most sensitive organisms to many pollutants (Kapoor and Solanki, 1992). Type of algae growing in the water body and its abundance helps in assaying the quality of water and its nutrients status, respectively. As, they are sensitive indicators of water quality; therefore, any changes in their population may provide much information about water characteristic, e.g., water chemistry and nutrient status. To evaluate the factors responsible for the growth of algae, the study of physicochemical characteristics of water is also essential as both are interrelated.

In the present study, work has been carried out to study the occurrence of a variety of algal species and the physicochemical characteristics of water during the study period, i.e., from July, 2004 to December, 2005. Therefore, the objective of the present study is to study the algal flora and the physicochemical characteristics of Lake Jaisamand.

MATERIALS AND METHODS

For study, two sampling stations (1. near boating site and 2. near Shiva temple) were chosen. Algal samples were collected in small plastic broad mouthed bottles and about 1.5 to 2.0 litres of water sample was collected in polythene cans, from each station, which were then carried to laboratory, Department of Botany, M.L.S. University for analysis. Water temperature was recorded at the site and dissolved oxygen was also fixed with Winkler A and B on the spot itself. Water sampling was done monthly in the morning between 9.00 and 10.00 A.M. during the study period. Conductivity and TDS were measured using Systronics Conductivity-TDS meter 308.

Algal samples were identified with the help of compound microscope (Olympus, model no. 633374) and light microscope (Towa, model no. 735666) using keys given by Palmer (1980) for freshwater algae, by Desikachary (1959) for cyanophyceae and by Bold and Wynne (1978) for green algae. The water samples were analyzed for water temperature, pH, conductivity, total dissolved solids (TDS), total alkalinity, hardness, chloride, sulphate, nitrate, phosphate, dissolved oxygen (DO) and biochemical oxygen demand (BOD) as per standard methods (Manivasakam, 1996 and APHA, 1995).

RESULTS

During the entire study period, a variety of algal species have been reported which belong to Cyanophyceae, Chlorophyceae, Euglenophyceae, Xanthophyceae, Dinophyceae and

Bacillariophyceae (Table 1). Cyanobacteria represent both the greatest diversity and biomass of the primary producers. Among Cyanophyceae, different species of *Oscillatoria* and *Phormidium* were common in July, '04. In October '04 *Chroococcus* and *Gloeocapsa* were found attached to stone (stairs). In the mid of December, '04, *Gloeothece palea* were reported at station 1, which formed bloom-like conditions in February, '05 and finally, disappeared completely in March, '05. Between December, '04 and February, '05, other species reported were *Oscillatoria anguina, Lyngbya* sp., *Phormidium cevennense, Stigeoclonium lubricum, Cholorococcum multinucleatum, Cladophora aegagropila, Pediastrum tetras, Cymbella ventricosa, Gomphonema parvulum, Fragilaria capucina, Nitzschia palea* and *Navicula gracilis* and *Synedra ulna*. In February, '05 *Fragilaria capucina* were also abundant and found as epiphytic on *Cladophora aegagropila*, but in March, '05 they were reported free floating and not attached to the green alga. In March, '05

Table 1 Algal flora of Lake Jaisamand during July, 2004-December, 2005

Cyanophyceae	Chlorophyceae	Bacillariphyceae	Xanthophyceae
Gloeothece palea	*Chlorococcum*	*Nitzschia palea*	*Tribonema*
	Multinucleatum^{+}		*bombycinum*
Microcystis aeruginosa		*Navicula gracilis*	
Kütz $^{++}$	*Pediastrum tetras*		
	P. simplex	*Fragilaria capucina***	**Dinophyceae**
Chroococcus sp.			
	Chlorella vulgaris	*Gomphonema parvulum*	*Cosmarium* sp.
Gloeocapsa sp.			
	Tetraedron muticum	*Cymbella ventricosa*	
Entophysalis sp.			**Euglenophyceae**
	Chlamydomonas reinhardi	*Achnanthes* sp.	
Oscillatoria anguina			*Euglena viridis*
O. agardhii	*Tetraspora gelatinosa*	*Synedra ulna*	*E. sangvlnea*
O. chlorina			*E. stellata*
O. lauterborni	*Planktosphaeria gelatinosa*		
O. limosa			
	Scenedesmus quadricauda		
Phormidium anomala			
P. cevennense	*Cladophora aegagropila*$^{+}$		
	C. glomerata		
Lyngbya digueti^{+}	*C.crispata*		
L.gracilis			
L. polysiphonia	*Oedogonium suecium*		
	O. boscii		
Anabaena planktonica			
	Spirogyra fluviatilis^{+}		
Stigeoclonium lubricum^{+}			
	Microspora wittrockii^{+}		

++ bloom forming algae
+ commonly found
** abundant and epiphytic

Lyngbya (*L. gracilis, L. polysiphoniae*), *Phormidium anomala, Oscillatoria* sp., *Pediastrum* (*P. simplex* and *P. tetras*), *Scenedesmus quadricauda* and *Cosmarium* were reported. In April and May, '05, *Oscillatoria* spp. were common. One species of *Xanthophyceae*, i.e., *Tribonema bombycinum*, a clean water green alga was also found in May, '05. In October, '05 just after rains, *Microcystis aeruginosa* Kütz. were abundant and formed blooms at both the stations. *Spirogyra fluviatilis* was reported in May, November and December, '05. In November, '05 *Oscillatoria* sp., *Entophysalis lemaniae, Tetraspora gilatinosa, Planktosphaeria gelatinosa, Microspora wittrockii, Cladophora crispata* and *Cymbella ventricosa* were reported. In December, '05 different species of *Euglena* (*E. viridis, E. sangvlnea* and *E. stellata*), *Cladophora* (*C. aegagropila* and *C. glomerata*), *Chlamydomonas reinhardi, Chlorella vulgaris, Pyrobotrys stellata, Tetraedron muticum, Anabaena planktonica, Oscillatoria* (*O. agardhii* and *O. limosa*), *Nitzschia palea* and *Navicula gracilis* were reported.

The results of physicochemical characteristics of Lake Jaisamand in the form of bimonthly averages from July, '04 to December, '05 are summarized in Table 2. The values of total alkalinity, hardness, sulphate, nitrate and phosphate were quite high in July-Aug., '05 which showed gradual reduction in concentration except for nitrate, which showed an increase in the next months. In September-October, '05 the values of pH were high (8.70 and 8.80) at both the stations. The nitrate concentration was high in November-December., '04 (5.93 ppm) and in November-December, '05 (7.53 ppm) at station 1 and 2, respectively. During January '05 to April, '05 TDS was found to cross the permissible limit of 500 ppm. The values of conductivity were high during the study period.

DISCUSSION

Algae represent an essential and basic part of the cycle of living organisms. Small aquatic organisms feed on algae and form the base of food chain. Algae may prove to be nuisance, if it starts forming bloom or mats. Blooms may cause problems of tastes and odours, filter and screen clogging and slime accumulation in pipes.

Blooming conditions developed twice during the study period. Water temperature, pH and the available nutrients play an important role in the growth of particular alga. The pH between 8.5-9.0 favours the growth of blue-green algae. According to Kodarkar et al., (1991), blooming has a direct correlation with pH, temperature, DO and available phosphates contributed by sewage and sedimentary cycle. High nitrate level (5.93 ppm) in November – December, '04 was responsible for increased growth of *Gloeothece palea*, which was then favoured by low water temperature (19.5°C) and high values of pH (8.6), hardness (162.50 ppm) and phosphate (0.049 ppm) in January-February '05 and resulted in blooming condition in Febreary '05 at station 1. The flow of rainwater carried the nutrients along with it and enriched the water, which in combination with suitable temperature and pH favoured the growth of *Microcystis aeruginosa* Kütz and resulted in blooming condition in October, '05 at both the stations. *M. aeruginosa* is a toxic alga. It secretes toxins (microcystins) on degradation, which may cause fish kills. At that time the average concentrations of nitrate (1.94 ppm and 3.88 ppm at station 1 and 2, respectively) and phosphate (0.04 ppm and 0.06 ppm at station 1 and 2, respectively) were found to be high. The abundance of *M. aeruginosa* might be responsible for increased concentration of alkalinity and hardness in November-December, '05.

Table 2 Physicochemical characteristics of lake Jaisamand during July, 2004 and December, 2005

S. No.	Parameters		July-Aug., 2004	Sep.-Oct., 2004	Nov.-Dec., 2004	Jan.-Feb., 2005	Mar.-Apr., 2005	May.-June, 2005	July-Aug., 2005	Sep.-Oct., 2005	Nov.-Dec., 2005
1.	Water	I	26.0	27.5	20.5	19.5	28.0	29.25	-	27.5	21.0
	temperature (°C)	II	25.0	27.5	20.5	19.0	27.5	29.75	-	28.0	21.25
2.	pH	I	8.40	8.36	8.12	8.69	8.67	8.40	-	8.76	8.05
		II	8.35	8.38	7.99	8.65	8.69	8.59	-	8.80	8.06
3.	Conductivity	I	-	0.538	0.498	1.03	1.06	0.60	-	0.468	0.471
	(mS/cm)	II	-	0.520	0.499	1.05	1.08	0.60	-	0.473	0.470
4.	TDS	I	-	256.0	248.0	517.0	533.0	300.0	-	233.0	232.0
	(ppm)	II	-	244.0	249.0	527.0	543.0	300.0	-	235.0	235.0
5.	Total	I	430.0	128.5	144.0	166.0	189.0	198.0	-	134.5	161.5
	alkalinity (ppm)	II	406.0	126.5	142.0	164.0	189.0	196.0	-	136.0	163.5
6.	Hardness (ppm)	I	250.0	80.0	155.0	162.5	161.5	158.5	-	123.5	143.5
		II	232.0	76.0	152.0	202.5	163.0	158.5	-	123.5	142.5
7.	Chloride	I	14.69	13.99	10.99	22.99	23.99	37.98	-	17.99	22.99
	(ppm)	II	14.49	12.49	12.99	19.99	27.99	35.98	-	18.99	22.99
8.	Sulphate	I	583.1	247.5	60.0	150.0	170.0	178.7	-	102.5	377.5
	(ppm)	II	430.0	399.7	80.0	101.0	150.0	115.0	-	212.5	345.0
9.	Nitrate	I	1.329	2.436	5.93	1.33	0.76	3.65	-	1.94	3.54
	(ppm)	II	1.329	3.262	3.21	6.31	2.75	4.93	-	3.88	7.53
10.	Phosphate	I	0.039	0.031	0.031	0.049	0.019	0.028	-	0.04	0.03
	(ppm)	II	0.098	0.033	0.049	0.043	0.017	0.030	-	0.06	0.03
11.	D.O.	I	8.16	8.67	8.97	7.95	7.35	6.63	-	6.41	4.89
	(ppm)	II	9.38	10.81	9.79	8.05	7.65	6.73	-	7.01	5.10
12.	B.O.D.	I	16.33	11.53	1.63	1.43	3.78	5.41	-	3.04	1.33
	(ppm)	II	18.37	9.79	2.04	1.96	2.09	5.10	-	4.05	1.48

I-station 1
II-station 2
(-) not taken

According to Alikunhi (1957), the highly productive waters generally have an alkalinity above 100 ppm. The vigorous growth of algae have the ability to reduce hardness, as much as, one-third (Palmer, 1980). In the present study also, the reduction in hardness can be noticed very clearly after the first blooming in January-February '05 (Table 2.) Although algae play an important part in aquatic life, but their abundance may lead to high BOD and low DO in a water body, which may even cause fish kills due to suffocation. Finally, in the end we may conclude that the growth of algae depends on the physicochemical characteristics of water. In turn, the abundance of algae is also capable of altering these water characteristics. Therefore, a regular monitoring of water should be done from time to time to maintain the quality of water and also the beauty of the lake. The abundance of nuisance planktonic forms, e.g., *Microystis aeruginosa* can be controlled by the use of copper sulphate or other algaecides. Copper sulphate is toxic to many algae at low concentration.

ACKNOWLEDGEMENTS

The presenting author is thankful to University Grant Commission, New Delhi for providing fellowship and to Department of Botany, M.L.S. University, Udaipur (Rajasthan) for providing facilities. I am also thankful to Dr. K. Kapoor, Professor, Department of Botany, M.L.S. University, Udaipur for his valuable guidance and support. Encouragement and help from Shri P. D. Vikal (father) and Ghanshyam Singh and Rokesh Kumar Singh (brothers) are also gratefully acknowledged.

REFERENCES

Alikunhi, K. H. (1957) Fish culture in India. Bull. 20, ICAR, New Delhi.

APHA (1995) Standard Methods for the Examination of Water and Waste Water, 19th edn. Washington, D. C.

Bold Harold, C. and Wynne Michael, J. (1978) Introduction to the Algae. Prentice-Hall of India Pvt. Ltd., New Delhi.

Desikachary, T. V. (1959) Cyanophyta. Indian Council of Agricultural Research, New Delhi.

Kapoor, K. and Solanki, P. S. (1992) In: Water Management Problems Related to Water Resources in Udaipur. *Nat. Sem. Strat. Env. Prot. Industrialization*, M.L.S. University, Udaipur. pp. 119-126.

Kodarkar, M. S., Muley, E. V. and Rao, V. (1991) Toxic algal blooms in the Lake Hussainsagar, Hyderabad. *J. Aquatic Bio*. 6 (1 & 2): 13-18.

Manivasakam, N. (1996) Physico-chemical examination of water, sewage and industrial effluents. Pragati Prakashan, Meerut (India).

Palmer, C. M. (1980) Algae and water pollution. Castle House Publication Ltd., England, pp. 123.

Whitton, B. A , Rott, E. and Friedrick, G. (1991) Use of Algae for monitoring rivers. Institute of Botanik, Universitat Innsbruck, Germany.

Wong, S. L. (1989) Algal assays to interpret toxicity guidelines for natural waters. *J. Env. Sci. Health*. 24:1001-1010.

Index